Lecture Notes in Computer Science   10198

Commenced Publication in 1973
Founding and Former Series Editors:
Gerhard Goos, Juris Hartmanis, and Jan van Leeuwen

More information about this series at http://www.springer.com/series/7407

João Correia · Vic Ciesielski
Antonios Liapis (Eds.)

Computational Intelligence in Music, Sound, Art and Design

6th International Conference, EvoMUSART 2017
Amsterdam, The Netherlands, April 19–21, 2017
Proceedings

 Springer

Editors
João Correia
University of Coimbra
Coimbra
Portugal

Vic Ciesielski
RMIT University
Melbourne, VIC
Australia

Antonios Liapis
University of Malta
Msida
Malta

ISSN 0302-9743   ISSN 1611-3349 (electronic)
Lecture Notes in Computer Science
ISBN 978-3-319-55749-6   ISBN 978-3-319-55750-2   (eBook)
DOI 10.1007/978-3-319-55750-2

Library of Congress Control Number: 2017934454

LNCS Sublibrary: SL1 – Theoretical Computer Science and General Issues

Printed on acid-free paper

This Springer imprint is published by Springer Nature
The registered company is Springer International Publishing AG
The registered company address is: Gewerbestrasse 11, 6330 Cham, Switzerland

Preface

EvoMUSART 2017—the 6th International Conference and the 15th European event on Biologically Inspired Music, Sound, Art and Design—took place during April 19–21, 2017, in Amsterdam, The Netherlands. It brought together researchers who use biologically inspired computer techniques for artistic, aesthetic, and design purposes. Researchers presented their latest work in the intersection of the fields of computer science, evolutionary systems, art, and aesthetics. As always, the atmosphere was fun, friendly, and constructive.

EvoMUSART has grown steadily since its first edition in 2003 in Essex, UK, when it was one of the Applications of Evolutionary Computing workshops. Since 2012 it has been a full conference as part of the Evo* co-located events.

EvoMUSART 2017 received 29 submissions. The peer-review process was rigorous and double-blind. The international Program Committee, listed herein, was composed of 58 members from 20 countries. EvoMUSART continued to provide useful feedback to authors: among the papers sent for full review, there were on average 3.9 reviews per paper. It also saw an increase in the number of accepted papers, and thus dissemination of knowledge, than previous years: 12 papers were accepted for oral presentation (41% acceptance rate), and 12 for poster presentation (41% acceptance rate).

This volume of proceedings collects the accepted papers. As always, the EvoMU-SART proceedings cover a wide range of topics and application areas, including: generative approaches to music, graphics, game content, and narrative; music information retrieval; computational aesthetics; the mechanics of interactive evolutionary computation; computer-aided design; and the art theory of evolutionary computation.

We thank all authors for submitting their work, including those whose work was not accepted for presentation. As always, the standard of submissions was high, and good papers had to be rejected.

The work of reviewing is done voluntarily and generally without official recognition from the institutions where reviewers are employed. Nevertheless, good reviewing is essential to a healthy conference. Therefore, we particularly thank the members of the Program Committee for their hard work and professionalism in providing constructive and fair reviews.

EvoMUSART 2017 was part of the Evo* 2017 event, which included three additional conferences: EuroGP 2017, EvoCOP 2017, and EvoApplications 2017. Many people helped to make this event a success.

We thank the SPECIES society for its involvement in the event. We thank the local organizing team of Evert Haasdijk and Jacqueline Heinerman, from VU University Amsterdam.

We thank Marc Schoenauer (Inria Saclay, Île-de-France), for continued assistance in providing the MyReview conference management system. We thank Pablo García Sánchez (University of Cádiz) for the Evo* publicity and website service.

In particular, we would like to acknowledge our invited speakers: Kenneth De Jong and Arthur Kordon.

Last but certainly not least, we especially want to express a heartfelt thanks to Jennifer Willies, as her dedicated work and continued involvement in Evo* has been invaluable for its progress and current success.

April 2017

João Correia
Vic Ciesielski
Antonios Liapis

Organization

EvoMUSART 2017 was part of Evo* 2017, Europe's premier co-located events in the field of evolutionary computing, which also included the conferences EuroGP 2017, EvoCOP 2017, and EvoApplications 2017.

Organizing Committee

Conference Chairs

João Correia	University of Coimbra, Portugal
Vic Ciesielski	RMIT University, Australia

Publication Chair

Antonios Liapis	University of Malta, Malta

Program Committee

Dan Ashlock	University of Guelph, Canada
Peter Bentley	University College London, UK
Daniel Bisig	University of Zurich, Switzerland
Tim Blackwell	Goldsmiths College, University of London, UK
Andrew Brown	Griffith University, Australia
Adrian Carballal	University of A Coruña, Spain
Amilcar Cardoso	University of Coimbra, Portugal
Peter Cariani	University of Binghamton, USA
Vic Ciesielski	RMIT, Australia
João Correia	University of Coimbra, Portugal
Palle Dahlstedt	Göteborg University, Sweden
Hans Dehlinger	Independent Artist, Germany
Eelco den Heijer	Vrije Universiteit Amsterdam, The Netherlands
Alan Dorin	Monash University, Australia
Arne Eigenfeldt	Simon Fraser University, Canada
José Fornari	NICS/Unicamp, Brazil
Marcelo Freitas Caetano	IRCAM, France
Philip Galanter	Texas A&M College of Architecture, USA
Andrew Gildfind	Google, Inc., Australia
Gary Greenfield	University of Richmond, USA
Carlos Grilo	Instituto Politécnico de Leiria, Portugal
Andrew Horner	University of Science and Technology, Hong Kong, SAR China
Patrick Janssen	National University of Singapore, Singapore
Colin Johnson	University of Kent, UK

Daniel Jones   Goldsmiths College, University of London, UK
Anna Jordanous   University of Kent, UK
Amy K. Hoover   Northeastern University, USA
Maximos   Aristotle University of Thessaloniki, Greece
   Kaliakatsos-Papakostas
Matthew Lewis   Ohio State University, USA
Yang Li   University of Science and Technology Beijing, China
Antonios Liapis   University of Malta, Malta
Alain Lioret   Paris 8 University, France
Louis Philippe Lopes   University of Malta, Malta
Roisin Loughran   University College Dublin, Ireland
Penousal Machado   University of Coimbra, Portugal
Tiago Martins   University of Coimbra, Portugal
Jon McCormack   Monash University, Australia
Eduardo Miranda   University of Plymouth, UK
Nicolas Monmarché   University of Tours, France
Marcos Nadal   University of Vienna, Austria
Gary Nelson   Oerlin College, USA
Michael O'Neill   University College Dublin, Ireland
Philippe Pasquier   Simon Fraser University, Canada
Somnuk Phon-Amnuaisuk   Brunei Institute of Technology, Malaysia
Jane Prophet   City University, Hong Kong, SAR China
Douglas Repetto   Columbia University, USA
Juan Romero   University of A Coruña, Spain
Brian Ross   Brock University, Canada
Jonathan E. Rowe   University of Birmingham, UK
Antonino Santos   University of A Coruña, Spain
Marco Scirea   IT University of Copenhagen, Denmark
Daniel Silva   University of Coimbra, Portugal
Benjamin Smith   Purdue University, Indianapolis, USA
Gillian Smith   Northeastern University, USA
Stephen Todd   IBM, UK
Paulo Urbano   Universidade de Lisboa, Portugal
Anna Ursyn   University of Northern Colorado, USA
Dan Ventura   Brigham Young University, USA

Contents

X   Contents

Algorithmic Songwriting with ALYSIA

Margareta Ackerman[1]([✉]) and David Loker[2]

[1] Department of Computer Science, San Jose State University, San Jose, CA, USA
margareta.ackerman@sjsu.edu
[2] Orbitwerks, San Jose, CA, USA
dloker@gmail.com

Abstract. This paper introduces ALYSIA: Automated LYrical Song-wrIting Application. ALYSIA is based on a machine learning model using Random Forests, and we discuss its success at pitch and rhythm prediction. Next, we show how ALYSIA was used to create original pop songs that were subsequently recorded and produced. Finally, we discuss our vision for the future of Automated Songwriting for both co-creative and autonomous systems.

Keywords: Algorithmic composition · Machine learning · Songwriting

1 Introduction

Music production has traditionally relied on a wide range of expertise far exceeding the capabilities of most individuals. Recent advancements in music technology and the parallel development of electronic music has brought the joy of music-making to the masses. Among the countless amateurs who enjoy expressing their creativity through music-making, there has emerged a newly-leveled playing field for professional and semi-professional musicians who are able to independently produce music with no more than a personal computer.

Technological advances have greatly simplified music production through systems such as Logic Pro[1], Live[2], or even the freely available Garage Band[3]. In addition to placing a wide range of instruments at our fingertips, without requiring the corresponding training, these systems provide a variety of mastering tools that enable the creation of high quality music with relative ease. Yet, despite this paradigm shift, the writing of songs - comprised of both melody and lyrics - remains an exclusive domain. A careful examination of electronic music reveals these limitations. The genre relies heavily on remixing existing music and much attention is paid to sound design, which does not require original composition or its juxtaposition with lyrics.

The creation of a new song continues to rely on diverse skills rarely possessed by one individual. The score alone calls for a combination of poetic ability to

---

[1] http://www.apple.com/logic-pro/.
[2] https://www.ableton.com/en/live/.
[3] http://www.apple.com/mac/garageband/.

© Springer International Publishing AG 2017
J. Correia et al. (Eds.): EvoMUSART 2017, LNCS 10198, pp. 1–16, 2017.
DOI: 10.1007/978-3-319-55750-2_1

create the lyrics, musical composition for the melody, and expertise in combining lyrics and melody into a coherent, meaningful and aesthetically pleasing whole. Once the score is complete, vocal skill and production expertise are needed to turn it into a finalized musical piece. While some may be satisfied with remixing existing music, the creation of truly original songs is often desired, in part due to restrictions associated with performing and remaking existing works. Can technology assist in the creation of original songs, allowing users to express their creativity through the skills they possess while letting technology handle the rest?

We introduce ALYSIA, a songwriting system specifically designed to help both musicians and amateurs write songs, with the goal of producing professional quality music. Our system enables the human creator to discover novel melodies to accompany lyrics. To the best of our knowledge, ALYSIA is the first songwriting system to utilize formal machine learning methodology. This lets us evaluate the success of our model based on its ability to correctly predict the pitch and rhythm of notes in the test set (which was not used to train the model). On a corpus of pop music, the system achieves an accuracy of 86.79% on rhythms and 72.28% on scale-degrees.

In the generation phase, ALYSIA is given short lyrical phrases, for which it can then provide numerous melody options in the style of the given corpus. As it stands today, ALYSIA is a co-creative system that allows the user to retain a sense of creative control while eliminating one of the most essential skills traditionally required for songwriting: the ability to effectively explore the space of potential accompanying melodies. As a co-creative songwriting partner, ALYSIA assists human creators in the song-making progress by filling in gaps in the user's expertise.

Helping people be creative, particularly in a manner that aids their social engagement both on and offline, can have substantial positive impact on their sense of well-being and happiness [2]. Technological advancement has led to the radical expansion of those who are able to engage in music making, and made possible today's large online communities for sharing musical works. Widening access to original songwriting stands not only to ease the path to professional musicianship, but also makes it possible for more people to enjoy the psychological benefits of engaging in this creative and socially rewarding activity.

Without any changes to the system, ALYSIA could be used to create complete melodies without human interference by simply selecting her first suggestion for each line of lyrics. However, a little human interaction can result in better music and gives the user creative freedom that would be lacking from even the best autonomous system.

We begin with an examination of previous work and how it relates to our contributions, followed by a technical description of our system. We then move onto the application of ALYSIA to songwriting, starting with a detailed discussion of our proposed co-creative process. Finally, we showcase three musical pieces created with ALYSIA, and conclude with a discussion of our vision for the future of algorithmic songwriting.

2   Previous Work

Algorithmic composition has a rich history dating back to the 1950s. A wide range of techniques have been applied to music generation (with no lyrics), spanning from expert systems to neural networks. See a survey by Fernández and Vico [1] for an excellent overview.

Our focus here is on the creation of songs, which introduces a lyrical component to algorithmic composition. Algorithmic songwriting is relatively new, and so far the state of art addresses it using Markov chains. For example, M.U. Sucus-Apparatusf [8] by Toivanen et al., is a system that was used for the generation of Finnish art songs. Rhythm patterns are randomly chosen from among those typically found in art songs. Chord progressions are subsequently generated by using second order Markov chains. Lastly, pitches are generated using a joint probability distribution based on chords and the previous note. M.U. Sucus-Apparatusf integrates the entire songwriting process, from lyric generation to melody and accompaniment. Yet, as with most previous systems, the major weakness is a lack of clear phrase structure [8].

Another full-cycle automated songwriting system comes from the work of Scirea et al. [7]. Their songwriting system SMUG was used to write songs from academic papers. The methodology relies on the evolution of Markov chains. It uses a corpus of mixed-genre popular songs, and integrates several rules, such as the high probability of rests at the ends of words.

Monteith et al. [4] study the generation of melodic accompaniments. While utilizing a corpus, the system works by generating hundreds of corpus-driven random options guided by a few rules, and then chooses among them with an evaluation criteria that incorporates some musical knowledge. This process is applied for both the rhythm and melody generation. For example, for generating rhythms, the system produces 100 possible downbeat assignments. The text of each line is distributed across four measures, so four syllables are randomly selected to carry a downbeat.

In other related works, Nichols [5] studies a sub-component of the problem we address here, particularly lyric-based rhythm suggestion. He considers the set of all accompanying rhythms, and defines a fully rule-based function to evaluate them, via considerations such as rare word emphasis through strong beats. As future work, Nichols suggests solving the rhythm generation processes through machine learning techniques, which is one of our contributions here. Lastly, in the work of Oliveira [6], the pipeline is inverted and lyrics are generated based on the rhythm of the provided melody.

Our approach does not encode any musical rules and uses Random forests instead of Markov chains. The lack of directly encoded musical knowledge gives our system the potential for genre independence. Musical genres are incredibly diverse, and rules that make sense for one genre can make it impossible to generate songs in another. For example, penalizing melodies with many repeating consecutive pitches is often a sensible choice, but it is unsuited to pop music, where such sequences are common. A machine learning model that does not explicitly encode musical expertise not only applies to existing genres, but may also be adapted to all future musical genres, which can differ from prior music in unpredictable ways.

Random forests offer some advantages over using Markov chains. One drawback of Markov chains is that while they can model likelihoods of sequences, they do not encapsulate the "why" of one note being chosen over another after the same initial sequence. For example, if note A is the first note, why is the next note B most of the time? Perhaps it is only B most of the time if the key signature is C-major, and it is actually F# most of the time under a different context. Such context can be discovered and used in systems like random forests given a suitable feature set. Additionally, we rely on machine learning methodology to formally evaluate our model, which has not been done in previous work on automated songwriting. Another difference is that we rely on a co-creative process, which aids with our goal of creating high quality music.

3   The Making of ALYSIA

ALYSIA is an entirely data-driven system that is based upon a prediction model. We chose to construct two models. The first predicts note duration, which we refer to as the *rhythm model*. The second predicts the scale degree of a note, with possible accidentals, which we refer to as the *melody model*. Both models rely on a similar set of features, though depending on which model is run first, one model will have access to extra information. In our case, the melody model benefits from the predictions made by the rhythm model. Note that combing the models would increase the number of output classes, requiring a larger corpus.

The most important component of our model is the set of features. The aim is to construct features that will allow the model to learn the mechanics and technicalities behind the structure of songs and their composition.

Since we provide melodies for text provided by the user, our training set consists of vocal lines of music, and not arbitrary melodies. The corpus consists of Music-XML (MXL) files, with a single instrument corresponding to the vocal line and the accompanying lyrics. As such, each note has a corresponding syllable.

The system consists of five distinct components.

1. The corpus
2. Feature extraction
3. Model building and evaluation
4. User lyrics to features
5. Song generation

We now discuss each of these in detail.

3.1   Feature Extraction

In order to build a model, we first extract a set of features from the Music-XML (MXL) files.

For each note, we extract the following features:

- First Measure - A boolean variable indicating whether or not the current note belongs to the first measure of the piece
- Key Signature - The key signature that applies to the current note
- Time Signature - The time signature that applies to the current note
- Offset - The number of beats since the start of the music
- Offset within Measure - The number of beats since the start of the measure
- Duration - The length of the note
- Scale Degree - The scale degree of the note (1–7)
- Accidental - The accidental of the note (flat, sharp, or none)
- Beat Strength - The strength of beat as defined by music21. We include the continuous and categorical version (Beat Strength Factor) of this variable.
- Offbeat - A boolean variable specifying whether or not the note is offbeat
- Syllable Type* - Classifies the current syllable in one of the following four categories: Single (the word consists of a single syllable), Begin (the current syllable is the first one in its word), Middle (the current syllable occurs in the middle of its word), End (the current syllable is the last one in its word).
- Syllable Number* - The syllable number within its word
- Word Frequency* - The word frequency of the word which the current note/syllable is part of. This value is obtained through the dictionary frequency as calculated by Pythons NLTK (http://www.nltk.org/) library using the Brown corpus[4].
- Word Rarity* - A function of word frequency, as defined by [5]. $WordRarity = 2(1 - \frac{\log_{10}(WordFrequency)}{7})$.
- Word Vowel Strength* - The word vowel strength (primary, secondary, or none) as indicated by the CMUDict v0.07.[5]
- Number of Vowels in the Word* - The number of vowels found in the word corresponding to the current note/syllable.
- Scale Degree, Accidental, and Duration of previous 5 notes

The featured marked with an * are generated from lyrics found within the vocal line. These same features must be generated using the user's lyrics when ALYSIA is used in generation mode. The features marked with an * were used by [5] as part of a rule-based evaluation criteria for rhythm suggestion.

3.2 Model Building and Evaluation

Using R, we train two models using random forests. Random forests were chosen for several reasons. They are well-suited for large numbers of categorical variables. Additionally, they allow non-linearity in combining features, without an explosion in the required size of the training set, in order to avoid overfitting. Non-linearity makes random forests more powerful than linear models such as

---

[4] Brown corpus: http://www.hit.uib.no/icame/brown/bcm.html.
[5] CMUDict v0.07, Carnegie Mellon University.

logistic regression. We randomly split the data using stratified sampling into a training set (75% of the data) with the rest for testing.

The accuracy of the rhythm model on test data was 86.79%. Figure 1 shows the confusion matrix. The row labels correspond to the duration found in the test sets, and column labels are the predicted labels. The number after the underscore corresponds to the number of dots in the durations. So, for example, a dotted sixteenth note is 16th_1.

As you can see in the confusion matrix, the model tends to perform better on rhythms which occur more frequently in the data set. A larger data set will enable us to test whether the trend will continue.

	16th_0	16th_1	32nd_0	eighth_0	eighth_1	half_0	half_1	quarter_0	quarter_1	quarter_2	whole_0	class.error
16th_0	551	0	0	68	2	1	0	3	0	0	0	0.11840000
16th_1	0	4	0	0	0	0	0	0	0	0	0	0.00000000
32nd_0	2	0	6	0	0	0	0	0	0	0	0	0.25000000
eighth_0	25	0	0	1182	5	3	0	24	0	0	0	0.04600484
eighth_1	14	0	0	5	81	0	0	4	0	0	0	0.22115385
half_0	2	0	0	15	4	69	0	18	0	0	2	0.37272727
half_1	0	0	0	3	0	1	3	7	0	0	0	0.78571429
quarter_0	11	0	0	77	1	3	0	310	0	0	0	0.22885572
quarter_1	0	0	0	16	1	5	1	8	28	0	0	0.52542373
quarter_2	0	0	0	0	0	0	0	0	0	0	0	NaN
whole_0	1	0	0	4	0	3	0	2	0	0	20	0.33333333

Fig. 1. Confusion matrix for Rhythm Model. Cell (i,j) in the above table specifies how often a node with duration type i was classified as having during j. The last column shows the error rate for each duration type (NaN indicating that no notes of this duration were found in the corpus).

The accuracy of the melody model was 72.28%. Figure 2 shows the confusion matrix. The numbers correspond to the scale degrees, and the label after the underscore represents the accidentals. So, for example, 1_none represents the first scale degree without accidentals. As with the rhythm model, we see that accuracy is higher when there are more examples.

Lastly, our machine learning model allows us to evaluate the importance of each feature. Figure 3 depicts the mean decrease Gini (which measures node impurity in tree-based classification) of each feature in the rhythm model (left) and the melody model (right). In both models, all 5 previous notes' scale degrees are some of the most important features. This suggests that including features for additional previous notes may be helpful. This also explains why Markov chains seem to do well for this task. However, we see that other features are also prominent.

For example, the rhythm model is most heavily influenced by beat strength, revealing the potential of using a data-driven approach over Markov chains, which do not give access to such features. Finally, it is important to note that lyrical features such as word rarity and frequency also play an important role.

	1_none	1_sharp	2_none	3_flat	3_none	4_none	4_sharp	5_none	6_flat	6_none	7_flat	7_none	class.error
1_none	504	0	18	0	39	2	1	17	0	27	0	11	0.1857835
1_sharp	0	2	0	0	0	0	0	0	0	0	0	0	0.0000000
2_none	48	0	271	0	34	5	0	4	0	4	2	9	0.2811671
3_flat	4	0	1	11	1	0	0	0	0	1	0	0	0.3888889
3_none	47	0	21	2	375	3	0	42	0	14	0	2	0.2588933
4_none	5	0	7	0	28	101	0	2	0	5	0	1	0.3221477
4_sharp	0	0	0	0	0	0	3	1	0	6	1	0	0.7272727
5_none	24	0	6	0	46	2	0	254	0	25	0	9	0.3060109
6_flat	0	0	0	0	0	0	0	6	4	0	0	0	0.6000000
6_none	29	0	2	0	22	11	3	21	0	237	0	13	0.2988166
7_flat	4	0	2	1	0	0	0	5	0	0	11	0	0.5217391
7_none	29	0	10	0	5	0	0	16	0	15	0	103	0.4213483

Fig. 2. Confusion matrix for Melody Model. Cell (i, j) in the above table specifies how often scale degree i was classified as scale degree j. The last column shows the error rate for each scale degree.

3.3 User Lyrics to Features

In order to generate music for a new set of lyrics using our models, we need to extract the same set of lyric features we were using for our models. This was done using a python script, which outputs a feature matrix for importing into R.

3.4 Song Generation

Using R, we loop through the lyric feature set one sentence at a time. For each line, we read the feature set from the lyrics given by the user and generate the rhythm, followed by the melody. Note that we generate one note at a time for each syllable in the lyrical phrases. We also keep track of 5 previous notes, since these are very important features in the model. Finally, we generate a file detailing the generated melodies for each line. There may be many depending on how many melodies per line were requested.

3.5 Corpus

Our corpus consists of 12,358 observations on 59 features that we extract from MXL files of 24 pop songs. From each song, only the melody lines and lyrics were used. For model building and evaluation, the data is split using stratified sampling along the outcome variable, which is scale degree with accent for the melody model, and note duration for the rhythm model. Seventy-five percent of the data is used for training, while the remaining data is used for evaluation.

Please note that unlike MP3 files, MXL files containing lyrics are much more difficult to acquire, limiting the size of our data. Previous work on automated songwriting at times does not use any training (it is instead rule-based), or trains separately on lyrics and melodies - which eliminates the possibility of learning how they interact. The size of the training set in previous work is often omitted[6].

---

[6] Due to the difficulties in attaining MXL files, we are currently creating a new larger corpus of songs, which will be analyzed in future work.

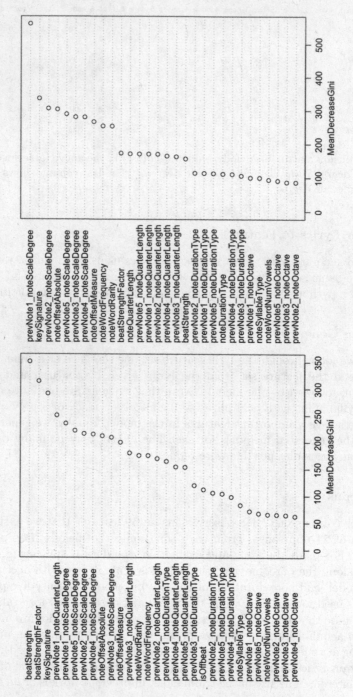

Fig. 3. Importance plot for the rhythm model (left) and the melody model (right), representing the significance of each feature in the respective models.

3.6   Parameters

When used in generation mode, ALYSIA allows the user to tune the following parameters:

Explore/Exploit. This 'parameter determines how heavily we lean on the model. Specifically, for each scale degree/note duration we generate, our model outputs a distribution over all possible outcomes. This parameter determines how many independent draws we make from this distribution. The final resulting note is the most common draw, with ties being broken by the original outcome distribution. That is, if scale degree 1 and 2 are tied after four draws, we take the scale degree that was originally more likely in the distribution output by the model for this data point. Given this definition, a higher explore/exploit parameter value means we exploit because we are almost always going to output the scale degree or duration that has the highest probability of occurring. This parameter allows us to favor the scale degrees and durations that are most likely, versus potentially taking a more varied approach and generating music that could be considered more experimental.

Melody Count. We generate a set number of melodies per line, outputting them to a file for later translation into midi and MXL files.

Rhythm Restriction. We allow the user to restrict the possible rhythmic outcome if they would like to omit certain note durations. For example, the user may want to disallow whole notes, or some faster notes such as a 32nd note.

4   The Co-Creative Process of Songwriting with ALYSIA

In this section, we describe the co-creative process of songwriting with ALYSIA. This is a new approach to writing songs that requires minimal to no musical training. ALYSIA makes it easy to explore melodic lines, reducing songwriting to the ability to select melodies based on one's musical taste.

The user provides ALYSIA with the lyrics broken down into separate lines. Then, the system gives the specified number of melodies (notes/rhythm combinations) to which the given lyrics can be sang. The number of melodies per line is specified by the user. ALYSIA outputs the melodies in both MXL and MIDI format. The MIDI form is particularly helpful, since it allows the user to play the variations, select the desired one, and directly utilize the MIDI to produce the song.

We now describe the workflow we used with ALYSIA. We typically ask for between 15 to 30 melodic variations per line of text. Among these, we select between three and ten melodic lines. It should be noted that nearly all of ALYSIA's suggestions are reasonable. For example, Fig. 4 lists the first 12 melody options provided by ALYSIA for the phrase *everywhere I look I find you looking*

Fig. 4. The first 12 melodies generated by ALYSIA to accompany the first line of the song *Everywhere* (the full song appears in the following section). All melodies are sensible and could be used towards the creation of a complete melody line. Having many reasonable options allows the user to incorporate their musical taste into songwriting with ALYSIA.

back. All of these melodies work fairly well with the underlying text and could be extended into full songs.

One may ask why we choose to look at 15 to 30 options if all are reasonable. Having a variety of options can lead to better quality songs while also enabling artists to incorporate their own musical preferences, without possessing the composition and text/melody juxtaposition skills traditionally required to engage in this art form.

When making our selections, we look for melodies that are independently interesting, and have intriguing relationships with the underlying text. We search for lines that match the emotional meaning of the text (happy or sad), as well as interesting word emphasis. For example, if the word "sunshine" appears in the text, we may select a melody that rises with this word (see Fig. 7). ALYSIA often suggests melodic variations for which we find interesting explanations. For example, in the original song *Why Do I Still Miss You* (see Fig. 5), ALYSIA suggested a melody where the phrase "went wrong" is on the lower end of scale, giving these words an appropriately dark interpretation.

The next step involves combining the melodic lines to form the complete song. This takes some trial and error, and several interesting variations can typically be attained. The most efficient approach we had found was to construct the song around a few melodic lines. Say, if there are particularly interesting melodies for lines 1 and 3, then lines 2 and 4 are chosen to fit with these. Having multiple options allows the artist to create several variations for verses and chorus repetitions, as often found in songs.

5   Songs Made with ALYSIA

Three songs were created using ALYSIA. The first two rely on lyrics written by the authors, and the last borrows its lyrics from a 1917 Vaudeville song. The songs were consequently recorded and produced. All recordings are posted on http://bit.ly/2eQHado.

The primary technical difference between these songs was the explore/exploit parameter. *Why Do I Still Miss You* was made using an explore vs exploit parameter that leads to more exploration, while *Everywhere* and *I'm Always Chasing Rainbows* were made with a higher exploit setting.

5.1   Why Do I Still Miss You

Why Do I Still Miss You, the first song ever written with ALYSIA, was made with the goal of testing the viability of using computer-assisted composition for production-quality songs. The creation of the song started with the writing of original lyrics, written by the authors. The goal was to come up with lyrics that stylistically match the style of the corpus, that is, lyrics in today's pop genre (See Fig. 5).

Next, each lyrical line was given to ALYSIA, for which she produced between fifteen and thirty variations. In most cases, we were satisfied with one of the options ALYSIA proposed within the first fifteen tries. We then combined the melodic lines. See Fig. 5 for the resulting score. Subsequently, the song was recorded and produced. One of the authors, who is a professionally trained singer, recorded and produced the song. A recording of this song can be heard at http://bit.ly/2eQHado.

This experiment, intentionally spanning the entire songwriting process from the lyrics to the final production, demonstrates the viability of our approach for

Fig. 5. The first original song created with ALYSIA.

making production-worthy works. It also shows that ALYSIA makes songwriting accessible to those with no songwriting ability, as the song was written by one of the authors, who does not possess this skill unassisted by our computer system. Informal assessment by the authors and many others consistently point to the surprising quality of this song, gauging it as having quality on par with that of fully human-made songs. Future user studies will help to formally assess this claim.

For this song, we used an explore/exploit parameter of 1, meaning that the distribution of notes output by the model was sampled only once. With regard to improving the application of our system, the making of *Why Do I Still Miss You* suggested that we should increase the exploit parameter. This was particularly important for the rhythm suggestion, which we found to be too varied. As such, *Why Do I Still Miss You* fully utilized the melody model, while making some manual adjustment to the rhythm. We stress that even with this sub-optimal setting for the exploit-explore setting, ALYSIA made songwriting accessible to someone who previously could not engage in this artform.

The two subsequent songs discussed below exploit our model more heavily, using sampling four times for the rhythm model and two times for the melody model, selecting the majority, which corresponds to an explore/exploit parameter of 4 and 2, respectively. Heavier reliance on the model allowed us to use the rhythm suggestions without alterations and also raised the proportion of high quality scale degree sequences.

5.2 Everywhere

Everywhere is the first original song written with ALYSIA using a high setting of the exploit parameter. This made it even easier to discover new melodies. In fact, so many of ALYSIA's melodies were interesting that combining melodies became more time consuming (of course, one could simply consider a smaller set of melodies per line to overcome this challenge). Nevertheless, the writing of *Everywhere* from start to finish took only about two hours. Figure 6 shows the score co-created with ALYSIA. A recording appears on http://bit.ly/2eQHado.

5.3 I'm Always Chasing Rainbows

The final song made with ALYSIA so far relies on lyrics from the song *I'm Always Chasing Rainbows*. The original music is credited to Harry Carroll, although the melody was in fact adapted from Chopin's Fantaisie-Impromptu. Lyrics were written by Joseph McCarthy. The song's appearance in Broadway musicals and motion pictures contributed to its popularity. As it was published in 1917, *I'm Always Chasing Rainbows* is in public domain.

Due to its Vaudeville's style and classical roots, we thought it would be interesting to recreate *I'm Always Chasing Rainbows* as a pop song. ALYSIA's version is given in Fig. 7. A production of ALYSIA's pop version of this Vaudeville song has been posted on http://bit.ly/2eQHado, where you can also find the original score.

6 Discussion: Co-Creative and Autonomous Songwriting

Songwriting is the art of combining melodies and lyrics; It is not enough to have beautiful melodies and poetic lyrics, the music and words must fit together into a coherent whole. This makes Algorithmic Songwriting a distinct sub-field of Algorithmic Composition. ALYSIA is a machine-learning system that learns the relationship between melodies and lyrics, and uses the resulting model to create new songs in the style of the corpus.

The unique challenges of songwriting were observed during the creation of the first musical to rely on computational creativity systems, Beyond the Fence, which played in London in the early months of 2016 [3]. During the panel on Beyond the Fence held at the Seventh International Conference of Computational Creativity, it was noted that the juxtaposition of music and text posed a substantial challenge in the making of this musical. This further illustrates the need for systems that focus on uncovering the complex relationships between music and lyrics.

Algorithmic songwriting offers intriguing challenges as both an autonomous and a co-creative system. An autonomous songwriting system producing works on par with those of expert human songwriters would mark a significant achievement. Yet, we can go beyond the score. What if, in addition to writing the song, an automated system could also perform and record its own compositions? A truly independent system would not only create a score, but incorporate the full spectrum of expertise required for the creation of a complete song, including the vocal performance, expressive rendition, and automated music production.

As we aspire to create autonomous songwriters, artists and hobbyists alike are thirsty for our help. Even if we had access to a fully autonomous songwriter,

Fig. 6. Another original song created with ALYSIA, using a higher setting for the exploit parameter.

it would not replace the need for a corresponding co-creative system. Whereas an autonomous songwriter could be used when complete works are desired, a co-creative variation would satisfy the human need for music making - much like the difference between the joys of eating and cooking. A co-creative algorithmic songwriter would expand our creative repertoire, making songwriting accessible to those who cannot otherwise enjoy this art-form, and be of great use to semi-professional and amateur musicians (particularly singers) who do have the luxury of hiring a songwriter.

In the development of a co-creative songwriting system, it is desired that the users retain creative control, allowing them to claim ownership of the resulting works, or at least experience the process as an expression of creativity. The goal is to relieve the burden of having to master all of the diverse skills needed for the creation of a song, giving users the freedom to focus on aspects of the creative process in which they either specialize or find most enjoyable.

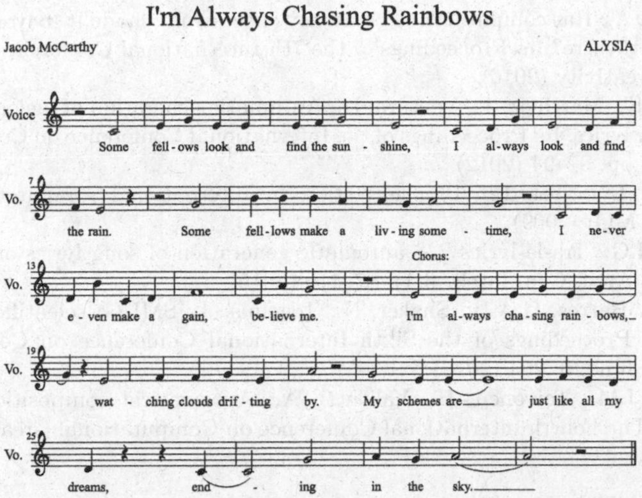

Fig. 7. Melody and rhythm by ALYSIA, lyrics by Joseph McCarthy from the popular Vaudeville song *I'm Always Chasing Rainbows*.

From a technical viewpoint, the goals of autonomous and co-creative algorithmic songwriters converge - the primary difference being that a co-creative system allows for more human interaction. ALYSIA is a long-term project in which we will simultaneously explore both of these paths as the system's technical foundation continues to expands.

Short-term goals for ALYSIA include the automation of chord progressions that would underlie the melody, as well as adding an evaluation phase that will allow ALYSIA to rank its own melodies. A couple of user studies are being planned where people at different levels of musical expertise will evaluate the songs made with ALYSIA. Another future direction is to integrate semantic-based features, which will allow the system to adjust melodies based on the meaning of the lyrics. To further our machine learning methodology, we are putting together a corpora of songs in Music XML format. For several different genres, a large corpus will be used to train a model using a neural network, which requires large data sets to avoid over-fitting. This will also let us explore differences in melodies resulting from different machine learning models. We are also experimenting with our system's potential for songwriting in different genres and languages, ranging from English pop songs discussed here to the creation of new arias in the style of classical Italian composer Giacomo Puccini.

References

1. Fernández, J.D., Vico, F.: Ai methods in algorithmic composition: a comprehensive survey. J. Artif. Intell. Res. **48**, 513–582 (2013)
2. Gauntlett, D.: Making is Connecting. Wiley, New York (2013)

3. Jordanous, A.: Has computational creativity successfully made it 'beyond the fence' in musical theatre? In: Proceedings of the 7th International Conference on Computational Creativity (2016)
4. Monteith, K., Martinez, T., Ventura, D.: Automatic generation of melodic accompaniments for lyrics. In: Proceedings of the International Conference on Computational Creativity, pp. 87–94 (2012)
5. Nichols, E.: Lyric-based rhythm suggestion. In: International Computer Music Conference (ICMC) (2009)
6. Oliveira, H.G.: Tra-la-lyrics 2.0: automatic generation of song lyrics on a semantic domain. J. Artif. Gen. Intell. **6**(1), 87–110 (2015)
7. Scirea, M., Barros, G.A.B., Shaker, N., Togelius, J.: SMUG: scientific music generator. In: Proceedings of the Sixth International Conference on Computational Creativity June, p. 204 (2015)
8. Toivanen, J.M., Toivonen, H., Valitutti, A.: Automatical composition of lyrical songs. In: The Fourth International Conference on Computational Creativity (2013)

On Symmetry, Aesthetics and Quantifying Symmetrical Complexity

Mohammad Majid al-Rifaie[1]([✉]), Anna Ursyn[2], Robert Zimmer[1],
and Mohammad Ali Javaheri Javid[1]

[1] Goldsmiths, University of London, New Cross, London SE14 6NW, UK
{m.majid,r.zimmer,m.javaheri}@gold.ac.uk
[2] University of Northern Colorado, Guggenheim 106A, Greeley, CO 80639, USA
anna.ursyn@unco.edu

Abstract. The concepts of order and complexity and their quantitative evaluation have been at the core of computational notion of aesthetics. One of the major challenges is conforming human intuitive perception and what we perceive as aesthetically pleasing with the output of a computational model. Informational theories of aesthetics have taken advantage of entropy in measuring order and complexity of stimuli in relation to their aesthetic value. However entropy fails to discriminate structurally different patterns in a 2D plane. In this work, following an overview on symmetry and its significance in the domain of aesthetics, a nature-inspired, swarm intelligence technique (Dispersive Flies Optimisation or DFO) is introduced and then adapted to detect symmetries and quantify symmetrical complexities in images. The 252 Jacobsen & Höfel's images used in this paper are created by researchers in the psychology and visual domain as part of an experimental study on human aesthetic perception. Some of the images are symmetrical and some are asymmetrical, all varying in terms of their aesthetics, which are ranked by humans. The results of the presented nature-inspired algorithm is then compared to what humans in the study aesthetically appreciated and ranked. Whilst the authors believe there is still a long way to have a strong correlation between a computational model of complexity and human appreciation, the results of the comparison are promising.

Keywords: Human aesthetic perception · Symmetry and complexity · Aesthetics · Swarm intelligence · Dispersive flies optimisation

1 Introduction

For decades, evolutionary computation enthusiasts and researches have been working on generating aesthetically pleasing images, which include Bimorphs of Dawkins [10], Mutator of Latham [39], and Virtual Creatures of Sims [37] and many more. Acknowledging the presence of some impressive outcome by researchers and digital artists, one of the remaining key questions is how to conduct the aesthetic selection. According to McCormack [25], the subjective

© Springer International Publishing AG 2017
J. Correia et al. (Eds.): EvoMUSART 2017, LNCS 10198, pp. 17–32, 2017.
DOI: 10.1007/978-3-319-55750-2_2

comparison process in the evolutionary process is slow and forms a bottleneck, even for a small number of phenotypes. Human users would take hours to evaluate many successive generations that in an automated system could be performed in a matter of seconds. Secondly, genotype-phenotype mappings are often not linear or uniform. That is, a minor change in genotype may produce a radical change in phenotype. Such non-uniformities are particularly common in tree or graph based genotype representations such as in Genetic Programming, where changes to nodes can have a radical effect on the resultant phenotype. In this study we approach the problem in the framework of dynamical systems and define a criterion for aesthetic selection in terms of its association with symmetry. The association of aesthetics and symmetry has been investigated from different points of view. This work is an extension of an earlier work [17], where the correlation between human aesthetic judgement and spatial complexity measure has been explored using information gain model.

In this paper, the authors present an overview of symmetry and its significance on aesthetics, giving examples of types of symmetries as well as providing an account on symmetric vs asymmetric analysis and their links to aesthetics and beauty. Then Dispersive Flies Optimisation[1] (DFO) [1], a swarm intelligence algorithm, is presented, followed by explanation on how the algorithm could be adapted to detect symmetries in images. This process is then further expanded to generate a quantitative figure representing the symmetrical complexity of an input image. The results of the swarm intelligence algorithm are then compared against a collection of images, which are ranked by humans for their aesthetics (the images are created by Jacobsen and Höfel, who have designed the dataset to study the human perception of aesthetics).

2   Symmetry and Aesthetics

The association of aesthetics and symmetry has been extensively investigated in the literature. A study to investigate the effect of symmetry on interface judgements, and the relationship between a higher symmetry value and aesthetic appeal for the basic imagery, showed that subjects preferred symmetric over non-symmetric images [4]. Further studies found that if symmetry is present in human face or body, the individual is judged as being relatively more attractive [14,33].

Symmetry plays a crucial role in theories of perception and is considered a fundamental structuring principle of cognition [23]. In the Gestalt school of psychology, things or objects are affected by where they are and by what surrounds them, with the aim of understanding the things or objects as more than the

---

[1] Despite the algorithm's simplicity, it is shown that DFO outperforms the standard versions of the well-known Particle Swarm Optimisation, Genetic Algorithm (GA) as well as Differential Evolution (DE) algorithms on an extended set of benchmarks over three performance measures of error, efficiency and reliability [1]. It is shown that DFO is more efficient in 84.62% and more reliable in 90% of the 28 standard optimisation benchmarks used; furthermore, when there exists a statistically significant difference, DFO converges to better solutions in 71.05% of problem set.

sum of their parts [5]. The Gestalt principles emphasise the holistic nature of perception where recognition is inferred (during visual perception) more by the properties of an image as a whole rather than its individual parts [18]. Thus, during the recognition process, elements in an image are grouped from parts to whole based on Gestalt principles of perception such as proximity, parallelism, closure, symmetry, and continuation [29]. In particular, symmetric objects are more readily perceived [8]. It is not surprising that humans find sensory delight in symmetry, given the world in which we evolved. In our world the animals that have interested us and our ancestors (as prey, menace, or mate) are overwhelmingly symmetric along at least one axis [32]. Evolutionary psychologists examine physical appearances such as symmetry as an indirect measure in mate selection [26, 27]. Additionally, symmetry is positively linked with both psychological and physiological health indicators [36]. In geometry, symmetrical shapes are produced by applying four operations of translations, rotations, reflections, and glide reflections.

Despite the above developing computational methods that generate symmetrical patterns is still a challenge; this is partially due to the difficulty of associating mathematics with the noisy, imperfect, real world, resulting in a small number of computational tools dealing with real-world symmetries [24].

Applying evolutionary algorithms to produce symmetrical forms leaves the formulation of fitness functions, which generate and select symmetrical phenotypes to be addressed. Lewis describes two strategies in evolutionary algorithms approach for generating and selecting symmetrical forms: "A common approach is to hope for properties like symmetry to gradually emerge by selecting for them. Another strategy is to build in symmetry functions which sometimes activate, appearing suddenly. However this leads to a lack of control, as offspring resulting from slight mutations (i.e. small steps in the solution space) bear little resemblance to their ancestors [22]".

2.1 Symmetry Examples

Symmetry exists not only in geometry but also in natural world and human works. For example, water, when in liquid state, has bilateral symmetry, with the symmetric stretch of the two O-H bonds and some molecular vibrations [19, 34]. When frozen, water becomes symmetrical in various ways (however, not always), usually developing the hexagonal crystals. Many molecules such as carbon dioxide, benzene, or carbontetrachloride are perfectly symmetrical. Ice, snowflakes, feather ice on the twigs, hail, sleet, icicles, glaciers, and polar caps, all develop their own order of symmetry and various arrangements of symmetry axes. Crystals such as table salt or copper contain not many kinds of atoms and have a simple structure. Proteins, for example those building teeth, usually have complex crystalline structure and many kinds of molecules. Average local molecular orientations in liquid crystals (fluids made of spontaneously aligning rod-like molecules) are described with a head-tail symmetry [35]. We cannot ascribe the crystalline structure to inanimate forms only. Crystalline material has been separated from the tobacco-mosaic virus protein in 1933 [38].

Some viruses can organize themselves into liquid crystals. Thus, for several reasons, "current knowledge about nanostructures makes difficult defining the distinction between organic and inorganic, living and inanimate, natural and artificial, or human and machine" [9].

2.2 Types of Symmetry

There are several types of symmetry, for example, the line or mirror symmetry, the radial, cylindrical, or spherical symmetry. Mirror symmetry of a symmetrical object is often defined as the correspondence in size, form, and arrangement of similar parts on the opposite sides of a point, line (axis), or plane.

Radial symmetry in an object occurs when it can be rotated around an imaginary line called the rotation axis and retain the same appearance as before rotating, repeating itself several times during a complete rotation. A centre of symmetry is equally distant from any point on the surface of a symmetrical object. For example, if a crystal has a centre of symmetry, then, when laid down on its face on a tabletop, it has at the top an inverted horizontal face of equal size and shape.

Several types of geometrical symmetry include bilateral (reflection or mirror), rotational (when an object looks the same after rotation), cylindrical, spherical, and helical symmetry (like in a drill bit), and also translational (where a particular translation - moving in a specified direction does not change the object), glide reflection (in a line or plane combined with a translation), or rotoreflection symmetry (which presents rotation about an axis, combined with reflection in a plane perpendicular to that axis).

Fractals, patterns, and symmetry exist in nature and in art projects. Fractals represent form of scale symmetry that appears when the objects magnified or reduced in size have the same properties. Fractal related concepts and processes are examined in selected fields of physics, biology, or computing, and to studies on astronomy. At the same time they pertain to our Planet's life and our own everyday experience. Mathematicians name some objects symmetrical with respect to a given mathematical operation applied to this object, when this operation preserves some property of the object. Such operations form a symmetry group of the object.

Claude Lévi-Strauss [21] recalled the opinion of a German philosopher Immanuel Kant (1724–1804) about an aesthetic judgement. Kant wrote about judgement of knowledge (conceptual judgement) and aesthetic judgement (nonconceptual judgement). In an aesthetic theory developed by Kant, judgements about beauty rest on feeling but they should be validated in harmony with mental structure, so they are not merely statements of taste or opinion. According to Kant, there are judgements of taste that are subjective and judgements of reason that are universally valid. Aesthetic judgement falls somewhere between these two kinds. Lévi-Strauss stated that, in this intermediary space, fractals are given the status of a work of art, because they are appealing and at the same time, objectively governed by reason.

2.3   Symmetry and Asymmetry Analysis

Scientists and artists see a purpose in symmetry investigations, for example, mathematicians, anthropologists, artists, designers, architects who conduct computer analysis of the facades, friezes, and some architectural details, as well as researchers in many fields of natural sciences, medicine, pharmacology, biology, geology, or chemistry. Many artists have created masterpieces this way. Artists used to transform patterns and repetitions to apply the unity or symmetry in their compositions (for example, by examining a Fibonacci sequence, prime numbers and magic squares, a golden section, or tessellation techniques).

Genetic algorithms and other evolutionary computing techniques are applied not only to the artistic areas but also many industries, including aeronautic and automotive design, electronic circuit design, routing optimization, modelling markets for investment, among other domains. Analysis of generative art systems may reveal an analogy with the natural systems; both systems maintain balance between order and disorder. While biological life takes on most of its forms in a spectrum between unstructured atmospheric gases and ordered crystals and minerals, generative art systems and artificial life (A-life) are placed somewhere in the middle of a continuum between disordered randomization, chaotic systems, and fractals or L-systems, and highly ordered forms such as symmetry and tiling [12, 13]. With generative approach, artists draw from natural phenomena observed in biology and physics, and their creative process may evolve into a sequence of iterative solutions and modifications transforming the artwork.

2.4   Aesthetics and Beauty

Natural objects displaying symmetry evoke wonder and surprise because their intricacy. For example, architecture and architectural details, such as stain windows, mosaics, and friezes, visual arts, pottery and ceramics, quilts, textiles, and carpets make a varied use of symmetry as an important principle in their design. Ferreira [11] examined architecture and cell biology in terms of biosemiotics, with architectural structures discussed as context-dependent semiotic objects with functional and/or aesthetic values. Both the natural and man-made environment can be perceived as locus, place, site, or a part of a mental map of a cultural framework. Maybe for that reason symmetry is so often seen not only beautiful but also conducive to visual communication. Possibly, artwork resulting from coding is able to convey the correctness of natural lines, symmetries, patterns, textures, light, and color, thus being something more than identification of natural objects.

In words of Andres Gaviria [15] (p.481), "aesthetics is concerned with the theory of sensual perception, while art is a social practice involved in certain forms of research and investigation processes and in the construction of particular types of artefacts." The field of aesthetics involves studies in the arts, philosophy of art, and our judgements about art works' qualities. Beautiful, harmonious, or emotionally pleasing objects have been traditionally considered aesthetic and thus valuable. The postmodern philosopher Jean-Francois Lyotard (1924–1998)

posed they would be sublime (Lyotard, Lessons on the Analytic of the Sublime,
1994). Then the criteria of beauty broadened, and judgements of aesthetic values
examined also social, political, moral, and many other aspects of the art objects.
Modern analytic approach in aesthetics is no longer restricted to an analysis of
natural beauty because, in opinion of cubists, dadaists, constructivists, conceptual artists, generative artists, and many others, beauty ceased to be central to
the definition of art. Computing science specialists examine aesthetics of electronic projects' usability, efficiency, and discuss aesthetics in terms of possible
applications to controlling computer products. Evaluation of tag clouds in terms
of the aesthetic quality obtained from an extensive user study confirms that
aesthetic values correlate with product usability.

The foundation for the digital exploitation would include the mathematical measure of aesthetics proposed by a mathematician George David Birkhoff
(2003/1933); the golden ratio (where the ratio of the smaller to the larger sub-
segment is the same as the ratio of the larger sub-segment to the whole segment);
the Zipf's law (studying the relation between the rank and frequency of natural
language utterances); fractal dimension (providing a ratio of the change in detail
to the change in the scale); basic gestalt design principles (telling about the
principles of figure-ground articulation, proximity, similarity, closure, symmetry,
continuity, past experience, common fate, and the good gestalt of perceptual
scenes), and the rule of thirds (advising that an image should be imaginary
divided into nine equal parts by two horizontal and two vertical lines, and that
important compositional elements should be placed along these lines or their
intersections). Computational aesthetic evaluation of evolutionary art systems
may refer to the empirical studies and psychological modelling of aesthetics and
neuroaesthetics.

Semir Zeki explored the domain of neuroaesthetics, a cross-disciplinary
research field related to the neural basis of artistic creativity and achievement.
Zeki examined brain activity associated with the perception of images. He arrived
at conclusion, the visual brain functions in search for dependable qualities to
obtain knowledge about the external world. Our survival in the world may thus
strongly depend on the accuracy and completeness of mental models used by
our mind to represent real life [28,41].

Birkhoff (1884–1944) proposed in his book entitled "Aesthetic Measure" a
mathematical theory of aesthetics intended for artistic purpose: in the equation
M=O/C, Aesthetic Measure (M) is a function of Order (O) divided by Complexity (C.) (see [7] and Mathematical poetry, 2015[2]). The Gestalt psychology theory
of mind postulated that brain has self-organizing tendencies and recognizes the
whole of a figure rather than its individual parts [7].

Some scientists use intuition as a guide in developing hypotheses, since laws
are reflection of symmetries, and there is a connection between beauty and
symmetry.

The theme of symmetry can certainly be considered inspirational to create biologically inspired art, because symmetrical forms and shapes possess an

---

[2] http://mathematicalpoetry.blogspot.com/2006/09/equation-for-aesthetic-measure-by.html.

aesthetic beauty and an order reflected by their geometry. We can appreciate these forms finding the importance of adaptations that animals develop as an answer to the conditions of life, examining mathematical order in natural forms, and re-creating it in our own artwork.

2.5 Graph and Visualisation Aesthetics

The optimal layout aesthetics has been investigated in the field of graph drawing and the aesthetics of graph drawing algorithms, to understand the effect of minimizing edge bends, minimising edge crossings, and improving symmetry. Node and edge shape, size, texture, and color are variables that could play a significant role in improving graph aesthetics [40]. Purchase [30] argued that Gestalt principles and neurophysiology can help explain which aesthetics might be important and why. The results obtained suggested that reducing the number of edge crossings, aligning nodes and edges to an underlying grid, and making the best use of symmetry, e.g. by maximizing symmetry of subgraphs were significant for graph aesthetics [20,31]. The overall layout and the spatial relationship between nodes and edges, including graph's symmetry, area, flow, and aspect ratio, determine the aesthetics of a graph [6].

The next section explains the swarm intelligence algorithm which will be used in detecting symmetries.

3 Dispersive Flies Optimisation

Dispersive Flies Optimisation (DFO) is an algorithm inspired by the swarming behaviour of flies hovering over food sources. DFO, which is a recently proposed algorithm, is one of the simplest yet robust continuous optimisation techniques with only two tunable parameters which makes it an-easy-to-implement yet strong optimiser. This algorithm has been applied to various fields including optimisation, medical imaging and digital art [2,3].

To describe this algorithm, as detailed in [1], the swarming behaviour of flies in DFO is determined by several factors and that the presence of threat could disturb their convergence on the marker (or the optimum value). Therefore, having considered the formation of the swarms over the marker, the breaking or weakening of the swarms is also noted in the proposed algorithm.

In other words, the swarming behaviour of the flies, in Dispersive Flies Optimisation, consist of two tightly connected mechanisms, one is the formation of the swarms and the other is its breaking or weakening. The algorithm and the mathematical formulation of the update equations are introduced below.

The position vectors of the population are defined as:

$$x_i^t = \left[x_{i1}^t, x_{i2}^t, ..., x_{iD}^t \right], \qquad i = 1, 2, ..., NP \tag{1}$$

where t is the current time step, D is the dimension of the problem space and NP is the number of flies (population size).

In the first generation, when $t = 0$, the ith vector's j^{th} component is initialised as:

$$x_{id}^0 = x_{min,d} + r\,(x_{max,d} - x_{min,d}) \tag{2}$$

where r is a random number drawn from a uniform distribution on the unit interval $U\,(0,1)$; x_{min} and x_{max} are the lower and upper initialisation bounds of the dth dimension, respectively. Therefore, a population of flies are randomly initialised with a position for each flies in the search space.

On each iteration, the components of the position vectors are independently updated, taking into account the component's value, the corresponding value of the best neighbouring fly (consider ring topology) with the best fitness, and the value of the best fly in the whole swarm:

$$x_{id}^t = x_{nb,d}^{t-1} + U\,(0,1) \times (x_{sb,d}^{t-1} - x_{id}^{t-1}) \tag{3}$$

where $x_{nb,d}^{t-1}$ is the value of the *neighbour's best* fly in the dth dimension at time step $t-1$; $x_{sb,d}^{t-1}$ is the value of the *swarm's best* fly in the dth dimension at time step $t-1$; and $U\,(0,1)$ is the uniform distribution between 0 and 1.

The algorithm is characterised by two principle components: a dynamic rule for updating the flies positions (assisted by a social neighbouring network that informs this update), and communication of the results of the best found fly to other flies.

As stated earlier, the swarm is disturbed for various reasons; one of the positive impacts of such disturbances is the displacement of the disturbed flies which may lead to discovering a better position. To consider this eventuality, an element of stochasticity is introduced to the update process. Based on this, individual components of flies' position vectors are reset if the random number, r, generated from a uniform distribution on the unit interval $U\,(0,1)$ is less than the *disturbance threshold* or *dt*. This guarantees a proportionate disturbance to the otherwise permanent stagnation over a likely local minima. Algorithm 1 summarises the DFO algorithm[3]. This algorithm has been applied to several others domains including medical imaging [2].

The next section details how DFO is instructed to detect symmetries in Jacobsen and Höfel [16] stimuli.

4   Symmetry Detection and Quantifying Symmetrical Complexity with DFO

This section details the process through which DFO is adapted for the purpose of symmetry detection. The algorithm is designed to detect symmetries around any focal points, these symmetries could be full or partially existing in part of the input image. The search space of the algorithm is 2D and consisting of

---

[3] The source code of DFO algorithm can be downloaded from the following web page: http://doc.gold.ac.uk/mohammad/DFO/.

Algorithm 1. Dispersive Flies Optimisation

1: **while** FE < function evaluations (FEs) allowed **do**
2:     **for** $i = 1 \rightarrow NP$ **do**
3:         $\boldsymbol{x}_i$.fitness $\leftarrow f(\boldsymbol{x}_i)$
4:     **end for**
5:     $sb \leftarrow \{sb, \forall\, f(\boldsymbol{x}_{sb}) = \min\left(f(\boldsymbol{x}_1), f(\boldsymbol{x}_2), ..., f(\boldsymbol{x}_{NP})\right)\}$
6:     $nb \leftarrow \{nb, \forall\, f(\boldsymbol{x}_{nb}) = \min\left(f(\boldsymbol{x}_{\text{left}}), f(\boldsymbol{x}_{\text{right}})\right)\}$
7:     **for** $i = 1 \rightarrow NP$ **do**
8:         **for** $d = 1 \rightarrow$ D **do**
9:             $\tau_d \leftarrow x_{nb,d}^{t-1} + \mathrm{U}\left(0,1\right) \times (x_{sb,d}^{t-1} - x_{id}^{t-1})$
10:             **if** $(r < dt)$ **then**
11:                 $\tau_d \leftarrow x_{min,d} + r\left(x_{max,d} - x_{min,d}\right)$
12:             **end if**
13:         **end for**
14:         $\boldsymbol{x}_i \leftarrow \tau$
15:     **end for**
16: **end while**

---

all the (x, y) coordinates of the pixels in the input image; as such the vector representing each fly is set to be its (x, y) coordinate.

The fitness of the each fly is determined by a random d_x and d_y, pointing to two equally distanced *areas* in the search space (see Fig. 1). In other words, the fly position is tested against being the point of symmetry in the image without having to evaluate all the possibilities. The rationale behind using a certain area (vs the whole image) lies in reducing the computational expense of running the algorithm; therefore, instead of comparing each pixel against its corresponding pixel, the symmetry of the point is only partially evaluated (by comparing to areas equally spaced from the position of the fly and calculating the difference between the two areas[4]).

In this work, the population size of the flies is empirically set to 100 (approximately quarter of the image side), the disturbance threshold is set to 0.01, the area of comparison is set to 10px (approximately checking $\frac{1}{500}$ of the search space in each evaluation) and 300 iterations are allowed. The size of original images are 453×453.

4.1   Experiments and Results

All the 252 experimental stimuli were adapted from an empirical study of human aesthetic judgement [16]. Jacobsen and Höfel [16] report on an empirical trial of human aesthetic judgement. Fifty-five young adults (15 males) participated in the experiment for course credit or partial fulfilment of course requirements. All were first or second year psychology students at the University of Leipzig.

---

[4] In this research, for simplicity, the difference of the sums of the two areas are considered. Other more sensitive or computationally expensive measures, such as the histogram of oriented gradients (HOG), could alternatively be used.

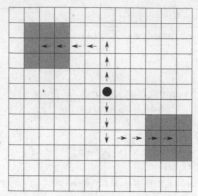

The fly's position is shown as a circle,
with $d_x = 4$ and $d_y = 3$; radius is set to 1.

Fig. 1. Partial symmetry evaluation

None of them had received professional training in the fine arts or participated in a similar experiment before. Participants reported normal or corrected-to-normal visual acuity. Subjects were asked to evaluate images from two groups; a group of asymmetrical images and a group of symmetrical images with at least one axis of reflection symmetry. The images consisted of a solid black circle showing a centred, quadratic, rhombic cut-out.

The black circular background was replaced by a square in order to reduce aliasing errors. The images in Fig. 2 show some of the images from Jacobsen and Höfel's research. The images are ordered by their mean aesthetic ratings (from the most aesthetically pleasing to the least).

Each image is fed into the DFO symmetry detection algorithm and the x, y coordinates of the best fly is logged, therefore each image will result in 300 coordinates showing where the the best fly in each iteration has mostly been located. This will clearly highlight the strongest point of symmetry in the image. In order to visualise the process, a few of the processed images are shown in Fig. 3. The best fly in each iteration leaves a mark in red (with transparency), therefore the areas will the strongest presence of red circle indicates the presence of symmetry.

In order to quantitatively calculate the performance of DFO over the input images, the collection of all best fly's positions are taken into account[5] and the *interquartile range* is calculated, highlighting the sparsity of the swarm concentration over the image well-depending on the nature of the patterns. In other words, the communications between the members of the swarm, result in identifying the best member. The communications between the members of the swarm are repeated in every iterations, and the best member of each iteration is identified and stored. It is the aggregation of the entire best members (throughout

---

[5] In other words, the (x, y) coordinate of the best fly in each iteration is logged, giving rise to 300 coordinates (the number of iterations allowed) for evaluating each image.

Fig. 2. Some randomly picked stimuli from Jacobsen and Höfel's dataset ordered in rows from the most beautiful pattern to the least beautiful one, starting in the upper left corner

the entire iterations) that forms the results. This gives every individual iteration a stronger say in dictating the overall outcome, representing every image in the dataset. Figure 4 shows the performance of the algorithm in one trial over all the images.

In order to verify the consistency of the results, the experiment is run 10 times for each of the 252 images and the results are shown in Fig. 5. In Jacobsen and Höfel's image dataset, the first image has the highest rank by the human observes, with the mean rating of 74.73 and the last has the lowest ranking with the mean rating of 28.58. The fitted line in the rating of these images are shown in Fig. 6.

As the results and the graphs illustrate, although there might be a long distance in having a close correlation between the output of a computational model and the human judgement of aesthetics, this paper aims to highlight the

Fig. 3. Images processed by DFO algorithm in one trial

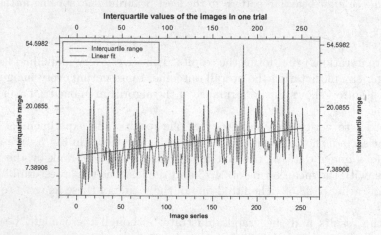

Fig. 4. Measuring symmetrical complexities in stimuli using interquartile range in one trial, running DFO over all the input images

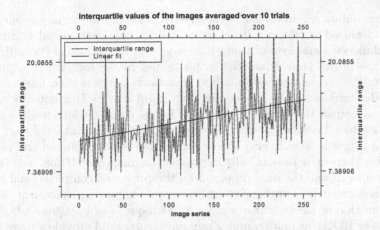

Fig. 5. Measuring symmetrical complexities in images using interquartile range in 10 trials with slope $= 0.021$, intercept $= 8.99$, $\chi = 8,032.65$, $R^2 = 0.069$

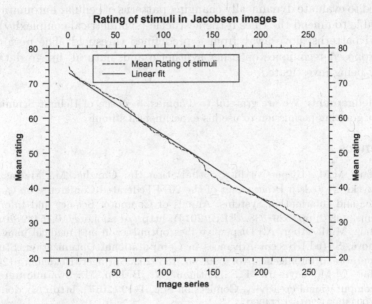

Fig. 6. Jacobsen and Höfel's image ratings: from the most beautiful (image #1) to the least beautiful (image #252)

result of using a simple swarm intelligence technique, which is sensitive to both global symmetry as well as the localise ones in the patterns.

5 Conclusion

Quantitative evaluation of order and complexity has always been the heart of computational aesthetics. In this work, in addition to presenting an overview on symmetry and its role in measuring aesthetics, a simple swarm intelligence technique is deployed to offer a method for measuring local and global symmetries. This techniques focuses on utilising the essence of the collective (vs individual) intelligence in the population; this is facilitated by aggregating the "core" of each iteration (i.e. each iteration's best member), which is then used in determining the outcome of the process for each input image. This outcome is then contrasted against the Jacobsen and Höfel image dataset which itself is ranked by humans based on their aesthetic judgements. It is demonstrated that whilst there is a long way from having a close correlation between human and machine aesthetics, there is a present, albeit not very strong, correlation between the human rankings and the ones proposed by the presented computational model.

Further research is ongoing to utilise more complex measurements for the fitness function of the swarm intelligence technique (e.g. histogram of oriented gradients or HOG), as more complex measurements could provide a more robust comparison between the areas. In addition to applying the proposed technique to the computational models of aesthetic evaluation, the result of this study could be applied to evaluate dynamically changing patterns of cellular automata and as such be able to amend the rules if the aesthetic (symmetrical complexity) of the generated patterns move away from a predefined threshold. The performance of the proposed techniques on other synthetic and 'natural' image datasets is currently being investigated.

Acknowledgement. We are grateful to Thomas Jacobsen of Helmut Schmidt University for granting permission to use his experimental stimuli.

References

1. al-Rifaie, M.M.: Dispersive flies optimisation. In: Ganzha, M., Maciaszek, L., Paprzycki, M. (eds.): Proceedings of the 2014 Federated Conference on Computer Science and Information Systems. Annals of Computer Science and Information Systems, vol. 2, pp. 529–538. IEEE (2014). http://dx.doi.org/10.15439/2014F142
2. al-Rifaie, M.M., Aber, A.: Dispersive flies optimisation and medical imaging. In: Fidanova, S. (ed.) Recent Advances in Computational Optimization. Studies in Computational Intelligence, vol. 610, pp. 183–203. Springer, Heidelberg (2016)
3. al-Rifaie, M.M., Leymarie, F.F., Latham, W., Bishop, M.: Swarmic autopoiesis and computational creativity. Connection Sci., 1–19 (2017). http://dx.doi.org/10.1080/09540091.2016.1274960
4. Bauerly, M., Liu, Y.: Computational modeling and experimental investigation of effects of compositional elements on interface and design aesthetics. Int. J. Man Mach. Stud. **64**(8), 670–682 (2006)
5. Behrens, R.: Design in the Visual Arts. Prentice-Hall, Upper Saddle River (1984)
6. Bennett, C., Ryall, J., Spalteholz, L., Gooch, A.: The aesthetics of graph visualization. Comput. Aesthetics **2007**, 57–64 (2007)
7. Birkhoff, G.D.: Aesthetic Measure. Harvard University Press, Cambridge (1933)
8. Carroll, J.M. (ed.): HCI Models, Theories, and Frameworks Toward a Multidisciplinary Science. Morgan Kaufmann Publishers, San Francisco (2003)
9. Cheetham, M.A.: The crystal interface in contemporary art: metaphors of the organic and inorganic. Leonardo **43**(3), 250–255 (2010)
10. Dawkins, R.: The Blind Watchmaker. Norton & Company, Inc., New York (1986)

11. Ferreira, M.I.A.: Interactive bodies: The semiosis of architectural forms. Biosemiotics **5**(2), 269–289 (2012)
12. Galanter, P.: Complexity, neuroaesthetics, and computational aesthetic evaluation. In: 13th International Conference on Generative Art (GA2010) (2010)
13. Galanter, P.: Computational aesthetic evaluation: past and future. In: McCormack, J., d'Inverno, M. (eds.) Computers and Creativity, pp. 255–293. Springer, Heidelberg (2012)
14. Gangestad, S.W., Thornhill, R., Yeo, R.A.: Facial attractiveness, developmental stability, and fluctuating asymmetry. Ethol. Sociobiol. **15**(2), 73–85 (1994)
15. Gaviria, A.R.: When is information visualization art? determining the critical criteria. Leonardo **41**(5), 479–482 (2008)
16. Jacobsen, T., Hofel, L.: Aesthetic judgments of novel graphic patterns: analyses of individual judgments. Percept. Mot. Skills **95**(3), 755–766 (2002)
17. Javid, M.A.J., Blackwell, T., Zimmer, R., al-Rifaie, M.M.: Correlation between human aesthetic judgement and spatial complexity measure. In: Johnson, C., Ciesielski, V., Correia, J., Machado, P. (eds.) EvoMUSART 2016. LNCS, vol. 9596, pp. 79–91. Springer, Cham (2016). doi:10.1007/978-3-319-31008-4_6
18. Jiang, H., Ngo, C.W., Tan, H.K.: Gestalt-based feature similarity measure in trademark database. Pattern Recogn. **39**(5), 988–1001 (2006)
19. Kettle, S.F.: Symmetry and Structure: Readable Group Theory for Chemists. Wiley, New York (2008)
20. Lau, A., Moere, A.V.: Towards a model of information aesthetics in information visualization. In: 2007 11th International Conference on Information Visualization, IV 2007, pp. 87–92. IEEE (2007)
21. Lévi-Strauss, C.: Look, Listen, Learn. Translated by Singer, B.C.J.: Basic Books, a division of Harper-Collins Publishers (1997). ISBN 0465068804
22. Lewis, M.: Evolutionary visual art and design. In: Romero, J., Machado, P. (eds.) The Art of Artificial Evolution. Natural Computing Series, pp. 3–37. Springer, Heidelberg (2008)
23. Leyton, M.: Symmetry, Causality, Mind. Bradford Books/MIT Press, Cambridge (1992)
24. Liu, Y.: Computational Symmetry. CMU Robotics Institute (2000)
25. McCormack, J.: Interactive evolution of L-system grammars for computer graphics modelling. In: Complex Systems: From Biology to Computation, pp. 118–130 (1993)
26. Møller, A.P., Cuervo, J.J.: Asymmetry, size and sexual selection: meta-analysis, publication bias and factors affecting variation in relationships, p. 1. Oxford University Press (1999)
27. Møller, A.P., Thornhill, R.: Bilateral symmetry and sexual selection a meta-analysis. Am. Nat. **151**(2), 174–192 (1998)
28. Onians, J.: Neuroarthistory: From Aristotle and Pliny to Baxandall and Zeki. Yale University Press, New Haven (2007)
29. Park, I.K., Lee, K.M., Lee, S.U.: Perceptual grouping of line features in 3-D space: A model-based framework. Pattern Recogn. **37**(1), 145–159 (2004)
30. Purchase, H.C.: Metrics for graph drawing aesthetics. J. Vis. Lang. Comput. **13**(5), 501–516 (2002)
31. Purchase, H.C., Plimmer, B., Baker, R., Pilcher, C.: Graph drawing aesthetics in user-sketched graph layouts. In: Proceedings of the Eleventh Australasian Conference on User Interface, vol. 106, pp. 80–88. Australian Computer Society, Inc. (2010)

32. Railton, P.: Aesthetic value, moral value and the ambitions of naturalism. In: Aesthetics and Ethics, chap. 3. University of Maryland (2001)
33. Randy, T., Steven, G.: Human facial beauty. Hum. Nat. **4**, 237–269 (1993)
34. Rankin, D.W., Mitzel, N., Morrison, C.: Structural Methods in Molecular Inorganic Chemistry. Wiley, New York (2013)
35. Senyuk, B., Liu, Q., He, S., Kamien, R.D., Kusner, R.B., Lubensky, T.C., Smalyukh, I.I.: Topological colloids. Nature **493**(7431), 200–205 (2013)
36. Shackelford, T.K., Larsen, R.J.: Facial symmetry as an indicator of psychological emotional and physiological distress. J. Pers. Soc. Psychol. **72**(2), 456–466 (1997)
37. Sims, K.: Evolving virtual creatures. In: Proceedings of the 21st Annual Conference on Computer Graphics and Interactive Techniques, pp. 15–22. ACM (1994)
38. Stanley, W.: Crystalline tobacco-mosaic virus protein. Am. J. Bot. **24**(2), 59–68 (1937)
39. Todd, S., Latham, W., Hughes, P.: Computer sculpture design and animation. J. Visual. Comput. Anim. **2**(3), 98–105 (1991)
40. Ware, C., Purchase, H., Colpoys, L., McGill, M.: Cognitive measurements of graph aesthetics. Inf. Visual. **1**(2), 103–110 (2002)
41. Zeki, S., Nash, J.: Inner Vision: An Exploration of Art and the Brain, vol. 415. Oxford University Press, Oxford (1999)

Towards Polyphony Reconstruction Using Multidimensional Multiple Sequence Alignment

Dimitrios Bountouridis[1]($\bowtie$), Frans Wiering[1], Dan Brown[2],
and Remco C. Veltkamp[1]

[1] Department of Information and Computing Sciences,
Utrecht University, Utrecht, Netherlands
d.bountouridis@uu.nl
[2] David R. Cheriton School of Computer Science,
University of Waterloo, Waterloo, Canada

Abstract. The digitization of printed music scores through the process of optical music recognition is imperfect. In polyphonic scores, with two or more simultaneous voices, errors of duration or position can lead to badly aligned and inharmonious digital transcriptions. We adapt biological sequence analysis tools as a post-processing step to correct the alignment of voices. Our multiple sequence alignment approach works on multiple musical dimensions and we investigate the contribution of each dimension to the correct alignment. Structural information, such musical phrase boundaries, is of major importance; therefore, we propose the use of the popular bioinformatics aligner MAFFT which can incorporate such information while being robust to temporal noise. Our experiments show that a harmony-aware MAFFT outperforms sophisticated, multidimensional alignment approaches and can achieve near-perfect polyphony reconstruction.

1 Introduction

Optical music recognition (OMR), has been one of the earliest applications of optical character recognition dating back to the late 1960s. The goal is to parse a printed music sheet, typically through scanning, and convert its elements (e.g. notes, clefs) to a digital format. From there, one can visualize, process or play back the digitized score. The OMR process typically comprises image recognition and machine learning components; however, despite the technology advancements, OMR has been imperfect with frequent pitch and temporal errors.

Recognition errors arise due to low quality printing, ambiguous music notation and written music's higher complexity than traditional text e.g. merged staff symbols. Temporal errors are those arising from predicting the wrong position (onset) or duration of a note. For example, incorrectly recognizing an eighth note as a quarter will result in all the following notes being shifted by an eighth. Multiple errors of this type tend to accumulate. The problem becomes more pronounced when dealing with polyphonic scores, since temporal note shifts can

© Springer International Publishing AG 2017
J. Correia et al. (Eds.): EvoMUSART 2017, LNCS 10198, pp. 33–48, 2017.
DOI: 10.1007/978-3-319-55750-2_3

result to an alignment of voices that besides being incorrect, can sound *inharmonious* as well. Interestingly, incorrect alignment of voices can occur even if the OMR is perfect, i.e. in printed musical sources from the 16th–18th centuries. The music from that period is generally typeset using a font consisting of a limited set of musical symbols. Because of this property, it is possible to attain recall rates between 85% and 100% on good-quality prints [19]. However, there are several complicating factors, two of which are particularly important for this research. One is that music was generally not printed in score format but in separate voices, each in its own "partbook". The other is that barlines were only introduced around 1600, so that an important mechanism for coordination of voices that might counterbalance rhythmical errors in the OMR, is missing.

Polyphony reconstruction can be defined as the task of restoring a polyphonic piece to its original temporal formation-arrangement. In the same manner, *polyphony construction* can be defined as the prediction of the temporal formation of a set of unaligned voices. Understanding the temporal aspect of polyphony, a requirement for both tasks, can find application to automatic music generation and musicological analysis as well. Surprisingly, there is a lack of related research, which can be partially attributed to the high complexity of the problem: cognitively, aligning music notes with each other to form "meaningful" polyphonic pieces, is a process involving many musical dimensions such as harmony, durations and structure (e.g. segments, phrases, repetitions).

Interestingly, the synchronization of music voices has many parallels to the well studied multiple sequence alignment problem in bioinformatics [28]. In biological sequence analysis, measuring the similarity of more than two sequences is performed by examining possible multiple sequence alignments (MSA) in order to find an optimum, given a "meaningful" distance measure [5]. Traditionally, MSA algorithms are applied on unidimensional sequences; however, similar to music, some biological sequences have multiple dimensions and consequently MSA approaches that can deal with such information have been investigated through the years; for example, MSA methods that are aided by proteins' secondary structure (an abstraction of the three-dimensional form of local segments) [18].

We adapt tools used in biological sequence analysis towards understanding the temporal aspect of polyphony and towards solving polyphony reconstruction in particular. We employ a multidimensional MSA approach that allows us to identify the contribution of each musical dimension to the correct reconstruction. Our first round of experiments shows that, besides harmonic relations, structural information is a crucial for the task. However, structural information is rarely available, and since most algorithms for automatic structure analysis rely on temporal information (e.g. onset positions, durations), their predictions can be highly unreliable when temporal corruption is present (e.g. due to OMR errors). To accommodate for such corruption, we propose the use of the MSA algorithm MAFFT (Multiple Alignment using Fast Fourier Transform) [14], which can guide the alignment by structural segments computed from non-temporal information, such as pitch. We show that a harmony-aware MAFFT can almost perfectly reconstruct a artificial dataset of temporally-corrupted polyphonic pieces; a fundamental step towards real-life applications.

The remainder is organised as follows. Section 2 presents a brief overview of the related literature. Section 3 defines the problem of polyphony reconstruction and explains the simplifications we make in order to reduce its inherent complexity. Section 4 investigates the importance of musical dimensions to polyphony reconstruction. Section 5 introduces MAFFT and proposes its use for the task. Discussion and conclusions are presented in Sect. 6.

2   Related Work

OMR is an area of active scientific research and as consequence, various solutions have been proposed through the years. However, the task is merely secondary to the scope of this paper. The reader is referred to [20] which provides a complete overview of the algorithms and related literature regarding OMR. We note that only few of the proposed systems address historical forms of music notation, such as mensural notation from the Renaissance or lute tablature. The best results on Renaissance polyphony (the notation that inspired this study) is attained with Aruspix[1].

Few relevant works are tangentially related to polyphony reconstruction. For example, Boulanger-Lewandowski *et al.* [2] use recurrent neural networks (RNN) to model temporal dependencies between polyphonic voices for the purposes of music generation and music transcription. Similarly, Lyu *et al.* [17] propose the fusion of a Long Short-Term Memory RNN and restricted Boltzmann machines (RBM) for the purpose of music generation. However, none of the approaches provide any insights regarding the cognitive process of polyphony construction.

A number of researches outside bioinformatics have adopted multidimensional multiple sequence alignment (MDMSA) through the years. For example, Joh *et al.* [13] use MDMSA to compute the similarity between activity patterns. Sanguansat [22] uses a multidimensional version of the Dynamic Time Warping (DTW) algorithm for the task of query-by-humming. Closer to our work, van Kranenburg [25] uses a multidimensional extension of the DTW-based pairwise alignment. In his work, the scoring function incorporates heuristics to accommodate for more dimensions and is applied on melody classification.

3   Problem Definition and Polyphony Representation

Monophonic melodies can be considered as sequences of three-dimensional objects known as notes. Each note n_i can be represented by its pitch, duration and onset components: $n_i = (p_i, d_i, o_i)$. Perceptually, these dimensions never appear in isolation but constantly interact with each other. For example, onsets and durations allow us to perceive the *rhythm* dimension. Pitch patterns and rhythm create melodic segments (e.g. phrases). In polyphonic pieces, where multiple melodic sequences (voices) sound in parallel, the dimension interaction is higher. For example, the polyphonic temporal organization of notes creates more

---

[1] www.aruspix.net.

complex rhythm patterns. In addition, notes of different pitches from different voices happening in similar onset times allow us to perceive *harmony*.

Pitch errors aside, errors in the duration and onset dimensions, such as those due to incorrect OMR, lead to both incorrect harmony and rhythm. The original piece can be temporally reconstructed, as soon as both components are corrected from errors. In this paper we are solely interested in reconstructing the original harmony as a first step towards the complete polyphony reconstruction.

The representation of a polyphonic piece as a tractable harmony structure is of major importance for the task. The sequential nature of certain music documents (e.g. melodies, chord progressions) has allowed for their representation as sequences of symbols in a wide range of pattern recognition and Music Information Retrieval tasks (MIR) tasks. It is therefore logical to apply this successful scheme to each voice in a polyphony. Our interest in harmony reconstruction solely, allows us to use a representation that does not consider the duration and onset dimensions. Each voice is represented as a sequence of pitch values folded into one octave and mapped into an alphabet.

The question now is how to encode the harmonic relations between the sequences. Multiple sequence alignment (MSA) is the arrangement of sequences (via the introduction of gaps "-") so that they have they have the same length, while keeping related symbols aligned. MSA seems like a perfect fit to encode harmony: gaps "-", that can be interpreted as rests, can be introduced to the pitch sequences such that original pitch alignments are retained (see Fig. 1).

The MSA representation of harmony allows us to reformulate the harmony reconstruction task: assume a polyphonic score S and a function $h : S \rightarrow A_{harm}$ that maps the score's harmony into a multiple sequence alignment (A_{harm}), for example by removing durations and focusing on simultaneous pitches. Given S^*, a corrupted version of S in terms of note duration and onsets, our goal is to find

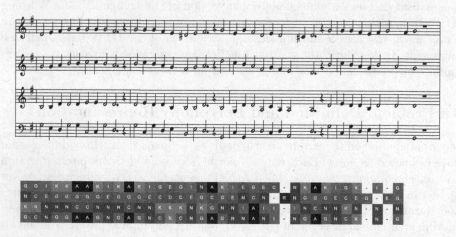

Fig. 1. A polyphonic score represented as a duration-onset-agnostic MSA (bottom). Colors are used for visualization purposes. Only the harmonic relations are retained from the original score. (Color figure online)

a score correction function f such that $h(f(S^*)) = h(S)$. To simplify, given a corrupted score in terms of onsets and durations, our goal is to realign the pitch sequences (voices) so that the original harmony is reconstructed.

4   Importance of Musical Dimensions

Our musical intuition suggests aligning music voices is a multidimensional process. Obviously, the relation between note pitches should meet the stylistic requirements for consonance and dissonance. However, one could posit that for example, it is more likely for notes of similar duration to be aligned together. This section aims to investigate to which extent various musical dimensions contribute to the correct harmony reconstruction.

Formally, we are interested in $\arg \min[d(h(f_D(S^*)), h(S))]$ where d is a distance function between two MSAs and D is a set of musical dimensions that can be incorporated in a function f. To achieve this we first need to have a reference set of correctly aligned polyphonic pieces, the S component (see Sect. 4.1). Secondly, we need an aligner of multiple sequences that can incorporate more than one dimensions, the f function (Sect. 4.2). We also need to establish a meaningful distance measure so that we can compare the polyphonic ground truth to the reconstruction created by the multidimensional MSA, the d function (Sect. 4.3). Finally, after we discuss the musical dimensions we consider in our work, the D component (Sect. 4.4), we put them to the test (Sect. 4.5) (Fig. 2).

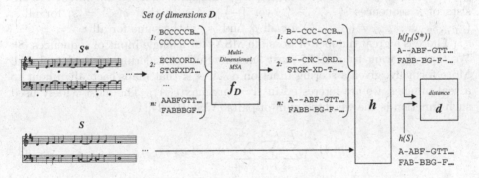

Fig. 2. The pipeline and components (S, S^*, D, f, h, d) for discovering the importance of musical dimensions in harmony reconstruction.

4.1   Dataset

Our dataset, called HYMNS[2], comprises of 153, sixteenth century 4-voice religious pieces. We picked this dataset due to its particular properties. First, all

---

[2] www.genevanpsalter.com/music-a-lyrics/2-complete-collections/
181-midi-collections.

pieces are transposed to the same key which allows us to learn a global harmony model. Secondly, the average length of the pieces is small enough to run different experiments with a reasonable time complexity. Finally, the number of notes over the length of each voice is almost the same, which avoids alignment ambiguities. For example, the sequence ABC can be aligned to both the beginning or end of ABCXXXABC, so any algorithm would have trouble distinguishing which alignment is better. This is an undesirable property in the context of this experiment. For more information regarding this problem, the reader is referred to [12].

4.2 Multidimensional Multiple Sequence Alignment (MDMSA)

We are interested in a function f that can incorporate more than one dimension to align the multiple voices of a polyphonic score. We use a multidimensional extension of the popular progressive alignment (PA) approach which was originally used for aligning multiple unidimensional sequences [9]. We selected PA due to its simplicity, both in terms of concept and development, and ease of adaptation to more than one dimension. In this section, we first present the concepts of unidimensional MSA and PA before their multidimensional extensions.

The unidimensional multiple sequence alignment is the output of a process that introduces gaps "-" to sequences of symbols so that they have the same length. Formally, given k sequences $s_1, s_2, ..., s_k$ over an alphabet $\mathcal{A}$, a gap symbol "-" $\notin \mathcal{A}$ and let $g : (\{-\} \cup \mathcal{A})^* \rightarrow \mathcal{A}^*$ a mapping that removes all gaps from a sequence containing gaps. A multiple sequence alignment A consists of k sequences $s'_1, s'_2, ..., s'_k$ over $\{-\} \cup \mathcal{A}$ such that $g(s'_i) = s_i$ for all i, $(s'_{1,p}, s'_{2,p}, .., s'_{k,p}) \neq (-, ..., -)$ for all p; and $|s'_i|$ is the same for all i.

There is a great number of possible MSAs for a single input of sequences [8]. We typically want to pick the most "meaningful" considering our task at hand. More formally: given a scoring function $c : A \rightarrow \mathbb{R}$ that maps each alignment to a real number, we are interested in $A' = \arg\max(c(A))$. The most widely used such function is the weighted sum-of-pairs (WSOP) [24]:

$$c(A) = \sum_{p=1}^{L} \sum_{i=1}^{k-1} \sum_{j=i+1}^{k} w_{i,j} v(s_{i,p}, s_{j,p}) \qquad (1)$$

where L is the length of the MSA, $w_{i,j}$ is a weight of the pair of sequences i, j and $v(a, b)$ is a "relatedness" score between two symbols $a, b \in \{-\} \cup \mathcal{A}$. The scores are typically stored in a matrix format called the substitution matrix. Literature suggests that A' would be "meaningful" as long as the substitution matrix captures "meaningful" relationships between symbols [8]. WSOP can also be extended to take into consideration affine gap scores (different scores for gap insertions and gap extensions).

The exact computation of A' is NP-hard [27], so it cannot be used in practice. Instead, the focus is on heuristic approaches that give good alignments not guaranteed to be optimal. The most popular approach is progressive alignment (PA) [11], which comprises three fundamental steps. At first, all pairwise alignments between sequences are computed to determine the WSOP similarity between each pair. In the second step, a similarity tree (guide tree) is constructed using a hierarchical clustering method. Finally, working from the leaves of the tree to the root, one aligns alignments, until reaching the root of the tree, where a single MSA is built. The drawback of PA is that incorrect gaps are retained throughout the process from the moment they are first inserted.

The unidimenional multiple sequence alignment can be extended to accommodate for multiple MSAs that we call "dimensions". More formally: a multidimensional multiple sequence alignment (MDMSA) consists of N multiple alignments $A_1, A_2, ..., A_N$. Each A_n consists of k sequences $s_1^n, s_2^n, ..., s_k^n$ over an alphabet $\{-\} \cup \mathcal{A}^n$ such that $|g(s_m^n)|$ is the same for all n and if $s_{m,p}^z$ is a gap at a dimension z, then $s_{m,p}^n$ is also a gap for all n. Figure 3 presents examples of simple MSA and MDMSA. In the same manner, WSOP can be extended to MDMSAs by summing over all dimensions:

$$c(A) = \sum_{n=1}^{N} W_n \sum_{p=1}^{L} \sum_{i=1}^{k-1} \sum_{j=i+1}^{k} w_{i,j} v_n(s_{i,p}^n, s_{j,p}^n) \qquad (2)$$

where W_n and v_n are the weight and scoring function of the nth dimension respectively. Extending the progressive alignment algorithm to accommodate MDMSAs is similarly straightforward; we extend pairwise alignment to multiple dimensions (through the multidimensional WSOP score).

Multiple Sequence Alignment		Multidimensional Multiple Sequence Alignment	
Input:	Output:	Input:	Output:
s_1: ABCABC	s'_1: ABCABC	s^1_1: ABCABC	s^1_1: ABCABC
s_2: ACC	s'_2: A-C--C	s^1_2: ACC	s^1_2: AC--C-
s_3: BCAC	s'_3: -BCA-C	s^1_3: BCAC	s^1_3: BC-AC-
		s^2_1: ZGGZGZ	s^2_1: ZGGZGZ
		s^2_2: GGG	s^2_2: GG--G-
		s^2_3: ZZZG	s^2_3: ZZ-ZG-

Fig. 3. Examples of: a multiple sequence alignment of three sequences (left), a multidimensional multiple sequence alignment of three sequences and two dimensions (right). Note that gaps in both dimensions are at the same positions.

4.3   Distance Between Two Multiple Sequence Alignments

Although the polyphonic scores S and S^* are multidimensional, both $h(f_D(S^*))$ and $h(S)$ are unidimensional MSAs reduced to only the pitch dimension. Therefore, we need to define a meaningful distance measure between two MSAs of the same sequences; the smaller the distance the more similar two alignments of the same voices should sound. The previous definition implies that any distance

measure we devise should correlate with the perceived distance between two harmony alignments A and B. For the sake of convenience, we make the following
intuitive assumption: the larger the portion of the voices that are misaligned,
the higher the perceived distance; any misalignment of voices leads to a bad
sounding polyphony, despite the fact that it might sound "nice" by pure chance.

Based on this definition we generate a synthetic set of corrupted HYMN polyphonic pieces: each note $n_i = (p_i, d_i, o_i)$ in a voice s_j is modified (doubled or
halved in duration) with a probability for modification $P = (l^2 \times o_i)/|s_j|$, where
l the misalignment degree. Every note with onset value larger than o_i is consequently shifted resulting to misalignment. Notes at the end of the piece (larger
o_i) have higher chance to be modified, since altering initial notes would result
to larger misaligned portions. We generate misalignments at different degrees
$l = 0.1, 0.2, ..., 0.8$. Figure 4 presents an example of two voices from the score of
Fig. 1 corrupted at two different degrees l.

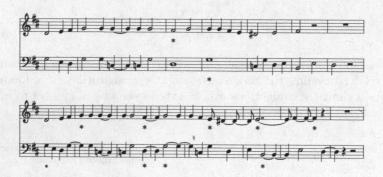

Fig. 4. Two (out of four) voices from the polyphonic score in Fig. 1 corrupted at different degrees of misalignment (top, bottom). Star signs "*" represent which notes were
modified in terms of duration (halved or doubled) to generate the misalignments.

Now we identify a good measure for two MSAs. A widely used distance
measure is based on the sum-of-pairs similarity, recoded as a dissimilarity: d_{SP}
represents the ratio of aligned symbols in A that could not be found in B over
$|A|$. Blackburne and Whelan [1] argue that d_{SP} is not a real metric because it
violates the core principles of symmetry and triangle inequality. They presented
four alternatives that differ in the way they treat gaps: *(a)* the "Symmetrized
SP", or d_{SSP} which aims to be a correction of the d_{SP} score by ignoring all
gaps *(b)* d_{seq} which incorporates raw gap information meaning that each gap
is simply recoded as G_i, indicating it occurred in sequence i, *(c)* d_{pos} which in
addition includes the position where gaps occur in a sequence, and *(d)* a metric
that incorporates information from the phylogenetic tree of the MSA; omitted
in our work since phylogeny information for our dataset is absent.

The relationship between the misalignment degree l and a ideal distance measure should be monotonic and linear; since as stated before, any distance measure, should correlate with the perceived distance between two polyphonic alignments. We measure these by calculating the Spearman and Pearson correlation coefficients respectively between l and the MSA pairwise distance. We also compute the coefficient of determination R^2. Table 1 presents those figures for all four distance measures considered in our work (all significance p values are smaller than 10^{70} and are omitted). The simple d_{SP} shows the highest correlation to the misalignment degree l, therefore it will be used from now on whenever we refer to a distance between two MSAs.

Table 1. The Spearman coefficient, the Pearson coefficient and the coefficient of determination R^2 for the d_{SP}, d_{SSP}, d_{pos} and d_{seq} measures.

	Spearman	Pearson	R^2
d_{SP}	0.826	0.801	0.641
d_{SSP}	0.800	0.765	0.586
d_{pos}	0.825	0.799	0.639
d_{seq}	0.817	0.796	0.633

4.4 Dimensions and Sequence Representation

We now explain the different dimensions of a polyphonic score that we consider in our work, and how they were represented as sequences, which is a prerequisite of the multidimensional PA algorithm.

Pitch. Pitch information is probably the most important dimension when it comes to harmony reconstruction. The representation of pitches into sequences is achieved by folding the pitch values into one octave and mapping them into an 12-sized alphabet of symbols.

Duration. We have also hypothesized that notes of similar duration might have higher chance to be sounded together in a polyphonic piece. It is also interesting to investigate to which extent duration corruption affects harmony reconstruction. We represent the duration dimension as a sequence by assigning an alphabetic symbol to each note value (e.g. thirty-second to "A", sixteenth to "B", eighth to "C" and so on).

Segment Boundaries. In musicology, "meaningful" units of notes are referred as "phrases", "segments", "sections" and so on, although the distinctions between them are vague. Music psychologists consider segmentation a fundamental listening function in terms of how humans perceive and structure music [16]. As such, information regarding segments has been frequently employed in MIR applications [21]. Given the reasonable assumption that humans generate a segment structure mentally as they listen to music, we hypothesize that the segment beginnings (or ends) have higher chance to be aligned between different polyphonic voices, i.e. segments boundaries are more likely to sound in parallel.

Our dataset does not include segment boundary information, therefore we use three automatic segmentation algorithms (applied on each voice seperately): $seg_{gestalt}$ by Tenney and Polansky [23] which is based on Gestalt principles, seg_{markov} which is based on Markov probabilities of segment boundaries derived from the Essen collection [15] and seg_{LBD} which is based on the Local Boundary Detection Model by Cambouropoulos [4]. A number of segmentation algorithms exist beyond the ones considered [21], however those three should be sufficient for our task: understanding the importance of the segmentation dimension in harmony reconstruction. All three are based on onset information so any score corruption might have a major effect on their output. We represent the segmentation dimension of each voice as a sequence by binning the space of values into 26 bins so that each note is assigned an alphabetic character corresponding to its bin index. For example, if the segmentation output $\in [0, 1]$ for a melody consisting of ten notes is $1.0, 0.2, 0.0, 0.0, 0.9, 0.2, 0.2, 0.5, 0.2, 1.0$, then the sequence representation would be ZFAAWFFMFZ.

Metric Weights. The importance of a note in the temporal domain can be represented by its metric weight (not to be confused with distance metrics). We use the Inner Metric Analysis (IMA) [26] to compute the metric weights based on the note onsets. Two different variations of the IMA algorithm are computed: $IMA_{spectral}$ and $IMA_{metrical}$. We represent both IMA dimensions into sequences by binning the space of values into 26 bins. Each note is assigned an alphabetic character corresponding to its bin index.

Figure 5 presents an example of four MSAs corresponding to different dimensions of a polyphonic score.

4.5   Experiment

Settings. We aim to find which dimension(s) are most important for the reconstruction of the harmony of a polyphonic piece S after it has been corrupted to S^*, i.e. the set D that minimizes $d(h(f_D(S^*)), h(S))$. Besides the distance of the reconstruction to the ground truth, we are also interested in the difference in distance before and after reconstruction $\delta = d(h(f_D(S^*)), h(S)) - d(h(S^*), h(S))$. We perform the experiment on the HYMNS dataset corrupted with misalignments at different degrees l (see Sect. 4.3).

Regarding the substitution matrix v_n for each dimension, we express the probabilities of symbols appearing in pairs in the so called log-odds scores. This means the substitution matrix for each dimension is learned from the ground truth in a similar manner as in [3, 7, 10]. Particularly for the pitch dimension, the substitution matrix can be considered a rough *harmony model*, since it encodes which pairs of pitch values are frequently sounded together. All substitution matrices are normalised to have zero mean and unit variance. All dimensions are assigned equal weights W_i for the sake of simplicity, although different weight settings may result to differences in performance. All pairs of sequences are

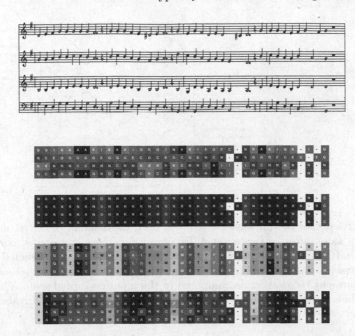

Fig. 5. Four properly aligned MSAs corresponding to the following dimensions of a polyphonic score (top): pitch, durations, $IMA_{spectral}$ and seg_{LBD}. Colors are used for visualization purposes. (Color figure online)

assigned equal weights $w_{i,j}$. Gap open and gap extend scores are set to -0.8 and -0.2 respectively. Although gap settings have great effect on alignments in general [6], preliminary results have showed that the core findings of our experiments are not affected.

In theory, given seven dimensions, we need to investigate the performance of $2^7 = 128$ different D set combinations. In practice, knowing that the pitch dimension is essential we can reduce that number to $2^6 = 64$, which is still impractical considering the time complexity of the PA. We therefore decided to combine dimensions empirically, starting from fewer dimensionalities to more.

Results. We start by combining the pitch dimension with any of the remaining six, i.e. $D = \{pitches, y\}$ $\forall y \in \{$ durations, $seg_{gestalt}$, seg_{markov}, seg_{LBD}, $IMA_{spectral}$, $IMA_{metrical}\}$. Figure 6 presents the $d(h(f_D(S^*)), h(S))$ and δ values achieved at different misalignment degrees. Considering the $d(h(f_D(S^*)), h(S))$ figure, three observations become immediately obvious: First, any addition to the pitch dimension makes the reconstruction more accurate. Second, duration is the dimension contributing the most. Third, all dimensions' positive contribution is weakened as the misalignment degree increases. It seems that a harmony model (pitch dimension) by itself is not sufficient to reconstruct the original harmony but the incorporation of more dimensions leads to a better reconstruction in comparison. However, all dimensions besides pitch are onset, duration-based and their reliability weakens as the degree of misalignment increases.

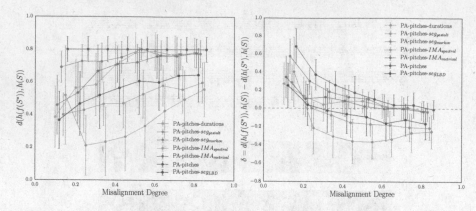

Fig. 6. The results for two dimensions, $|D| = 2$. Left: the distance of the reconstructed alignment (y axis), after corrupted at different misalignment degrees (x axis) to the ground truth. Right: the difference in distance (δ) between the reconstructed and the corrupted version to the ground truth. δ values above 0 mean that the method results in a worse harmony reconstruction compared to the input corrupted score. The results for the unidimensional MSA using only pitch information (PA-pitches) are also plotted as a baseline.

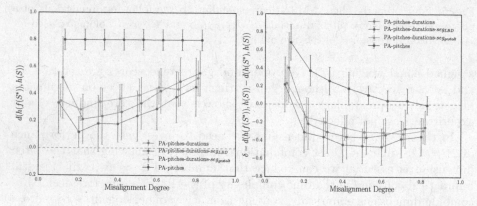

Fig. 7. The results for three dimensions, $|D| = 3$. The results for the one-dimensional MSA using only pitch information (PA-pitches) and the two dimensional PA using pitch and duration (PA-pitches-duration) are also plotted as baselines.

Considering the δ figure (right part of Fig. 6), it become obvious that only the duration, $seg_{gestalt}$ and seg_{LBD} dimensions result in an improvement on the reconstruction (any δ value above 0 means we make the alignment worse). Also the improvement happens only when the misalignment degree is above 0.2. In other words, even though the fusion of the pitch with any other dimension results in a better harmony reconstruction than the pitch dimension solely, only a couple of dimension combinations actually have a positive effect. Also, given an almost perfect polyphony we are more likely to make it worse than fix it.

In general, the results show that the pitch dimension by itself is inadequate for harmony reconstruction as is the incorporation of only one extra dimension. However, there is a strong indication that the incorporation of more dimensions will lead to a better reconstruction. Based on the previous findings we proceed into combining three dimensions: $D = \{pitches, durations, y\}\ \forall y \in \{\ seg_{gestalt},\ seg_{LBD}\}$. Results are presented in Fig. 7. We can make the following observations. First, the pitches-durations-seg_{LBD} performs better than just the pitches-durations-$seg_{gestalt}$ and the pitches-durations approaches. Second, similar to the previous experiment, given a near-perfect initial score (misalignment degree less than 0.2), any combination of dimensions will result to a worse reconstruction.

5   Harmony Reconstruction Using MAFFT

The previous experiments revealed that the duration and segmentation dimensions (beside pitches) are the most important for harmony reconstruction: notes of similar duration and notes at segment boundaries are more likely to be sounded together in a polyphonic piece. Unfortunately, as soon as the note durations are altered drastically (highly corrupted), both dimensions become unreliable to be used for harmony reconstruction via multidimensional MSA. Segmentation particularly, degrades since the algorithms used in our work rely on musical heuristics applied on onset and duration information. It becomes clear that harmony reconstruction requires a segmentation technique that is impervious to duration-onset errors and according to our knowledge, such an algorithm for single voices, does not exist. It is also clear that the only reliable information for our task is pitches. Consequently, the question becomes whether useful structural information can be extracted from multiple pitch sequences corresponding to the different voices of a polyphonic piece.

Interestingly, locating very similar sub-regions (segments) between large sequences has been in important task in bioinformatics. Such segments can efficiently reduce MSA runtimes and as a consequence, MSA solutions that incorporate segmentation, such as DIALIGN [18] and MAFFT [14], have found successful application. MAFFT in particular, is a unidimensional progressive alignment method at its core, but uses the fast Fourier transform to identify short sub-regions that are high-scoring matches between the sequences in the alignment.

We hypothesize that MAFFT's pipeline can be a viable solution to the harmony reconstruction problem. Therefore, we apply MAFFT solely on the pitch dimension and we incorporate a harmony model similarly to the PA approaches, i.e. a learned log-odds substitution matrix from the pitch dimension of the ground truth. Figure 8 presents the results for MAFFT compared to the best performing PA-pitches-durations-seg_{LBD} approach from the previous experiment. For the sake of completeness we also include a five-dimensional approach PA-pitches-durations-seg_{LBD}-$seg_{gestalt}$, although we know in advance its performance will not be robust to high misalignments degrees. For this dataset, the results show that MAFFT achieves almost perfect harmony reconstruction and performs better than any multidimensional PA approach. In addition, given an

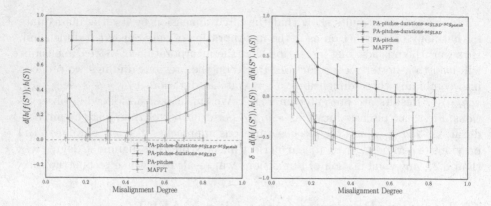

Fig. 8. The results for various dimensionalities and MAFFT.

almost perfect polyphonic score, MAFFT is more likely to fix it than worsen it. More importantly, MAFFT's performance is stable and invariant to any duration and onset noise. Consequently, given our particular task (harmony reconstruction), a harmony-aware MAFFT approach is the most reliable solution.

6   Conclusions

In this paper we introduced the problem of polyphony reconstruction and tackled its harmonic component: how to align pitches from different voices after they have been corrupted in terms of durations and onsets. By using a multidimensional version of MSA we showed that structural information, namely segment boundaries, are essential for the correct polyphony reconstruction. Since most segmentation algorithms are based on duration and onset information, we proposed the use of the bioinformatics MSA aligner MAFFT extended with a harmony model. We additionally showed its superiority and perfect fit for the task.

However, polyphony reconstruction is far from being considered solved, while we cannot claim that we now understand the cognitive process behind aligning voices. Besides excluding the crucial rhythm component, our work made a set of simplifications; first, we employed progressive alignment which is heuristic rather than an exact MSA algorithm. Secondly, each musical dimension was considered equally important, although the literature contradicts this. Thirdly, we have not yet investigated how pitch errors could be dealt with. And finally, the repertoire we have chosen is rhythmically simple in comparison to most polyphony from the 16th century. Despite those facts, our work set strong foundations for understanding polyphony and towards a complete solution to the polyphony reconstruction problem.

Aknowledgments. The authors would like to thank Meinard Müller and Hendrik Vincent Koops for comments that greatly improved the manuscript.

References

1. Blackburne, B.P., Whelan, S.: Measuring the distance between multiple sequence alignments. Bioinformatics **28**(4), 495–502 (2012)
2. Boulanger-Lewandowski, N., Bengio, Y., Vincent, P.: Modeling temporal dependencies in high-dimensional sequences: application to polyphonic music generation and transcription. arXiv preprint arXiv:1206.6392 (2012)
3. Bountouridis, D., Koops, H.V., Wiering, F., Veltkamp, R.C.: A data-driven approach to chord similarity and chord mutability. In: International Conference on Multimedia Big Data, pp. 275–278 (2016)
4. Cambouropoulos, E.: The local boundary detection model (LBDM) and its application in the study of expressive timing. In: International Computer Music Conference, pp. 17–22 (2001)
5. Carrillo, H., Lipman, D.: The multiple sequence alignment problem in biology. SIAM J. Appl. Math. **48**(5), 1073–1082 (1988)
6. Carroll, H., Clement, M.J., Ridge, P., Snell, Q.O.: Effects of gap open and gap extension penalties. In: Biotechnology and Bioinformatics Symposium, pp. 19–23 (2006)
7. Dayhoff, M.O., Schwartz, R.M., Orcutt, B.C.: 22 a model of evolutionary change in proteins. In: Atlas of Protein Sequence and Structure, vol. 5, pp. 345–352. National Biomedical Research Foundation Silver Spring, MD (1978)
8. Durbin, R., Eddy, S.R., Krogh, A., Mitchison, G.: Biological Sequence Analysis: Probabilistic Models of Proteins and Nucleic Acids. Cambridge University Press, Cambridge (1998)
9. Feng, D.-F., Doolittle, R.F.: Progressive sequence alignment as a prerequisitetto correct phylogenetic trees. J. Mol. Evol. **25**(4), 351–360 (1987)
10. Henikoff, S., Henikoff, J.G.: Amino acid substitution matrices from protein blocks. Proc. Natl. Acad. Sci. **89**(22), 10915–10919 (1992)
11. Hogeweg, P., Hesper, B.: The alignment of sets of sequences and the construction of phyletic trees: an integrated method. J. Mol. Evol. **20**(2), 175–186 (1984)
12. Hudek, A.K.: Improvements in the accuracy of pairwise genomic alignment (2010)
13. Joh, C.-H., Arentze, T., Hofman, F., Timmermans, H.: Activity pattern similarity: a multidimensional sequence alignment method. Transp. Res. Part B Methodol. **36**(5), 385–403 (2002)
14. Katoh, K., Misawa, K., Kuma, K., Miyata, T.: Mafft: a novel method for rapid multiple sequence alignment based on fast fourier transform. Nucleic Acids Res. **30**(14), 3059–3066 (2002)
15. Lartillot, O., Toiviainen, P., Eerola, T.: A matlab toolbox for music information retrieval. In: Preisach, C., Burkhardt, H., Schmidt-Thieme, L., Decker, R. (eds.) Data Analysis, Machine Learning and Applications, pp. 261–268. Springer, Heidelberg (2008)
16. Lerdahl, F., Jackendoff, R.: An overview of hierarchical structure in music. Music Percept. Interdisc. J. **1**(2), 229–252 (1983)
17. Lyu, Q., Wu, Z., Zhu, J., Meng, H.: Modelling high-dimensional sequences with LSTM-RTRBM: application to polyphonic music generation. In: International Conference on Artificial Intelligence, pp. 4138–4139. AAAI Press (2015)
18. Morgenstern, B., Dress, A., Werner, T.: Multiple dna and protein sequence alignment based on segment-to-segment comparison. Proc. Natl. Acad. Sci. **93**(22), 12098–12103 (1996)

19. Pugin, L., Crawford, T.: Evaluating omr on the early music online collection. In: International Society on Music, Information Retrieval, pp. 439–444 (2013)
20. Rebelo, A., Fujinaga, I., Paszkiewicz, F., Marcal, A.R.S., Guedes, C., Cardoso, J.S.: Optical music recognition: state-of-the-art and open issues. Int. J. Multimedia Inf. Retrieval 1(3), 173–190 (2012)
21. Rodríguez López, M.E.: Automatic Melody Segmentation. Ph.D. thesis, Utrecht University (2016)
22. Sanguansat, P.: Multiple multidimensional sequence alignment using generalized dynamic time warping. WSEAS Trans. Math. 11(8), 668–678 (2012)
23. Tenney, J., Polansky, L.: Temporal gestalt perception in music. J. Music Theory 24(2), 205–241 (1980)
24. Thompson, J.D., Higgins, D.G., Gibson, T.J.: Clustal w: improving the sensitivity of progressive multiple sequence alignment through sequence weighting, position-specific gap penalties and weight matrix choice. Nucleic Acids Res. 22(22), 4673–4680 (1994)
25. van Kranenburg, P.: A computational approach to content-based retrieval of folk song melodies. Ph.D. thesis (2010)
26. Volk, A., Garbers, J., Van Kranenburg, P., Wiering, F., Veltkamp, R.C., Grijp, L.P.: Applying rhythmic similarity based on inner metric analysis to folksong research. In: International Society on Music Information Retrieval, pp. 293–296 (2007)
27. Wang, L., Jiang, T.: On the complexity of multiple sequence alignment. J. Comput. Biol. 1(4), 337–348 (1994)
28. Wang, S., Ewert, S., Dixon, S.: Robust joint alignment of multiple versions of a piece of music, pp. 83–88 (2014)

Melody Retrieval and Classification Using Biologically-Inspired Techniques

Dimitrios Bountouridis[1]([✉]), Dan Brown[2], Hendrik Vincent Koops[1], Frans Wiering[1], and Remco C. Veltkamp[1]

[1] Department of Information and Computing Sciences,
Utrecht University, Utrecht, The Netherlands
d.bountouridis@uu.nl
[2] David R. Cheriton School of Computer Science,
University of Waterloo, Waterloo, Canada

Abstract. Retrieval and classification are at the center of Music Information Retrieval research. Both tasks rely on a method to assess the similarity between two music documents. In the context of symbolically encoded melodies, pairwise alignment via dynamic programming has been the most widely used method. However, this approach fails to scale-up well in terms of time complexity and insufficiently models the variance between melodies of the same class. Compact representations and indexing techniques that capture the salient and robust properties of music content, are increasingly important. We adapt two existing bioinformatics tools to improve the melody retrieval and classification tasks. On two datasets of folk tunes and cover song melodies, we apply the extremely fast indexing method of the Basic Local Alignment Search Tool (BLAST) and achieve comparable classification performance to exhaustive approaches. We increase retrieval performance and efficiency by using multiple sequence alignment algorithms for locating variation patterns and profile hidden Markov models for incorporating those patterns into a similarity model.

1 Introduction

The retrieval and classification of music documents is a fundamental problem in Music Information Retrieval (MIR), and vital to the music recommendation task that underpins the ever-growing digital music industry. Assessing the similarity between two musical recordings or scores is at the core of both tasks, with direct ties to musicological analysis and music cognition. The proliferation of large music collections has placed efficiency at the center of MIR research therefore, accurate and efficient similarity methods, applied on various retrieval scenarios, are currently of vital importance.

Retrieval and classification, in general, requires building representations of previously seen classes. Two popular metaphors to this problem derive from cognition models that explains how humans categorize a concept or object [27]. The *exemplar* metaphor assumes that a new object is compared to all examples

© Springer International Publishing AG 2017
J. Correia et al. (Eds.): EvoMUSART 2017, LNCS 10198, pp. 49–64, 2017.
DOI: 10.1007/978-3-319-55750-2_4

in a class. In computational systems this corresponds to the nearest-neighbor method. The alternative posits that a new object is compared with a representation called a *prototype* or a profile model, an abstraction created through experience that contains the most common features of the members of the class.

In the exemplar metaphor, the most common similarity method comprises converting the musical documents into sequential strings, and comparing them using alignment via dynamic programming. This approach is mainly driven by a pre-defined set of relationships between symbols, encoded as a fixed scoring matrix, and pre-defined gap penalties for changes made to one sequence or the other. Theoretically, the final similarity score between the sequences is as "meaningful" as the relationships themselves [13]. Alignment via dynamic programming has been proven very useful for identification or classification tasks where strong similarities are present [34,38]. However, the dynamic programming technique is slow and fails to efficiently scale to large datasets and long sequences. As a consequence, scalability and efficiency in MIR have been investigated by a range of researchers [4,9,23,29,30].

Most methods generally agree on the idea that representations that capture salient and robust properties of musical content are a more intuitive approach to speed up the retrieval and classification tasks. Consequently, we argue that profiles or prototypes for each class of musical documents are more appropriate than exemplars. However, applying alignment via dynamic programming on a prototype-retrieval metaphor is not straight forward. This relates to another drawback of the alignment method: its inappropriate way of modelling similarity. Studies suggest that music similarity is a complex process [32,41]. Intuitively, certain salient parts of a melody or a chord progression are less likely to change than others in a folk melody rendition or a cover song. Additionally, musicologists have argued that similarity raises from a listening process that involves altered, but not beyond recognition musical patterns called *variations*. The simplistic nature of the substitution matrix clearly cannot accommodate for the proper handling of music uncertainty and variance (or stability), because the matrix focuses only locally on individual notes.

Interestingly, in bioinformatics and biological sequence analysis, a similar situation has emerged. Various solutions have been proposed to both reduce time-complexity and properly model similar evolutionary relationships. For example, in an exemplar retrieval metaphor, the widely popular indexing framework BLAST (Basic Local Alignment Search Tool) [1,2] uses heuristics to filter out unnecessary comparisons from the database while it only explores a small part of the dynamic programming space by identifying high-matching substrings. On the other hand, in a prototype retrieval metaphor, profile hidden Markov models (profile HMMs) [14,26] have been successfully used to summarize alignments of multiple biologically-related sequences. Such profiles capture the variance in a class of sequences and can be used for accelerated database searches.

We propose the adaptation of BLAST and profile HMMs to increase the efficiency and performance of the melody classification and retrieval tasks. Although BLAST has already found application in the other music domains [22,29], we are interested in reestablishing its practical benefits in melodies. On the other hand, in a prototype retrieval metaphor, we consider multiple sequence alignment and profile HMMs appropriate for capturing the shared salience between music variations. We conduct an empirical study and evaluation on two symbolic datasets of folk tunes and cover song melodies.

The remainder of the paper is organised as follows: Sect. 2 presents the general background and previous related research on MIR and bioinformatics. Sections 3 and 4 describe our proposed tools for the exemplar and prototype retrieval metaphors. Sections 5 and 6 present our experimental setup and results respectively. Finally, concluding remarks are in Sect. 7.

2   Background and Related Work

Sequence alignment via dynamic programming is common in a variety of domains, including bioinformatics and medicine [33]. Music documents of sequential format, such chord transcriptions or melodies, can also be compared using the same method; gaps "–" are introduced in the sequences, until they have the same length and the amount of "relatedness" between symbols at the same index position is maximized. Given that the quality of an alignment between two sequences $s_1 : c_1, c_2, .., c_n$ and $s_2 : c'_1, c'_2, .., c'_m$ is the sum of alignment scores of the individual symbols, most pairwise alignment methods use a dynamic programming method credited to Needleman and Wunsch [31]. The optimal (highest scoring) alignment can be generated by filling a cost matrix D recursively:

$$D(i, j) = max \begin{cases} D(i - 1, j - 1) + sub(c_i, c'_j) \\ D(i - 1, j) - \gamma \\ D(i, j - 1) - \gamma \end{cases}$$

where $sub(c_i, c'_j)$ is the substitution scoring function (typically encoded as a matrix) and γ is the gap penalty which is assigned when a symbol is aligned to gap. The score of the optimal alignment is stored in $D(n, m)$, while the alignment itself can be obtained by looking for adjacent maxima backwards from $D(n, m)$ to $D(0, 0)$. The Needleman and Wunsch approach is a global alignment method, since it aims to find the best score among alignments of full-length sequences. On the other hand, local alignment introduced by Smith and Waterman [35], aims to find the highest scoring alignments of partial sequences by tracking back from $max(D(i, j))$ instead of $D(n, m)$, and by forcing all $D(i, j)$ to be non-negative. Local alignment allows for the identification of substrings (patterns) of high similarity which can be mapped to the concept of musical segments (e.g. verses, choruses, chord progressions).

We consider pairwise alignment via dynamic programming to be unfit for music classification and retrieval and particularly for melodic sequences. First, the $O(nm)$ time complexity is impractical when it comes to large databases. Although various heuristics have been introduced to reduce the time complexity of dynamic programming [33], in practice, for k sequences in the database the system would run in $O(k^2 nm)$ time [29]. Müller et al. [30] reduce the computational cost by first computing alignment at a coarse resolution level. The alignment is then projected into a finer level for refinement. Hu *et al.* [19] proposed an approach that relies on locating and extending promising matches incrementally on the D cost matrix to find the "best" match. Inspired by bioinformatic heuristics, Martin *et al.* [29] adapted BLAST, which reduces the cost of local alignment, for cover song identification on a large dataset of audio recordings. Martin *et al.* argue that substantial decrease in retrieval performance (compared to pairwise alignment) is compensated by the gain in computation times.

The second drawback of alignment via dynamic programming is its scoring function (or substitution matrix), which is the only component where domain knowledge can be incorporated. Such a simple, and globally applied, structure cannot accommodate for the particularities of certain classes of music documents and music similarity in general. Nevertheless, an expert-annotation study [41] of a folk-song melody dataset highlighted the importance of melodic contour, rhythm, lyrics and motifs in melodic similarity. Based on these findings, van Kranenburg [38] extended the scoring function of the typical pairwise alignment to include multiple musical dimensions (e.g. inner-metric analysis, phrase boundaries). On a melody classification task, he showed that expert-based heuristics could achieve an almost perfect 99% accuracy. At placing songs into families, Hillewaere *et al.* [17] showed that a simple all-versus-all adaptation of standard edit distance using only pitch information gives 94% accuracy for the same task. In a related work, Boot *et al.* [5] investigated the importance of repeated patterns on melodic similarity. Based on the assumption that shared patterns between variations carry more salience, they showed that pattern-based compression can achieve almost state-of-the-art classification accuracy. However, their approach failed to show similarly promising results in a retrieval scenario, presumably due to the quality of the automatic pattern extraction algorithms.

The problem of sequence similarity is well studied in bioinformatics. In a prototype retrieval metaphor, profile methods are used by most protein class identification pipelines. Sequences that include the active site in a protein (the position in the 3-dimensional structure of the protein where it binds to other molecules) are aligned using multiple alignment pipelines, and then the common region is identified using a variety of summarization methods. The summarization methods can range from the very simple PROSITE patterns (which are simple ungapped regular expressions) [3] to position-specific scoring matrices (essentially a probabilistic extension of regular expressions where there is a distribution of symbols at each position) [2]. Profile hidden Markov models (profile HMMs) [14] are also architectures that summarize multiple sequence alignments and have made major contributions to the field of computational molecular biology [13]. They are probabilistic automata that take a multiple sequence alignment and convert it into a position-specific scoring system that can be used for database searching. These models still often enable complex indexing strategies [15] that speed up the assignment process, which is necessary when studying millions of newly-generated sequences.

Despite their success, profile HMMs have found little application in MIR. Most notably, Chai and Vercoe [10] have used them to model and classify folk songs into their corresponding country of origin. Although their results showed that folk tunes from different countries share commonalities, they suggested applying profile HMMs to more discriminable data sets. Wang *et al.* [43] have used them to align different performances of the same musical piece. Results showed that profile HMMs can improve alignment accuracy and robustness over state-of-the-art pairwise methods but not significantly compared to other approaches (e.g. progressive alignment). However, they have shown to be faster, rendering them ideal for large music databases. Bountouridis *et al.* [6] have shown that profile HMMs perform better that other MSA summarization methods, in a inlier-outlier separation scenario on datasets of folk melodies, chord transcriptions and musical audio.

3   Exemplar Retrieval

The exemplar metaphor, manifested as the nearest-neighbor method, is widely used in alignment-based music retrieval. In the previous section we saw that near-perfect melody classification accuracies can be achieved by extending the typical alignment scoring with musical heuristics. The question therefore becomes whether similarly high performance can be achieved while reducing the comparison times. Boot *et al.* [5] already established the practical benefits of melodic patterns for efficiently tackling the task, but the retrieval results were not very promising. Martin *et al.* [29], on the other hand, already established that the audio-derived music sequences can be compared in near-linear time using the pattern-based BLAST. Intuitively, BLAST can find application to melodic sequences.

3.1   BLAST

BLAST [1,2] uses a local alignment method at its core, which implies that similarity is modelled as a simple substitution matrix. Its efficiency mainly relies on the idea that a good alignment contains highly similar sub-strings (called seeds). This allows for the substantial reduction of the number of database comparisons needed for a single query, since target sequences with no matching sub-strings can be filtered out. After the initial seeds are located, BLAST extends them in both directions to find longer regions of similarity above a certain threshold. Those regions are later used as anchor points that aid the local alignment. The efficiency of BLAST also relies on its indexing strategy; the whole database is divided into words, of equal size to the seed length, which are stored in a look-up table for fast accessing.

The balance between sensitivity (true positive rate) and specificity (true negative rate), in a retrieval scenario using BLAST, is mainly determined by the size of seed length s_L. Although, the work by Martin *et al.* [29] has investigated that particular relationship, their findings cannot be generalized to other datasets, as they were considering digitizations of audio recordings with specific settings.

4   Prototype Retrieval

Volk *et al.* [40] presented a comprehensive study that shed light on the importance of variations for the perceived music similarity. They argue that a variation-based computational model of music similarity requires two steps: first, the detection of variation (or stability) patterns and second, the development of an overall similarity measure that incorporates the patterns. This is a challenging task, mostly because musicology and music cognition have yet to provide us with sufficient knowledge regarding the nature of musical variations [40].

Both components of a variation-based model of music similarity can be mapped to ideas and tools from the field of computational biology. As Krogh states: "the variation in a class of sequences can be described statistically, and this is the basis for most methods used in biological sequence analysis" [25]. This agrees with the sequential, and widely supported probabilistic [36], nature of certain music representations (e.g. melodies, chord progressions) which has allowed for the successful application of sequential alignment and analysis algorithms.

Multiple sequence alignment (MSA) is an extension of pairwise alignment to more sequences, and it is a widely used sequence-analysis tool in bioinformatics. It is a way to organize sequences such that relevant features are aligned together [21]. Although "relevancy" depends on the context, the major benefit of MSA is that it allows us to identify regions of low stability or high variation. In a musical context, this can potentially translate to locating variation patterns; a prerequisite of any model of music similarity according to Volk *et al.* [40].

Profile HMMs, briefly described in Sect. 2, are theoretically more appropriate for assessing the similarity between sequences than the typical pairwise alignment because they provide a powerful framework for dealing with uncertainty and randomness in sequential data. The scoring and penalizing at each position is dependent on the MSA and not on some arbitrary chosen fixed values (e.g. scoring matrix, gap penalties). This agrees with our music intuition that certain parts of a melody are more likely to change in a variation than others.

Our proposed pipeline to enhance melody retrieval, in a prototype retrieval metaphor, comprises two steps. First, we reveal the salient parts of the variations belonging to a class, by building an MSA. Second, we encode them in a structure that allows for comparison and database searching, by building a profile HMM on top of the MSA. These can be mapped to the notions of variation and similarity modelling respectively. Both are further explained in the following sections.

4.1   Multiple Sequence Alignment (MSA)

Our approach firstly requires us to identify salient patterns between the variations belonging to a class. We propose representing melodies as sequences of symbols (from a finite alphabet) before aligning them using an MSA algorithm, such that shared patterns are revealed.

Multiple sequence alignment is the output of a process that introduces gaps "–" to sequences of symbols so that they have the same length. Formally, given k sequences $s_1, s_2, ..., s_k$ over an alphabet $\mathcal{A}$, a gap symbol "–" $\notin \mathcal{A}$ and let

$g : (\{-\} \cup \mathcal{A})^* \rightarrow \mathcal{A}^*$ a mapping that removes all gaps from a sequence containing gaps. A multiple sequence alignment A consists of k sequences $s_1', s_2', ..., s_k'$ over $\{-\} \cup \mathcal{A}$ such that $g(s_i') = s_i$ for all i, $(s_{1,p}', s_{2,p}', .., s_{k,p}') \neq (-, ..., -)$ for all p; and $|s_i'|$ is the same for all i.

There is a great number of possible MSAs for a single input of sequences [13]. We typically want to pick the most "meaningful" considering our task at hand. More formally: given a scoring function $c : A \rightarrow \mathbb{R}$ that maps each alignment to a real number, we are interested in $A' = \arg\max(c(A))$. The most widely used such function is the weighted sum-of-pairs (WSOP) [37]:

$$c(A) = \sum_{p=1}^{L} \sum_{i=1}^{k-1} \sum_{j=i+1}^{k} w_{i,j} v(s_{i,p}, s_{j,p}) \tag{1}$$

where L is the length of the MSA, $w_{i,j}$ is a weight of the pair of sequences i, j and $v(a, b)$ is a "relatedness" score between two symbols $a, b \in \{-\} \cup \mathcal{A}$. The scores are typically stored in a matrix format called the substitution matrix. Literature suggests that A' would be "meaningful" as long as the substitution matrix captures "meaningful" relationships between symbols [13]. WSOP can also be extended to take into consideration affine gap scores (different scores for gap insertions and gap extensions).

Representation. Related literature [21] and the previous definitions suggest that, in order to achieve a "meaningful" alignment we need to carefully select the music features that we will represent as sequences. Consequently, the representation of melodies into sequences of symbols and their relationship are of major importance. The works of van Kranenburg [38] and Hillewaere et al. [17] revealed the importance of the pitch dimension, so our work considers melodies as pitch-contours, meaning series of relative pitch transitions constrained to the region between $+11$ and -11 semitones (folded to one octave). Besides their simplicity and key-invariance, pitch contours have been found to be more significant to listeners for assessing melodic similarity than alternative representations [16].

Scoring. The next step towards building a "meaningful" MSA is to define the a relationship between symbols which translates to a similarity scoring matrix $\in \mathbb{R}$ and gap open and extend penalties. It is typical to vary the gap values and investigate their effect, since they have shown to affect the quality of an MSA [8]. We use the simplest scoring matrix: $v(i, i) = 1$ if $i = j$ and $v(i, j) = -1$ if $i \neq j$.

The previous paragraphs briefly explained the alignment score function c and the representations in $A' = \arg\max(c(A))$. The following paragraphs explain the two different MSA algorithms (the heuristics for the $\arg\max$ function) that are investigated in our work. Remember that the literature suggests that A' would be meaningful as long as c is meaningful too. Considering the simple sequence representation and the naive scoring model, we hypothesize that the intrinsic properties of different music-agnostic MSA algorithms will lead to better and more meaningful alignments.

Progressive Alignment. The exact computations of A' is NP-hard [42], so it cannot be used in practice. Therefore, heuristic approaches that give good alignments not guaranteed to be optimal have been developed. The most popular approach is progressive alignment (PA) [18], which comprises three fundamental steps. At first, all pairwise alignments between sequences are computed to determine the WSOP similarity between each pair. At the second step, a similarity tree (guide tree) is constructed using a hierarchical clustering method. Finally, working from the leaves of the tree to the root, one aligns alignments, until reaching the root of the tree, where a single MSA is built. The drawback of PA is that incorrect gaps are retained throughout the process since the moment they are first inserted.

MAFFT. MAFFT [20] is a progressive alignment method that uses the fast Fourier transform (FFT), or an FFT approximation, to identify short subregions of one sequence or intermediate alignment that are high-scoring matches with sub-regions from another sequence or alignment. This pre-processing allows MAFFT to guide its alignment by an initial "anchoring" phase, thus reducing the overall runtime. Besides being an efficient alignment method, MAFFT's FFT approach is in theory more well-suited to music sequences: according to Margulis [28], the phrase structure of a melody is of major importance for the human perception of variation patterns. By treating the located sub-regions as gap-free segments, MAFFT can be the closest to partitioning melodies into perceptually meaningful units. In addition, Bountouridis et al. [6] have shown that MAFFT performs better than PA in a inlier-outlier separation task on music datasets of different nature.

4.2 Profile Hidden Markov Models (Profile HMMs)

An introduction and literature review of profile HMMs can be found at [14,25]. Here we briefly describe them. Profile HMMs are linear, left-to-right models comprised of three types of states: match, delete and insert states. Match states correspond to columns in the MSA with low variability and capture the distribution of symbols in that column. Insert states model columns of high variability, and capture the likelihood of a column being extended with the insertion of symbols. Delete states simply allow for the skipping of match states.

An example will better explain the creation of a profile HMM. Figure 1 shows a profile HMM generated from a small MSA. The structure has six match states (besides the "start" and "end" states), although the length of the MSA is eight columns. That is because we have set a gap-to-symbol ratio $\theta_{ms} = 0.3$ below which a column can be modeled as a match state. For n match states there are n delete states and $n + 1$ insert states. The transition probabilities from state to state are initially set to 0 but are modified as we parse the MSA. In our example, the columns with index 6 and 7 are modelled as an increase in the transition probability from the match state M5 corresponding to column 5 to its insert state I5. Each match state M_i has an emission distribution probability P_{M_i} corresponding to symbol distribution of that column. In order to avoid

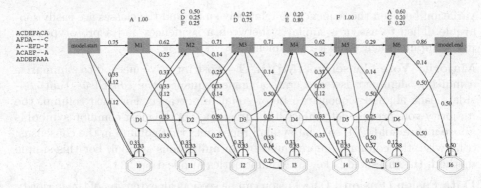

Fig. 1. An example of a profile HMM (right) created from a multiple sequence alignment (left). From top to bottom: probability distribution of the match states (only the most frequent symbols are shown) and match, delete, insert states. Not all columns have corresponding match states since $\theta_{ms} = .3$ in that particular example.

over-fitting, such as in the case of M1 where only the symbol "A" is found in the column, it is common to use pseudo-counts; where we manually increase the counts of every symbol in the alphabet. Insert states have also an emission probability distribution, but in contrast to the match states, they are assigned a fixed background symbol distribution P_I. Pseudo-counts can also be applied to the transitions between states. Altering the gap-to-symbol ratio θ_{ms}, emission pseudo-counts c_{em} and transition pseudo-counts c_{tr}, affects the flexibility of the profile HMM, meaning the allowed variation from the MSA sequences.

Profile HMMs can be trained by unaligned sequences too, using algorithms such as the Baum-Welch expectation maximization or gradient descent. A more reliable technique is to estimate a draft structure from a satisfactory MSA and then re-estimate the model's parameters using the Baum-Welch algorithm [14].

Comparing or aligning a sequence to a profile HMM is performed by finding the most likely path of states given the sequence. The Viterbi or Forward algorithms are typically used for this task, with time complexity $O(NM)$ similar to other dynamic programming methods, where N and M are the sequence length and number of states of the model respectively.

4.3 Alternative MSA Summarizations

Profile HMMs are not the only method for summarizing an MSA. In contrast to profile HMMs, all of the methods described below aim to represent an MSA as a single sequence rather than a probabilistic model. We denote this representation as "prototype" in order to avoid confusion with profile HMMs.

Random Exemplar (Random). The most naive approach to prototype modeling is picking a random sequence from the class and considering it the prototype: no MSA is required. This approach works in practice only when the variation between the sequences is relatively small and the classes are easily separable. Then to assign a similarity between a sequences and a prototype, one just computes their pairwise alignment.

Majority-Vote Consensus (MjV). The most intuitive method to summarize a multiple alignment is to generate a single sequence, the *consensus*, that considers each aligned sequence to be of equal importance. For each column, the majority vote process determines if the frequency of the most common symbol is above a threshold, θ. If so, that symbol represents that column in the consensus; otherwise, the column is represented by an ambiguous symbol. For this simple method, the threshold is key in bioinformatics applications [11].

Data Fusion (Fusion). Data Fusion can be seen as an extension of the majority vote approach. In addition to finding the most common symbol per column, it also uses the agreement between rows as a weight to favor values of rows with higher agreement [12]. Data Fusion has already found successful application in music, i.e. in the task of automatic chord recognition from audio [24]. We omit the details due to lack of space, however the reader is forwarded to the aforementioned publications.

5   Experiments

We have proposed using bioinformatics-inspired techniques for both exemplar and prototype based retrieval metaphors. In both cases, we design a classification and a retrieval experiment. For the first task the goal is to assign individual melodies to their corresponding class. For the latter given a class, the goal is to rank higher the melodies that belong to it.

5.1   Datasets

Our experiments use two datasets of symbolically represented melodies of varying size and nature. Variations of the same melody are grouped into classes. Summary statistics are presented in Table 1.

The Annotated Corpus of the Meertens Tune Collections [39] is a set of 360 Dutch folk songs grouped into 26 "tune families" and annotated by Meertens Institute experts. Each contains a group of melody variations related through an oral transmission process. For this TUNEFAM-26 data set, expert annotators assessed the perceived similarity of every melody over a set of dimensions (contour, rhythm, lyrics, *etc.*) to a set of 26 prototype "reference melodies".

The Cover Song Variation data set [7], or CSV-60, is a set of expert-annotated, symbolically-represented vocal melodies derived from matching structural segments (such as verses and choruses) of different renditions of sixty pop and rock songs. CSV-60 is inherently different from TUNEFAM-26 in two ways. First, the grouping of melodies into classes is certain: the songs were pre-chosen as known covers of songs of interest. Secondly, cover songs are typically not a byproduct of an oral transmission process: cover artists have access to the original version. All melodies in both datasets, are represented as sequences as described in Sect. 4.1.

Table 1. Summary statistics for the two datasets of our experiments.

	TuneFam-26	Csv-60
Number of classes	26	60
Number of sequences	360	243
Class size	13.0 *(4.0)*	4.0 *(1.1)*
Sequence length	43.0 *(14.9)*	53.0 *(21.4)*

5.2 Evaluation Framework

Exemplar Retrieval Settings. We are interested in evaluating whether BLAST can achieve comparable performance to the typical pairwise alignment in a nearest-neighbor retrieval scenario. Each target sequence in the dataset is compared to the query and ranked according to its similarity. For each query, the 1st-nearest neighbor is used to predict its class (e.g. tune family). Mean Average Precision (MAP) of the correct class in the ranked list is also computed. Since pairwise alignment (denoted as "PW-NN") can be highly dependant on gap open and extend settings, we experiment with different configurations (0.3–0.1, 0.5–0.1, 0.7–0.1, 0.7–0.3, 0.9–0.1, 0.9–0.3). Since BLAST is dependent on the seed size s_L, we experiment with different values $(3, 4, 5)$. BLAST's substitution matrix is set to the simple $v(i, i) = 1$ if $i = j$ and $v(i, j) = -1$ if $i \neq j$, while gap open and extend setting are set to default (1.1–0.1).

Prototype Retrieval Settings. We are interesting in evaluating whether our proposed profiling approach, i.e. MSA and profile HMMs, enables internal variation while still characterizing a class. We set up a leave-50%-out cross-validation retrieval system that randomly partitions each class into two equal-size training and testing sets. The split datasets are denoted TuneFam-26-H and Csv-60-H respectively. The *training* set is used to generate the prototypes. This includes building the MSA and summarizing with different prototype methods. For the MSA, we experiment with different algorithms (see Sect. 4.1) and different gap open and extent settings (0.3–0.1, 0.5–0.1, 0.7–0.1, 0.7–0.3, 0.9–0.1, 0.9–0.3). The match and mismatch scores are set to 1 and -1 respectively. Regarding the profile HMM training settings, we set $\theta_{ms} = 0.4$, $c_{em} = 1$ and $c_{tr} = 1$ which are considered typical. We re-estimate the model parameters by using the Baum-Welch algorithm. For the majority vote method, we experiment with three different threshold settings: 0.3, 0.5 and 0.7.

Each prototype is compared to all the sequences in the *test* set, and the MAP of the correct class in this ranked list over ten runs is computed. For the classification task, each sequence in the *test* set is compared to all the prototypes and the highest ranked class is used for prediction. For PW-NN, each sequence in the *training* set is compared to the *test* set.

6   Results

The MAP and mean classification accuracy (Acc) results of the PW-NN method
(our baseline) for the Csv-60-(H) and TuneFam-26-(H) datasets are presented
in Table 2. Our results for the TuneFam-26 set, agree with the findings of van
Kranenburg [38]. Interestingly, gap settings have little effect on the overall performance in both datasets.

Exemplar Retrieval Results. The BLAST results are presented in Table 3.
Clearly, the retrieval performance of BLAST cannot compare with that of
PW-NN. However for TuneFam-26, the highest classification accuracy (0.85),
achieved with $s_L = 5$, is only 0.08 lower than the best baseline performance,
0.93 (0.7–0.1 gap settings). It should be noted that Boot *et al.* [5] achieved
0.89 accuracy on the same dataset by using expert annotations of salient patterns. BLAST, achieves comparable performance without the incorporation of
any musical heuristics. With the Csv-60 set, the BLAST results show similar
behavior; 0.77 accuracy ($s_L = 5$) while the baseline is at 0.83 (0.7–0.1 gap settings). Due to the previous findings and its high efficiency, it is safe to state that
BLAST can be a reliable and fast solution for melody classification.

Prototype Retrieval Results. The MAP and mean classification accuracy
(Acc) results for the Csv-60-H and TuneFam-26-H datasets are presented
in Tables 4 and 5 respectively. Our profile HMM-based model shows the highest
retrieval performance in general. More specifically MAFFT-pHMM achieves MAP
scores of 0.76 and 0.83 for the Csv-60-H and TuneFam-26-H respectively. Both

Table 2. The baseline MAP and mean classification accuracy (Acc) results for the Csv-
60-(H) and TuneFam-26-(H) datasets using pairwise alignment run with different gap
settings (0.3–0.1, 0.5–0.1 and so on).

	.3–.1		.5–.1		.7–.1		.7–.3		.9–.1		.9–.3	
	MAP	Acc	MAP	Acc	MAP	Acc	MAP	Acc	MAP	Acc	MAP	Acc
Csv-60	.64	.82	.64	.82	.65	.83	.64	.83	.65	.83	.64	.82
TuneFam-26	.61	.92	.62	.93	.58	.92	.60	.93	.57	.92	.61	.93
Csv-60-H	.67	.77	.66	.79	.65	.78	.65	.76	.67	.78	.65	.76
TuneFam-26-H	.57	.86	.59	.85	.61	.87	.60	.86	.62	.88	.63	.88

Table 3. The MAP and mean classification accuracy (Acc) results for the Csv-60 and
TuneFam-26 datasets using BLAST run with different seed lengths s_L (3, 4, 5).

	3		4		5	
	MAP	Acc	MAP	Acc	MAP	Acc
Csv-60	.33	.65	.37	.66	.46	.77
TuneFam-26	.50	.85	.47	.84	.51	.85

Table 4. The MAP and mean classification accuracy (Acc) results for the Csv-60-H dataset algorithms are compared on different MSAs created by different algorithms (Mafft, PA) run with different gap settings (0.3–0.1, 0.5–0.1 and so on).

	.3–.1		.5–.1		.7–.1		.7–.3		.9–.1		.9–.3	
	MAP	Acc	MAP	Acc	MAP	Acc	MAP	Acc	MAP	Acc	MAP	Acc
Mafft-fusion	.70	.63	.68	.64	.66	.63	.70	.66	.68	.66	.66	.62
Mafft-pHMM	**.73**	**.73**	**.74**	**.76**	**.73**	**.71**	**.76**	**.74**	**.72**	**.72**	**.74**	**.71**
Mafft-MjV-.3	.60	.49	.59	.51	.59	.51	.60	.52	.57	.51	.58	.51
Mafft-MjV-.5	.60	.48	.59	.51	.59	.51	.60	.52	.57	.51	.59	.50
Mafft-MjV-.7	.53	.38	.51	.38	.49	.38	.46	.35	.46	.38	.48	.39
Mafft-random	.67	.60	.69	.64	.67	.63	.70	.64	.65	.60	.68	.60
PA-fusion	.64	.55	.66	.59	.63	.56	.68	.64	.67	.62	.67	.59
PA-pHMM	.68	.62	.66	.69	.65	.71	.70	**.74**	.64	.65	.68	.65
PA-MjV-.3	.67	.54	.67	.55	.69	.60	.68	.55	.67	.58	.65	.55
PA-MjV-.5	.67	.54	.67	.55	.69	.60	.68	.55	.67	.58	.65	.55
PA-MjV-.7	.64	.49	.65	.51	.69	.54	.61	.44	.63	.52	.59	.45
PA-random	.64	.49	.66	.52	.70	.53	.71	.60	.65	.51	.67	.56

Table 5. The MAP and mean classification accuracy (Acc) results for the TuneFam-26-H dataset algorithms are compared on different MSAs created by different algorithms (Mafft, PA) run with different gap settings (0.3–0.1, 0.5–0.1 and so on).

	.3–.1		.5–.1		.7–.1		.7–.3		.9–.1		.9–.3	
	MAP	Acc	MAP	Acc	MAP	Acc	MAP	Acc	MAP	Acc	MAP	Acc
Mafft-fusion	.71	.53	.75	.57	.74	.62	.77	.58	.74	.66	.77	.62
Mafft-pHMM	**.81**	**.72**	**.80**	**.70**	**.83**	**.70**	**.79**	**.69**	**.82**	**.71**	**.81**	**.70**
Mafft-MjV-.3	.67	.55	.70	.60	.76	.63	.74	.63	.73	.65	.75	.64
Mafft-MjV-.5	.66	.51	.67	.52	.72	.55	.72	.56	.71	.57	.72	.54
Mafft-MjV-.7	.50	.28	.48	.29	.46	.32	.45	.28	.48	.30	.42	.25
Mafft-random	.57	.42	.58	.41	.60	.47	.60	.43	.60	.49	.66	.50
PA-fusion	.65	.37	.66	.37	.70	.38	.76	.53	.69	.44	.77	.55
PA-pHMM	.70	.63	.72	.64	.72	.64	.73	.67	.71	.66	.77	.66
PA-MjV-.3	.51	.44	.57	.47	.64	.54	.70	.63	.69	.60	.74	.64
PA-MjV-.5	.51	.44	.57	.47	.64	.54	.70	.64	.69	.59	.74	.63
PA-MjV-.7	.47	.39	.55	.40	.60	.49	.68	.56	.65	.52	.70	.54
PA-random	.50	.30	.55	.28	.63	.34	.65	.40	.61	.32	.67	.41

are significantly better ($p < 0.005$ using Wilcoxon signed-rank test) than the second highest MAP performed by random exemplars and Data Fusion. Mafft-pHMM is also significantly better than PW-NN; 0.67 and 0.63 for the Csv-60-H and TuneFam-26-H respectively. For classification, PW-NN performs generally better than any prototype methods. Most notably in the TuneFam-26-H set, PW-NN achieves an accuracy of 0.88 (0.9–0.1 gap settings), while the second best (0.72) is achieved by the much faster MAFFT-pHMM. The same pattern holds at smaller extent for Csv-60-H; 0.79 (0.3–0.1 gap settings) and 0.76 for PW-NN and MAFFT-pHMM respectively.

7 Discussion and Conclusions

In this paper we proposed the adaptation of two popular tools of computational biology in the context of melody classification and retrieval. In a nearest-neighbor experiment, we showed that the pattern-based indexing tool BLAST can achieve high classification accuracy, comparable to music-aware sequence-compression methods that use global alignment via dynamic programming. As expected though, due to the improper modelling of similarity using a substitution matrix, the retrieval performance did not show similar behaviour. Nevertheless, we showed that MSA and profile HMMs are a great fit for modelling the randomness and uncertainty of variations and incorporating them into class profiles. The results showed that our proposed similarity model can outperform the global alignment method in a prototype retrieval metaphor. Profile HMMs were also shown to outperform alternative MSA summarization methods. Considering the previous findings and the high efficiency of profile HMMS when it comes to sequence comparison, we find that profile HMMs are a reliable and fast solution for melody retrieval. Interestingly, their performance is notably higher when the MSA comes from MAFFT instead from progressive alignment. This suggests that MAFFT's FFT-based internal heuristics are more appropriate than the typical progressive alignment for melodic sequences.

In general, BLAST and profile HMMs (trained on MAFFT MSAs) can be reliable and efficient solutions for large-scale melody classification and retrieval respectively, without the incorporation of musical heuristics. Future work should investigate their performance on music documents of more popular sequential formats (e.g. chords). Interestingly, profile HMMs, offer the possibility of generating variations based on the training MSA which can find applications in the field of automatic music generation. From a music-cognition perspective, it would be interesting to investigate whether listeners can distinguish generated from real variations. Our work can be considered as the first step towards more exciting research and applications.

References

1. Altschul, S.F., Gish, W., Miller, W., Myers, E.W., Lipman, D.J.: Basic local alignment search tool. J. Mol. Biol. **215**(3), 403–410 (1990)
2. Altschul, S.F., Madden, T.L., Schäffer, A.A., Zhang, J., Zhang, Z., Miller, W., Lipman, D.J.: Gapped blast and psi-blast: a new generation of protein database search programs. Nucleic Acids Res. **25**(17), 3389–3402 (1997)
3. Bairoch, A.: Prosite: a dictionary of sites and patterns in proteins. Nucleic Acids Res. **19**(Suppl), 2241 (1991)

4. Bertin-Mahieux, T., Ellis, D.P.W.: Large-scale cover song recognition using hashed chroma landmarks. In: Applications of Signal Processing to Audio and Acoustics, pp. 117–120 (2011)
5. Boot, P., Volk, A., de Haas, W.B.: Evaluating the role of repeated patterns in folk song classification and compression. J. New Music Res. 1–16 (2016)
6. Bountouridis, D., Koops, H.V., Wiering, F., Veltkamp, R.C.: Music outlier detection using multiple sequence alignment and independent ensembles. In: Amsaleg, L., Houle, M.E., Schubert, E. (eds.) SISAP 2016. LNCS, vol. 9939, pp. 286–300. Springer, Cham (2016). doi:10.1007/978-3-319-46759-7_22
7. Bountouridis, D., Van Balen, J.: The cover song variation dataset. In: The International Workshop on Folk Music Analysis (2014)
8. Carroll, H., Clement, M.J., Ridge, P., Snell, Q.O.: Effects of gap open and gap extension penalties. In: The Biotechnology and Bioinformatics Symposium, pp. 19–23 (2006)
9. Casey, M., Slaney, M.: Fast recognition of remixed music audio. In: Acoustics, Speech and Signal Processing, vol. 4, p. IV-1425 (2007)
10. Chai, W., Vercoe, B.: Folk music classification using hidden Markov models. In: International Conference on Artificial Intelligence, number 6 in 4. Citeseer (2001)
11. Day, W.H.E., McMorris, F.R.: Threshold consensus methods for molecular sequences. J. Theor. Biol. 159(4), 481–489 (1992)
12. Dong, X.L., Berti-Equille, L., Srivastava, D.: Integrating conflicting data: the role of source dependence. Proc. VLDB Endowment 2(1), 550–561 (2009)
13. Durbin, R., Eddy, S.R., Krogh, A., Mitchison, G.: Biological Sequence Analysis: Probabilistic Models of Proteins and Nucleic Acids. Cambridge University Press, Cambridge (1998)
14. Eddy, S.R.: Profile hidden Markov models. Bioinformatics 14(9), 755–763 (1998)
15. Finn, R.D., Clements, J., Eddy, S.R.: Hmmer web server: interactive sequence similarity searching. Nucleic Acids Res. gkr367 (2011)
16. Gómez, E., Klapuri, A., Meudic, B.: Melody description and extraction in the context of music content processing. J. New Music Res. 32(1), 23–40 (2003)
17. Hillewaere, R., Manderick, B., Conklin, D.: Alignment methods for folk tune classification. In: Spiliopoulou, M., Schmidt-Thieme, L., Janning, R. (eds.) Data Analysis, Machine Learning and Knowledge Discovery, pp. 369–377. Springer, Cham (2014)
18. Hogeweg, P., Hesper, B.: The alignment of sets of sequences and the construction of phyletic trees: an integrated method. J. Mol. Evol. 20(2), 175–186 (1984)
19. Hu, N., Dannenberg, R.B., Tzanetakis, G.: Polyphonic audio matching and alignment for music retrieval. Computer Science Department, p. 521 (2003)
20. Katoh, K., Misawa, K., Kuma, K., Miyata, T.: Mafft: a novel method for rapid multiple sequence alignment based on fast fourier transform. Nucleic Acids Res. 30(14), 3059–3066 (2002)
21. Kemena, C., Notredame, C.: Upcoming challenges for multiple sequence alignment methods in the high-throughput era. Bioinformatics 25(19), 2455–2465 (2009)
22. Kilian, J., Hoos, H.H.: Musicblast-gapped sequence alignment for MIR. In: International Society for Music Information Retrieval Conference, pp. 38–41 (2004)
23. Kim, S., Narayanan, S.: Dynamic chroma feature vectors with applications to cover song identification. In: Multimedia Signal Processing, pp. 984–987 (2008)
24. Koops, H.V., de Haas, W.B., Bountouridis, D., Volk, A.: Integration and quality assessment of heterogeneous chord sequences using data fusion. In: International Society for Music Information Retrieval Conference, pp. 178–184 (2016)

25. Krogh, A.: An introduction to hidden Markov models for biological sequences. New Compr. Biochem. **32**, 45–63 (1998)
26. Krogh, A., Brown, M., Saira Mian, I., Sjölander, K., Haussler, D.: Hidden Markov models in computational biology: applications to protein modeling. J. Mol. Biol. **235**(5), 1501–1531 (1994)
27. Malt, B.C.: An on-line investigation of prototype and exemplar strategies in classification. J. Exp. Psychol. Learn. Mem. Cogn. **15**(4), 539 (1989)
28. Margulis, E.H.: Musical repetition detection across multiple exposures. Music Percept. Interdisc. J. **29**(4), 377–385 (2012)
29. Martin, B., Brown, D.G., Hanna, P., Ferraro, P.: Blast for audio sequences alignment: a fast scalable cover identification. In: International Society for Music Information Retrieval Conference, pp. 529–534 (2012)
30. Müller, M., Mattes, H., Kurth, F.: An efficient multiscale approach to audio synchronization. In: International Society for Music Information Retrieval Conference, pp. 192–197. Citeseer (2006)
31. Needleman, S.B., Wunsch, C.D.: A general method applicable to the search for similarities in the amino acid sequence of two proteins. J. Mol. Biol. **48**(3), 443–453 (1970)
32. Pampalk, E.: Computational models of music similarity and their application in music information retrieval. na (2006)
33. Ratanamahatana, C.A., Keogh, E.: Everything you know about dynamic time warping is wrong. In: Third Workshop on Mining Temporal and Sequential Data, pp. 1–11. Citeseer (2004)
34. Serra, J., Gómez, E., Herrera, P., Serra, X.: Chroma binary similarity and local alignment applied to cover song identification. IEEE Trans. Audio Speech Lang. Process. **16**(6), 1138–1151 (2008)
35. Smith, T.F., Waterman, M.S.: Identification of common molecular subsequences. J. Mol. Biol. **147**(1), 195–197 (1981)
36. Temperley, D.: Bayesian models of musical structure and cognition. Musicae Sci. **8**(2), 175–205 (2004)
37. Thompson, J.D., Higgins, D.G., Gibson, T.J.: Clustal w: improving the sensitivity of progressive multiple sequence alignment through sequence weighting, position-specific gap penalties and weight matrix choice. Nucleic Acids Res. **22**(22), 4673–4680 (1994)
38. van Kranenburg, P.: A computational approach to content-based retrieval of folk song melodies. Ph.D. thesis, Utrecht University (2010)
39. van Kranenburg, P., de Bruin, M., Grijp, L., Wiering, F.: The Meertens tune collections. In: Meertens Online Reports (2014)
40. Volk, A., Haas, W.B., Van Kranenburg, P.: Towards modelling variation in music as foundation for similarity. In: Proceedings of the 12th International Conference on Music Perception and Cognition (2012)
41. Volk, A., Van Kranenburg, P.: Melodic similarity among folk songs: an annotation study on similarity-based categorization in music. Musicae Sci. **16**, 317–339 (2012). page 1029864912448329
42. Wang, L., Jiang, T.: On the complexity of multiple sequence alignment. J. Comput. Biol. **1**(4), 337–348 (1994)
43. Wang, S., Ewert, S., Dixon, S.: Robust joint alignment of multiple versions of a piece of music. In: International Society for Music Information Retrieval, pp. 83–88 (2014)

Evolved Aesthetic Analogies to Improve Artistic Experience

Aidan Breen$^{(\boxtimes)}$, Colm O'Riordan, and Jerome Sheahan

National University of Ireland, Galway, Republic of Ireland
{a.breen2,colm.oriordan,jerome.sheahan}@nuigalway.ie

Abstract. It has been demonstrated that computational evolution can be utilised in the creation of aesthetic analogies between two artistic domains by the use of mapping expressions. When given an artistic input these mapping expressions can be used to guide the generation of content in a separate domain. For example, a piece of music can be used to create an analogous visual display. In this paper we examine the implementation and performance of such a system. We explore the practical implementation of real-time evaluation of evolved mapping expressions, possible musical input and visual output approaches, and the challenges faced therein. We also present the results of an exploratory study testing the hypothesis that an evolved mapping expression between the measurable attributes of musical and visual harmony will produce an improved aesthetic experience compared to a random mapping expression. Expressions of various fitness values were used and the participants were surveyed on their enjoyment, interest, and fatigue. The results of this study indicate that further work is necessary to produce a strong aesthetic response. Finally, we present possible approaches to improve the performance and artistic merit of the system.

Keywords: Computational evolution · Genetic algorithms · Genetic Programming · Grammatical evolution computational analogy · Aesthetic analogy

1 Introduction

Analogy making is a cognitive tool that humans begin to use from an early age with children as young as six demonstrating a clear understanding and use of spacial analogy in problem solving tasks [7,8,10,22]. It has been said that analogy may even be "the most important cognitive mechanism" [10] that we use to make sense of the world around us. Computer science researchers, recognising the value of analogies, have explored the possibilities of computational analogy making in problem solving for quite some time. However, problem solving is not the only application of analogy. Analogy is used heavily in the artistic world as both an inspiration and subject, as we discuss in more detail in Sect. 2.2 below. It is the combination of computational analogy making and artistic analogy that we explore in this paper.

© Springer International Publishing AG 2017
J. Correia et al. (Eds.): EvoMUSART 2017, LNCS 10198, pp. 65–80, 2017.
DOI: 10.1007/978-3-319-55750-2_5

Recent work we have conducted demonstrates the design of a system capable of making aesthetic analogies between two artistic domains. The proposed system, described in detail in Sect. 3.1 below, makes use of *mapping expressions* to make artistic analogies. We have shown that these *mapping expressions* can be successfully evolved using Grammatical Evolution, discussed in Sect. 2.4, to estimate a mapping between two empirically gathered aesthetic data sets. While gathering aesthetic data sets in a reliable manner, and evolving *mapping expressions* is not trivial, implementing the proposed system to generate real-time visual displays based on a live music input is also a challenging task. The work presented in this paper describes an implementation of that system.

Beyond this implementation, an exploratory study was conducted to test the effectiveness of this approach and to guide the development of similar systems in the future. A number of music and visual displays were generated in real time, and recorded for reliable presentation to subjects. Subjects watched the displays and gave feedback on their enjoyment, the interestingness of each display, and their fatigue or boredom over time.

1.1   Contribution and Layout

The primary contribution of this paper is a description of the implementation of an aesthetic analogy system using *mapping expressions* evolved using Grammatical Evolution. This implementation demonstrates the value of a *mapping expression* as a real-time artistic tool and the validity of the system design proposed in previous work.

This primary contribution is composed of the following secondary contributions. First, we explore the use of live audio as an input to *mapping expressions* in order to generate visual displays. Second, we explore the use of MIDI messages in the place of a live audio signal as input. Third, we demonstrate that evolved *mapping expressions* of varying complexities can be evaluated in real-time using both audio signals and MIDI messages as input demonstrating the value of *mapping expressions* as artistic tools in a live performance. Finally, we demonstrate how a visual display can be generated using the output of a *mapping expression* with acceptable time delays.

The layout of the paper is as follows. First we present and briefly discuss related work in Sect. 2. In Sect. 3 we discuss the development of a system capable of creating real-time aesthetic analogies beginning with the evolution of *mapping expressions* using Grammatical Evolution in Subsect. 3.1 followed by the real-time system implementation in Sect. 4. In Sect. 5 we present the methodology of a study conducted to test the effect of the implemented system on the enjoyment, interest and fatigue of subjects. We present the results of this study in Sect. 6. Finally, we present a discussion and conclusion in Sects. 7 and 8.

2   Related Work

2.1   Computational Analogy

Computational Analogy (CA) focuses on the use of analogical problem solving as a computational approach to artificial intelligence. CA systems have been proposed since the 1960's [5]. As the field matured, analogical systems began to fit broadly into three main categories; Symbolic, Connectionist and Hybrid systems [6]. Symbolic systems made use of symbolic logic, means-ends analysis and classical logical techniques [1,5,21]. Connectionist systems made use of networks, with spreading-activation and back propagation building networks of similarity between domains [4,12]. Hybrid models often combined other models and made use of an agent based, distributed structure [16,17].

2.2   Artistic Analogies

Perhaps the most obvious examples of analogy in art might be found in the work of Wassily Kandisky. Kandisnky is credited as being one of the first purely abstract painters, with his accounts describing what we now call synaesthesia, a condition that involves the mixing of senses within the brain. The condition results in a person literally seeing sounds, or hearing colours. Kandisky held a life-long obsession with the connections between music and painting and also published a number of books on this topic [14].

Paul Klee produced and published similar work. Indeed, Klee and Kandinsky were colleagues at the Bauhaus school of art, design and architecture. Klee's handbooks [15] are cited as influential to most of modern abstract art with explicit notes on the representation of sounds as visuals.

In a more contemporary setting, lighting design has become ubiquitous with stage performance. Live music performance, especially popular, electronic, and rock music, all rely heavily on stage lighting design that enhances the feel of a performance. This can be seen as a practical analogy, where music is the source domain and lighting is the target domain. For some time, media player applications have displayed automatically generated visuals when playing music. These systems use simplistic generation techniques, and are not generally regarded as artistic, or artificially intelligent systems.

Further research has been conducted on the emotional connections between music and visuals. In her thesis, Behravan proposes a system which makes use of the detectable emotional aspects of music and visuals using psychological models of emotion together with an "artificial ear" and "artificial eye" [2]. Using a feedback loop and weighting mechanism tuned using Genetic Programming the system matches the emotional aspects of input music and generated visuals. The proposed system is functionally similar to the system proposed in this work, however the approach proposed in this paper aims to avoid subjective emotional aspects in favour of objective physical attributes within the domains of music and visuals.

2.3 Aesthetic Models

There have been numerous attempts to define aesthetic experience and to model or predict aesthetic value. Birkhoff's over-simplified "aesthetic measure" formula $M = O/C$, the ratio of order (O) to complexity (C) [3] has been often criticised but it did begin an important discussion on the definition of aesthetics. More recently, this has been tackled by Ramachandran who defined 8 "laws of artistic experience" [20]. Of the factors outlined, many are measurable, such as contrast and symmetry, but others remain abstract such as "grouping". Recent discussion has continued in the area [9,11,19].

The work in this paper relies on the existence of aesthetic models which are used as the core of the fitness function in the evolution of *mapping expressions*. Our previous work uses models gathered from two similar studies surveying people on their aesthetic preferences for pairs of music notes and pairs of colours. We adopt the same aesthetic models here.

2.4 Grammatical Evolution

Grammatical Evolution (GE) as proposed by O'Neill and Ryan [18] is a computational evolution approach whereby a grammar is used in a genotype-phenotype mapping. The application of this genotype-phenotype mapping enforces a stronger resemblance to biological systems where DNA codons are used to create proteins of a particular shape. In both systems, a many-to-one relationship occurs where many genotypes may produce one particular phenotype, which introduces a natural robustness while still allowing crossover and mutation to take effect. By defining a grammar, we can take any genotype and guarantee a grammatically correct phenotype.

A gene consists of an array of integers, known as codons. The grammar defines a starting non-terminal expression, and a set of terminal and non-terminal expressions. Terminal expressions represent fixed pieces of the output *mapping expression*, such as an input variable, a constant value or a mathematical operator. Non-terminal expressions represent pieces of the output *mapping expression* that are recursively replaced by other terminal or non-terminal expressions. The specific replacement expression is dictated by successive codons, with legal replacement expressions as defined by the grammar, equally distributed across the potential values of the codon. The recursive replacement continues until either a complete legal expression (phenotype) is created, or a length threshold is reached. It is common for an expression to require more codons than are present in the gene array and in this case, we simply begin at the start of the array again.

GE provides a number of distinct advantages. Primarily, a GE system suits the creation of executable expressions which can be easily defined by a relatively small grammar. In comparison to other Evolutionary Programming systems, the implementation is relatively straightforward and simple to implement. The output of the GE system, in this case a lisp like s-expression, can be parsed and executed simply and efficiently. This is of particular value in this work as a real-time execution of the expression is necessary when generating visuals with live music input. Using a grammar provides the ability to include useful 'pre-baked' expressions like sin, cos and log functions, as well as application-specific expressions like *plus90*, used in this implementation to represent offsetting a variable by 90 degrees on the colour wheel. Finally, the output *mapping expression* is a textual, human readable expression which can be stored in a text file for later evaluation, analysis or debugging.

3   System Methodology

3.1   Evolving Aesthetic Analogies

We have previously described an analogical approach based on a metalanguage constructed of *Mapping Expressions*. Each *Mapping Expression* maps one measurable aesthetic attribute in a source domain to a similar measurable aesthetic attribute in a separate target domain. An analogy may contain multiple *Mapping Expressions*, each mapping the value of some aesthetic attribute in the source domain to an attribute in the target domain. The structure of the metalanguage is illustrated in Fig. 1. In previous work, the musical harmony and colour harmony were used as measurable aesthetic attributes. We adopt this structure here also.

The GE system used to generate output *mapping expressions* is implemented as follows. The fitness of an individual is a measure of the similarity between the input musical harmony and output visual harmony. An individual contains a single gene. Genes are represented as an array of 8-bit positive integer (0–255) codons. Gene arrays of 60 codons were used within a population of size 50 and a population is seeded with entirely random genes. To begin the evolutionary process, the fitness of each individual is calculated. Tournament selection is then used to select individuals for evolution. Both single-point and double-point crossover are used to build the succeeding generation. Elitism is employed to maintain the peak population fitness from one generation to the next. Mutation is applied at the gene level where each codon may have its value randomly reassigned based on the mutation rate shown in Eq. 1 where α represents the number of generations since a new peak fitness was reached.

$$Mut_1 = \left(\frac{0.02}{70}\alpha\right) + 0.01 \tag{1}$$

A hyper-mutation rate is applied after a threshold to increase variation further. This approach ensures we allow adequate exploration of the genotype landscape while allowing local optimisation to occur for a short period. Evolution is halted after a threshold of generations where the peak fitness has not increased. The parameters used are summarised in Table 1.

Our previous work has shown that *Mapping Expressions* evolved with this procedure produce increasingly more accurate mappings between domains over time, with evolved expressions showing a distinctly higher fitness than randomly generated expressions.

Table 1. Genetic algorithm parameters.

Parameter	Value
Population size	50
Chromosome length	60
Crossover rate	0.8
Standard mutation rate (Mut_1)	See Eq. 1
Hyper-mutation rate (Mut_2)	1.0
Mutation threshold	100
Halting threshold	200

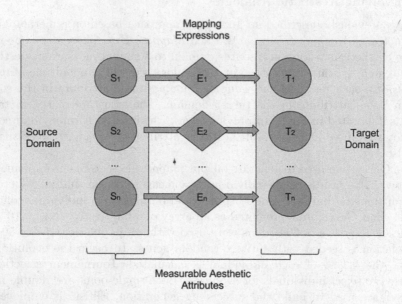

Fig. 1. Analogy structure overview with *Mapping Expressions* E_1 to E_n mapping source attributes S_1 to S_n to target attributes T_1 to T_n.

4 Real-Time Analogies

One of the major challenges of designing a system that receives live music input is making sense of the data being received. Our initial design intended to analyse live audio samples using signal processing techniques to detect salient information. While this approach has valuable applications, implementation is not reliable. Nevertheless, it still produces potentially useful results, presented in Sect. 4.1 below. The alternative approach, outlined in Sect. 4.2, uses a digital MIDI signal, removing most of the issues with live audio sampling in favour of a noise free, reliable input. This approach, however, is less flexible, limited to live input from digital instruments and synthesizers only.

4.1   Sampled Musical Input

A Fast Fourier Transform (FFT) was implemented to identify the frequencies being played in a live audio sample. In testing this approach we identified the strongest frequencies to be the fundamental frequencies of notes being played on an instrument. By taking the strongest N frequencies, the harmony of the audio could be calculated and sent on for *mapping expression* evaluation.

There are some obvious drawbacks to this approach. Chord recognition is a complex and actively researched field. Simply taking the strongest frequencies of a FFT is an unreliable, naive approach. The fundamental frequency of a note being played is not guaranteed to be the strongest frequency and the usefulness of this approach is diminished further as more instrumentation is included and the frequency spectrum becomes more crowded with overtones and noise. Another issue with this approach is the speed at which the calculated harmony changes. In testing, we found the harmony to fluctuate wildly as the frequencies identified changed. To combat this, a smoothing window was used to find the

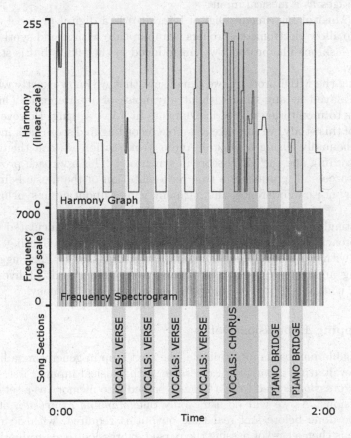

Fig. 2. Harmony calculated using a naive Fast Fourier Transform approach compared to the frequency spectrum and manually identified parts of a sample song.

mode harmony value. Using a short window of 10 samples was quite effective, improving the signal to noise ratio greatly without adding any significant delay.

Another major drawback of using a FFT with live audio is the timing of visual updates. Without a smoothing window, colour changes are rapid and distinctly unenjoyable. With a smoothing window, colours change at a more enjoyable pace, but do not seem to change in synchrony with any musical cadence.

Regardless of these drawbacks, using the FFT approach produced some interesting results as illustrated in Fig. 2. The figure compares the calculated harmony, the frequency spectrum and the manually identified sections of a sample song. Even using the naive FFT approach, it is clear that the harmony value is indicative of the part of the song, showing obvious differences between the verses, chorus and bridge.

4.2 MIDI Musical Input

To combat the downfalls of a FFT with live audio, we investigated the use of MIDI messages as a musical input.

MIDI (Musical Instrument Digital Interface) is a digital protocol originally designed to allow electronic controllers communicate with sound synthesizers in a modular fashion. The protocol was introduced in the 1980's but is still widely used today.

By using the MIDI protocol, we can ensure that we know exactly what notes are being played at any time without any noise or interference. This clearly restricts us to monitoring digital instruments in a live setting. However, for the purposes of this study, we can take a sample recording and score each instrument in MIDI manually. Using a Digital Audio Workstation we can then play the sample recording and the MIDI score in synchrony. This approach proved to be the most successful, producing a true representation of the music being played without any noise interference, and harmony values updating in synchrony with the music.

Interestingly, as illustrated in Fig. 3, the harmony value calculated using the MIDI approach is remarkably similar to the value calculated using the FFT approach, with parts of the section similarly distinguishable. This suggests that while using an FFT to detect the timing, notes and chords of live audio is unreliable, it may still be a useful approach to calculate harmony.

4.3 Mapping Expression Delay

With a reliable musical input available, the next step in generating a live visual output is evaluating a *mapping expression* with musical input. First, a chosen *mapping expression* is read from a file and parsed into memory as a set of nested sub-expressions. As we will be using only one *mapping expression* at a time, this can be done before any real time output is required, with no overhead. Evaluation is then a case of passing the parsed expression to an evaluation script which recursively evaluates each sub-expression.

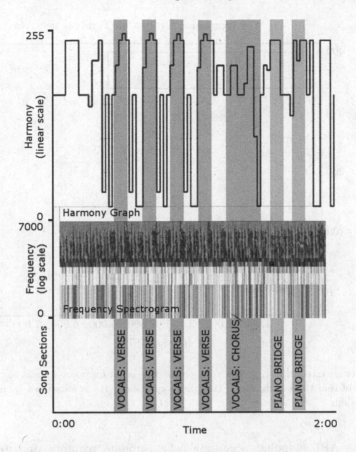

Fig. 3. Harmony calculated using a MIDI score compared to the frequency spectrum and manually identified parts of a sample song.

Evolved expressions vary in size between less than 10 sub-expressions, to over 7000, however, the vast majority of fit expressions contain less than 100 sub-expressions. The evaluation of larger expressions leads to a time delay. This is illustrated in Fig. 4, where we plot the average time taken to calculate the fitness of *mapping expressions*, by size, measured in number of sub-expressions. The figure displays the time taken to evaluate fitness rather than real-time evaluation times. This approach was chosen in order to obtain a fair evaluation delay across all input musical harmony values. The real-time evaluation delay can be accurately obtained by dividing the fitness evaluation time by 11, as shown in blue in Fig. 4.

4.4   Visual Display Generation

Finally, presenting and updating a visual display with minimal time delay is also a challenge. To achieve this, we implemented a *visualisation server*, a web

Expression Evaluation Time

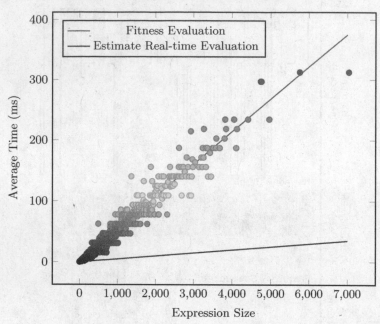

Fig. 4. Average execution time to evaluate the fitness of mapping expressions with an estimation of real-time evaluation. Mapping expression size is measured by number of sub-expressions.

server with API endpoints accepting colour update requests, and web pages served to render the display. Websockets were used to update the display as new colour update requests were received. The display could then be shown on a data projector, external monitors or any other display device. This approach was not ideal, but presented a cost effective and robust solution without the need for expensive stage lighting equipment and proprietary hardware for data transfer.

The overhead of evaluating a larger *mapping expression* as well as communicating with a visualisation tool could result in unacceptable delays, preventing suitable time synchronisation between audio and visuals. According to the ITU-R BT.1359-1 (1998) standard, a sound/vision timing difference of more than 90 ms is unacceptable, while anything less than 45 ms is undetectable. By plotting the time delay between a MIDI signal being received and a visual being rendered, as shown in Fig. 5, we can estimate this value. The figure displays a typical run with maximum delays of 79 ms, minimum delays of 11ms, and average delay of 24.357 ms indicating an acceptable delay.

Time Delay

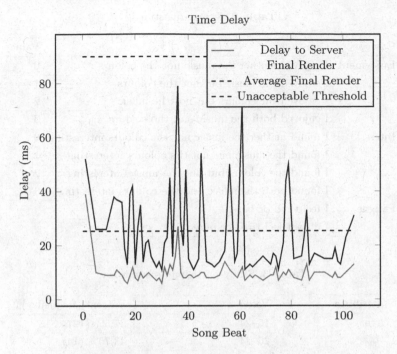

Fig. 5. Time delay in milliseconds from a MIDI message being sent to the Visualisation Server, and the visualisation being rendered for a single typical run

5   Survey Methodology

5.1   Experimental Setup

Using the approach outlined in Sect. 3.1, over 500 generations of *mapping expressions* were generated and stored. Three sample generations were selected and the lowest, median and highest fitness individual expressions were chosen for display. For each *mapping expression*, a visual display was generated using a sample music piece.

In the interest of displaying the same colours to all participants, and to avoid colour grading or exposure related loss of colour strength, colours were displayed on a screen which illuminated the room. The display consisted of a constant red background colour, and a varying foreground colour. The colours were arranged on screen as three vertical stripes, with the foreground colour in the center. This allowed the colours to illuminate the room distinctly.

The music piece, an arrangement of guitar, vocal harmony, piano, and light drums was used as it demonstrated a relatively strong periodic variation in musical harmony. The piece was composed by the authors to be enjoyable without focusing on any particularly polarising genre and also to avoid any intellectual property issues. An external speaker was used to ensure high audio quality throughout the study.

Table 2. Survey questions.

Category	Question	Value
Enjoyment	I enjoyed neither the music nor the colours	0
	I enjoyed the music but not the colours	1
	I enjoyed the colours but not the music	2
	I enjoyed both the music and the colours	3
Interest	I found neither the music nor the colours interesting	0
	I found the music but not the colours interesting	1
	I found the colours but not the music interesting	2
	I found both the music and the colours interesting	3
Fatigue	I feel tired or bored	
	I do not feel tired or bored	

Table 3. Study participants summary.

Attribute	Value	Percent (frequency)
Age	15–24	37.5% (12)
	25–44	43.75% (14)
	45–64	18.75% (6)
Gender	Male	56.25% (18)
	Female	43.75% (14)
Musical background	No experience	43.75% (14)
	Beginner or Intermediate	43.75% (14)
	Expert	12.5% (4)

Each of the 9 videos were of 2 min in length and were presented in a random order for each viewing. After each video, participants were asked to provide one answer to each of 3 multiple choice questions as shown in Table 2. In total, 32 participants were surveyed. A summary of the age, gender and musical background distributions are shown in Table 3[1].

6 Results

Each participant was presented with nine visualisations, divided into three fitness levels: low, medium and high. Each of the four mutually exclusive responses to Enjoyment and Interest questions were given a value, 0 to 3 as shown in Table 2.

The input variables include 3 categorical variables (or factors), namely Fitness, Gender, and Musical Background and 1 quantitative input (or covariate), Age. The response values within each fitness level were summed, obtaining

---

[1] Full results are available upon request. Please contact a.breen2@nuigalway.ie.

an ordered categorical response variable, taking values 0 to 9 representing the aggregate enjoyment or interest of a participant at each fitness level.

One complication in the analysis is that certain observations are correlated, namely those on the same participant at the different fitness levels.

In the software SPSS [13], suitable analysis can be conducted using Generalised Estimating Equations. This analysis provides no evidence of an effect of any of the three factors (Fitness, Gender, Musical Background), nor of the covariate (Age) on either of the responses (Enjoyment and Interest).

7   Discussion

The core goal of this work is to examine the possibility and practicality of implementing a system such as that proposed in our previous work. To this end, a system has been implemented that makes use of evolved *mapping expressions* and generated real-time aesthetic analogies by taking a live musical input, evaluating *mapping expressions* and generating a visual display. While this system was capable of creating aesthetic analogies, there is no evidence that these analogies have an effect on the aesthetic experience of a piece of music. While this is a negative aesthetic result, the implementation was successful. Given these results, a number of improvements could be made to improve the performance of the system, in terms of both optimizing the computational aspects of the system, and improving the aesthetic appeal of output visual displays.

7.1   Implementation

The grammatical evolution based system proved successful for evolving *mapping expressions*. There are obvious improvements that could be made, including increasing the gene length, increasing the size of the population and tweaking other parameters to obtain fitter individuals.

Obtaining a reliable musical input proved to be a major challenge in this work. The final system used a predefined MIDI score as musical input. While this approach might suit a live performance using digital instruments, it is clearly not practical for an acoustic performance. Without a predefined MIDI score, further work is needed for live audio input.

While live music performance may currently be beyond reach for analogue instruments, using a Fast Fourier Transform with a smoothing window does seem to be useful for identifying harmony with more granularity. This may have practical applications in song-part identification.

With a predefined MIDI input, it may be possible to look ahead at the score to dynamically change analogies. Meta-data MIDI messages may also be used for this purpose. Further, this may also be a practical use of the sampled music approach which may be used to identify the section of a song with more subtlety than the MIDI based approach.

The evaluation of *mapping expressions* is currently not a major factor in the input-output delay. However, if multiple attributes, longer gene length, or more complex expression selection approaches were put in place, this may become a bottleneck. Fortunately, the design of the *mapping expression* allows for some optimisation to be done. Firstly, with expressions and sub-expressions taking on a tree like structure, expression sub-trees that do not contain any dynamic input variables like music input may be pruned. Using this approach, large sub-trees could be replaced with constant values, greatly reducing the processing required for each evaluation of the expression. Secondly, the use of multiple aesthetic attributes lends itself to multi-threading or distributed evaluation which may further reduce any processing bottle-neck that occurs. Finally, it may be possible to build a single mapping expression using sub-trees built from separate genes. This approach would allow the distribution of processing even for the evaluation of single expression, though this approach would clearly not be required without substantially larger expressions.

There are a number of drawbacks with the approach taken for displaying visuals. Notably, the use of a web browser to generate a display presented on a screen or data projector is clearly inferior to the use of professional stage lighting hardware. While the current implementation was cost effective and quick to build, professional stage lighting hardware would create a far better atmosphere. The use of dedicated lighting hardware would also greatly reduce the input-output delay with current delays being mainly the result of rendering within the browser.

7.2   Survey

Our hypothesis assumed that an evolved *mapping expression* between the measurable attributes of musical and visual harmony would produce an improved aesthetic experience compared to a random *mapping expression*. The survey conducted did not verify this hypothesis. We believe this is primarily due to the effect of musical and visual harmony being too weak to have an observable effect. To remedy this, we believe the use of more rudimentary analogies, such as musical loudness to visual brightness, will have more obvious observable effects.

A second factor may be the simplistic structure of the analogy with just one *mapping expression*. Similar to the use of musical loudness to visual brightness suggested above, the use of these expressions in combination may also create more obvious observable effects allow a more dynamic analogy to be made, perhaps improving the aesthetic quality of the analogy. Using the implementation outlined in this paper, the use of multiple attributes should be possible without major architectural changes. However, gathering the required aesthetic response data may be prohibitive.

Our previous work suggests the use of a supervised fitness method which may encourage the creation of analogies that generate more recognisable effects. Using the music-visual analogy as an example, this process might complement a normal rehearsal, allowing human supervision without inducing a great deal of fatigue.

We have mentioned that the use of predefined MIDI lookahead and the sampled music approach may allow the use of multiple analogies for separate parts of a song. This approach may also improve the aesthetic quality of the system.

Finally, we hope to further investigate the consistency of feedback between judges using an output created by hand. Our survey is based on the assumption that an aesthetically pleasing visual can be created using the defined apparatus. It is possible that such an output is not possible or detectable given the constraints of the system and the inconsistency of responses.

8   Conclusion

It has been demonstrated that a system may be implemented to make use of Grammatical Evolution to make real-time aesthetic analogies by taking a live musical input, evaluating evolved *mapping expressions* and generating a visual display. *Mapping expressions* have been shown to have evaluation times fast enough to allow real time music to visual mapping. Furthermore, this mapping enables real time mapping within acceptable limits (90 ms), even considering the time delay resulting from input audio preprocessing and output visual rendering.

Statistical analysis of a survey conducted to investigate the effect of the system showed no evidence of any effect of any factors (Fitness of expressions, Gender, Musical Background) or the covariate (Age) on the enjoyment or interest of participants in a music and visual display.

Finally, a number of possible improvements have been proposed which may impact the aesthetic and artistic value of the system. The main proposed improvements are the use of multiple *mapping expressions*, the use of multiple measurable aesthetic attributes, and the use of a human supervised fitness method.

References

1. Becker, J.D.: The modeling of simple analogic and inductive processes in a semantic memory system. In: Proceedings of IJCAI-1969, pp. 655–668, Washington (1969). http://citeseerx.ist.psu.edu/viewdoc/download?doi=10.1.1.77.8346&rep=rep.1&type=pdf
2. Behravan, R.: Automatic mapping of emotion in music to abstract visual arts. Ph.D. thesis, University College London (2007)
3. Birkhoff, G.: Aesthetic Measure. Cambridge University Press, Cambridge (1933)
4. Eliasmith, C., Thagard, P.: Integrating structure and meaning: a distributed model of analogical mapping. Cogn. Sci. **25**(2), 245–286 (2001)
5. Evans, T.G.: A program for the solution of a class of geometric-analogy intelligence-test questions. Technical report, Air Force Cambridge Research Labs LG Hanscom Field (1964)
6. French, R.M.: The computational modeling of analogy-making. Trends Cogn. Sci. **6**(5), 200–205 (2002). http://www.ncbi.nlm.nih.gov/pubmed/11983582
7. Gentner, D.: Children's performance on a spatial analogies task. Child Dev. **48**, 1034–1039 (1977)

8. Gentner, D., Forbus, K.D.: Computational models of analogy. Wiley Interdisc. Rev. Cogn. Sci. **2**(3), 266–276 (2011). http://doi.wiley.com/10.1002/wcs.105
9. Goguen, J.A.: Art and the brain: editorial introduction. J. Conscious. Stud. **6**(6), 5–14 (1999)
10. Hofstadter, D.R.: Analogy as the core of cognition. In: Gentner, D., Holyoak, K., Kokinov, B. (eds.) The Analogical Mind: Perspectives from Cognitive Science, pp. 499–538. MIT Press, Cambridge (2001)
11. Huang, M.: The neuroscience of art. Stanford J. Neurosci. **2**(1), 24–26 (2009). http://www.stanford.edu/group/co-sign/huang.pdf
12. Hummel, J.E., Holyoak, K.J.: Distributed representations of structure: a theory of analogical access and mapping. Psychol. Rev. **104**(3), 427 (1997)
13. IBM Corp: IBM SPSS Statistics for Windows (2013)
14. Kandinsky, W., Rebay, H.: Point and Line to Plane. Courier Corporation, North Chelmsford (1947)
15. Klee, P.: Pedagogical Sketchbook. Praeger Publishers, New York (1925). http://monoskop.org/images/3/30/Klee_Paul_Pedagogical_Sketchbook_1960.pdf
16. Marshall, J.B., Hofstadter, D.R.: The metacat project: a self-watching model of analogy-making. Cogn. Stud. Bull. Japn. Cogn. Sci. Soc. **4**(4), 57–71 (1997)
17. Mitchell, M.: Analogy-Making as Perception: A Computer Model. MIT Press, Cambridge (1993)
18. O'Neil, M., Ryan, C.: Grammatical Evolution, pp. 33–47. Springer, Heidelberg (2003)
19. Palmer, S.E., Schloss, K.B., Sammartino, J.: Visual aesthetics and human preference. Ann. Rev. Psychol. **64**, 77–107 (2013)
20. Ramachandran, V.S., Hirstein, W.: The science of art: a neurological theory of aesthetic experience. J. Conscious. Stud. **6**(6), 15–35 (1999)
21. Reitman, W.R.: Cognition and Thought: An Information Processing Approach. Wiley, New York (1965)
22. Thibaut, J.P., French, R., Vezneva, M.: The development of analogy making in children: cognitive load and executive functions. J. Exp. Child Psychol. **106**(1), 1–19 (2010)

Deep Artificial Composer: A Creative Neural Network Model for Automated Melody Generation

Florian Colombo[(✉)], Alexander Seeholzer, and Wulfram Gerstner

School of Computer and Communication Sciences and School of Life Sciences,
Brain-Mind Institute, Ecole Polytechnique Fédérale de Lausanne,
Lausanne, Switzerland
florian.colombo@epfl.ch
http://lcn.epfl.ch

Abstract. The inherent complexity and structure on long timescales make the automated composition of music a challenging problem. Here we present the Deep Artificial Composer (DAC), a recurrent neural network model of note transitions for the automated composition of melodies. Our model can be trained to produce melodies with compositional structures extracted from large datasets of diverse styles of music, which we exemplify here on a corpus of Irish folk and Klezmer melodies. We assess the creativity of DAC-generated melodies by a new measure, the novelty of musical sequences, showing that melodies imagined by the DAC are as novel as melodies produced by human composers. We further use the novelty measure to show that the DAC creates melodies musically consistent with either of the musical styles it was trained on. This makes the DAC a promising candidate for the automated composition of convincing musical pieces of any provided style.

Keywords: Automated music composition · Deep neural networks · Sequence learning · Evaluation of generative models

1 Introduction

If music composition *simply* consists of applying rules, where does this leave musical creativity? Modern composers trying to escape the rules and structural constraints of traditional western music had to confront themselves with their inevitable presence [1] – by trying to flee from some limitations, we fall into others. While considered a drawback by some composers, others saw the very well defined structure of music as a possibility for its formalization [2].

The generation of music according to its inherent structure goes back as far as Ada Lovelace, who saw in the analytical engine of Charles Babbage, the ancestor of computers, the potential to *"compose elaborate and scientific pieces of music"* [3]. In the present age of machine intelligence, many approaches have been undertaken to algorithmically write musical pieces [4]. A large challenge

© Springer International Publishing AG 2017
J. Correia et al. (Eds.): EvoMUSART 2017, LNCS 10198, pp. 81–96, 2017.
DOI: 10.1007/978-3-319-55750-2_6

for music production algorithms is to be trainable, allowing them to learn the structural constraints of a given body of music, while at the same time producing original musical pieces adhering to these constraints without simply replaying learned motifs.

We present here the Deep Artificial Composer (DAC), a trainable model that generates monophonic melodies close to, but different from, tunes of a given musical style, which still carry well-defined musical structure over long timescales. To achieve this, the DAC employs multi-layer recurrent neural networks consisting of variants of Long Short-Term Memory (LSTM) units [5] to implement a generative model of note transitions. Networks based on LSTM units and their variants [6,7] have yielded impressive results on tasks involving temporal sequences, such as text generation [8,9] or translation [10,11]. Importantly, when used as generative models, neural networks have been able to produce artistic and nonartistic data close to real data or human art [12,13]. A prominent recent example was able to extract and apply arbitrary painting styles to any given picture [14].

Earlier attempts at extracting and generating the complex temporal structures of music with neural networks do not yet produce convincing musical pieces, mainly because of the lack of long timescale musical structure [15,16] and creativity/novelty of the produced pieces [17]. In particular, earlier work also based on LSTM networks proposed to improvise patterns of notes constrained on the pentatonic scale and harmony for blues improvisation [18], or to reproduce exactly a unique melody [17]. While both of these works are promising for automated melody generation, it remained unaddressed how these neural networks could generalize from arbitrary corpora to compose new songs in similar styles, carrying a shared structure with the melodies in the given training corpus.

The DAC presented here is an extension of the generative model of music presented in [19]. There, the authors introduced a neural network that could learn some of the temporal dependencies between a note in a melody and the history of notes in the same melody. Also, they introduced the separation of pitch and duration features of musical notes to explicitly model their relation with two different RNNs. Here, we introduce additional network layers, a more generalizable processing of melodies, as well as a different training protocol. Importantly, we demonstrate that we can train the model on a corpus containing two inherently different musical styles. By evaluating on a large scale the melodies generated by DAC networks with a measure of musical novelty, we show that the DAC can learn to compose consistent songs in different scales and styles.

We begin by introducing the model in Methods as part of a general formalization of iterative sequence generation, motivating and comparing the DAC with other generative models. In the Results section, we show the evolution of the DAC networks during training and present our measure of novelty that we then use to assess the melodies generated by the network. This allows us to show that the generated melodies are close to, but different from, melodies from the learned musical corpus, and classify the style of generated melodies. Finally, we present and analyze two example melodies that the DAC imagined.

2   Methods

Data Processing and Representation. Melodies in symbolic notation (`abc` notation [20]) are converted to `MIDI` sequences of notes, including the full note sequences resulting from symbolic repetitions with possibly different endings (e.g. "prima/seconda volta"). The Python package music21 [21] is then used to extract the corresponding sequences of notes, which are used to build integer alphabets of possible pitches and durations, into which melodies are translated (see Fig. 1**A**). In addition, silences are encoded as an additional alphabet entry in the pitch representation, while song endings are represented as additional entries in both the duration and pitch alphabets.

For a given corpus of melodies, we first construct the alphabets that contain all the pitches and durations values present in the entire dataset and map each of them to a unique integer value. We then represent the notes of each song with respect to these alphabets, thus producing two sequences of integers (pitches and durations) for each song. Songs containing very rare events (total occurrence in all songs < 10) in terms of either duration or pitch are removed from the dataset. The probability of very rare notes occurring in generated melodies (see below) will be closely related to the number of their occurrence in the training set – by removing songs containing these rare notes we reduce the alphabet sizes, which leads to a smaller number of parameters and a decreased time needed to train the model, without affecting the notes occurring in generated melodies much. Finally, we remove duplicate songs from the dataset (determined by unique pitch sequences).

The preprocessing of melodies presented here is more general than the one presented in an earlier model [19], where the `abc` notation was used directly. First, this makes the implementation of our model applicable to the more universal and more widely available `MIDI` format. Second, in contrast to the representation used there, we *do not transpose songs* into C Major/A minor, in order to retain and learn the actual key of songs. It is worth noting that this data normalization could be performed for smaller datasets to increase the number of examples of each pitch transition, at the cost of information about musical key.

To represent pitches and durations as input to DAC networks, we use a one-hot encoding scheme. This consists of mapping integer values i (for pitch or duration) to binary vectors where each entry is 0 except the one at dimension i, which is set to 1. This input vectors will later be denoted by $\mathbf{p}^s[n]$ (for pitch) and $\mathbf{d}^s[n]$ (for duration) of the nth note in song s.

General Statistical Formalism for Melody Generation. We considering a monophonic melody as a sequence of N notes $x[1 : N]$. The DAC is built under the assumption that each note $x[n]$ is drawn sequentially from a probability distribution (the "note transition distribution") $T[n]$ over all possible notes. Generally, for a melody s, the transition distribution of note at time step n will depend on song dependent inputs $\varPhi^s[n]$ (e.g. the history of notes up to note n)

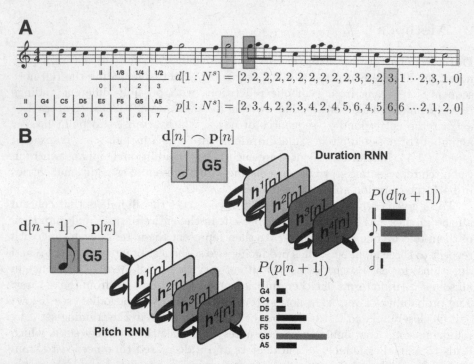

Fig. 1. Deep Artificial Composer networks: (**A**) Representation of the nursery rhyme *Brother John* in the DAC by two sequences of integers $d[1 : N^s]$ and $p[1 : N^s]$. Each integer in the top sequence corresponds to a single duration given by the duration alphabet to the left. Similarly, each integer in the bottom sequence corresponds to a single pitch given by the alphabet of pitches to the left. (**B**) The flow of information in DAC networks. The current duration $d[n]$ and pitch $p[n]$ are inputs to the duration network (bottom), while the pitch network (top) receives as inputs the current pitch $p[n]$ and the upcoming duration $p[n+1]$. Both RNNs learn to approximate the distribution over the next possible durations and pitches based on their inputs and their previous hidden state (recurrent connections). In order to generate melodies from the artificial composer, the upcoming duration and pitch are sampled from the estimated transition distributions (bars on the lower right).

and a set of fixed model parameters Θ. Then, the probability of a melody s could be expressed as the probability of all notes occurring in sequence:

$$P(x^s[1 : N^s]) = \prod_{n=1}^{N^s} T\big(x^s[n] \mid \Phi^s[n], \Theta\big) . \tag{1}$$

However, the inputs and parameters of this distribution have to be precisely defined – they could in principle encompass many different dimensions: the notes, the harmony, the tempo, the dynamic, and meta-parameters like for example the composer's musical and nonmusical experiences, or his general state of mind at the time of composition.

A simple approximation of the distribution T is given by Markov chain models of order k [22], where the probability of the next note depends only on the model parameters Θ_{MC} and history dependent input consisting of the last k notes:

$$\Phi^s_{MC}[n+1] = x^s[n-k:n].\qquad(2)$$

Markov chain models, however, can capture only a fraction of possible long-range temporal structure inherent in music [15]: due to the fixed restriction on the amount of past events that can influence the probability of occurrence of a note, this class of models will not capture all the complexity and temporal dependencies of music even with parameters trained to give the best possible approximation (the maximum likelihood estimate for a given set of data).

Rich temporal structure can in principle be captured by models that rely on recurrent neural networks (RNN). RNNs are probabilistic models that have potentially unlimited memory since their internal (so called "hidden") state H_{RNN} can depend on the entire history of prior notes in the song s. In this way, the input for each note x at time n can depend on the network representation of the full history $H_{RNN}(x^s[1:n-1])$ in addition to the current note $x^s[n]$ and the set of parameters Θ_{RNN}:

$$\Phi^s_{RNN}[n+1] = H_{RNN}(x^s[1:n-1], \Theta_{RNN}) \cup x^s[n].\qquad(3)$$

Splitting the Joint Probability of Duration and Pitch. For the transition distribution $T(\Phi^s_{DAC}[n], \Theta_{DAC})$ of the DAC we chose a conditional approach, where the note transition distribution T is split into two transition distribution: T_D for *duration* transitions and T_P for *pitch* transitions.

This splitting of the transition distributions into pitch and duration is an important difference of this work to previous attempts at symbolic music modeling with RNNs [16–18]. By this splitting, the DAC explicitly encodes and learns the duration (and pitch) of each note. An important design choice is that we train two separate RNNs separately to approximate the two transition distributions while *providing the upcoming duration as an additional input to the pitch network*. This architecture conditions the distribution of upcoming pitches on the duration of the upcoming note, effectively allowing the DAC to take into account the relation between a note duration and its pitch. Conditioning the pitch predictions on the upcoming duration is motivated by the fact that in standard western music theory there exists a relation between the duration of a note and its pitch. For example, a short note followed by a longer note has a high chance to be at the end of a musical phrase: long notes act as a tension release, typically exemplified by a short leading-tone which resolves to a longer note a semitone higher or lower. Additionally, rhythm is generally a more stable feature of western music. Indeed, in a corpus of songs written in a common style, examples of similar rhythmical patterns are more frequent than examples of similar melodic patterns, i.e. the rhythm is more shared across songs than the melody. With a song's rhythm being less informative than its melody [23], we chose to condition the melody on the rhythm in order to optimize the *expressiveness* of our model: for a given rhythmical (melodic) pattern, many (few) examples of different melodies will be seen by the model, from which the DAC could extract transition distributions.

By representing a note in song s by its duration and pitch as $note^s[n] = (p[n], d[n])$, we can formalize the splitting of the joint probability in the DAC model in the general formalism introduced above:

$$T(note^s[n] \mid \Phi^s_{AC}[n], \Theta_{AC}) = T(p[n], d[n] \mid \Phi^s_{AC}[n], \Theta_{AC}) \qquad (4)$$
$$= T_D(d[n] \mid \Phi^s_D[n], \Theta_D)$$
$$\times T_P(p[n] \mid \Phi^s_P[n], \Theta_P), \qquad (5)$$

where Θ_D and Θ_P are the parameters of the duration and pitch networks, respectively. The inputs are given by

$$\Phi^s_D[n+1] = H_D(d^s[1:n-1], p^s[1:n-1], \Theta_D) \cup d^s[n] \cup p^s[n], \qquad (6)$$
$$\Phi^s_P[n+1] = H_P(p^s[1:n-1], d^s[1:n], \Theta_D) \cup d^s[n+1] \cup p^s[n], \qquad (7)$$

where H_D and H_P are the internal (hidden) network states of the duration and pitch networks, respectively.

Summarizing, the DAC is constituted of two RNNs with similar architecture (see Fig. 1**B**). On the one hand, the duration network is trained to approximate the probability distribution over possible upcoming durations given the *current pitch and duration* and the network internal representation H_D of previous notes. On the other hand, the pitch network is trained to approximate the probability distribution over possible upcoming pitches given the *duration of the upcoming note* in addition to the current pitch and the network internal representation H_P (which contains the current duration).

Network Architecture. The DAC model was implemented in `python` using the `theano` library [24]. The pitch and duration RNNs are each composed of a binary input layer, four recurrent layers each composed of 128 gated recurrent units (GRU) and an output layer (see Fig. 1B). GRUs are a recent variant of the popular LSTM unit [5] that is particularly suited for sequence learning tasks and requires fewer parameters at similar performance [7,25].

The input layers are one-to-one mappings of the inputs to binary input units, in agreement with the splitting of the conditional probability (see the previous section). The input to the duration network $\mathbf{x}_D[n]$ is a concatenation of the duration and pitch vectors $\mathbf{d}[n]$ and $\mathbf{p}[n]$. The input to the pitch network $\mathbf{x}_P[n]$ is a concatenation of the current pitch and *upcoming duration* vectors $\mathbf{p}[n]$ and $\mathbf{d}[n+1]$. Each layer is then fully connected to every downstream layer.

In contrast to [19], we also increased the number of hidden layers to 4, which enables us to learn more complex structure and longer timescale dependencies.

The update equations for the activations of the recurrent layer $\mathbf{h}^i[n]$ at time step n for layer $i \in \{1, 2, 3, 4\}$ is given by the update equations of GRUs [10]. They are provided here with an adaptation to our architecture:

$$\mathbf{h}^i[n] = \mathbf{z}^i[n] \odot \mathbf{h}^i[n-1] + (\mathbf{1} - \mathbf{z}^i[n]) \odot \widetilde{\mathbf{h}}^i[n] \,, \tag{8}$$

$$\widetilde{\mathbf{h}}^i[n] = \tanh\left(\mathbf{w}_{y^i h^i}\mathbf{y}^i[n] + \mathbf{r}^i[n] \odot \mathbf{w}_{h^i h^i}\mathbf{h}^i[n-1]\right), \tag{9}$$

$$\mathbf{z}^i[n] = \sigma\left(\mathbf{w}_{y^i z^i}\mathbf{y}^i[n] + \mathbf{w}_{h^i z^i}\mathbf{h}^i[n-1] + \mathbf{b}^i_z\right), \tag{10}$$

$$\mathbf{r}^i[n] = \sigma\left(\mathbf{w}_{y^i r^i}\mathbf{y}^i[n] + \mathbf{w}_{h^i r^i}\mathbf{h}^i[n-1] + \mathbf{b}^i_r\right), \tag{11}$$

where $\mathbf{z}^i[n]$ and $\mathbf{r}^i[n]$ are the activations of the update and reset gates, respectively, $\mathbf{w}$ and $\mathbf{b}$ are the network parameters, $\sigma(x) = (1 + \exp(x))^{-1}$ is the logistic sigmoid function, $\odot$ denotes the element-wise product and $\mathbf{y}^i$ is the feed-forward input to layer i, which consists of both the global inputs $\mathbf{x}[n]$ as well as hidden layer activations $\mathbf{h}^{j<i}[n]$.

The update equation for the output unit k activation $\mathbf{o}[n, k]$ at note n is

$$\mathbf{o}[n, k] = \Theta\big(\mathbf{w}_{y^o o}\mathbf{y}^o[n] + \mathbf{b}^o\big)[k] \,, \tag{12}$$

where $\mathbf{y}^o$ is the feed-forward input of the output layer, which consists of the global inputs $\mathbf{x}[n]$ and all hidden layer activations $\mathbf{h}^{\{1,2,3,4\}}[n]$, and $\Theta(\mathbf{x})[k] = \frac{e^{\mathbf{x}[k]}}{\sum_j e^{\mathbf{x}[j]}}$ is the Softmax function. The Softmax normalization ensures that the values of the output units sum to one, which allows us to interpret the output of the two RNNs as probability distributions. In particular,

$$\mathbf{o}_D[n, d] \approx P(d[n+1]) \,, \tag{13}$$

$$\mathbf{o}_P[n, p] \approx P(p[n+1]) \,. \tag{14}$$

Training. To train the DAC, it is presented with melody examples from a training corpus and its parameters are updated in order to produce a shape of the estimated transition distributions T_P and T_D closer to the transition distributions of the training data.

A training epoch consists of feeding both networks 100 randomly selected melodies from a fixed partition of 80% of the melodies in the corpus (training sct). After a whole song has been fed to the model, the parameters of the pitch and duration networks are updated with the ADAM optimizer [26] in order to maximize the log likelihood of the networks given the song s:

$$\mathcal{L}^s(\theta) = \frac{1}{N_s - 1} \sum_{n=1}^{N_s-1} \log\big(P(x^s[n+1])\big) \,, \tag{15}$$

where $P\big(x^s[n+1]\big)$ is given by the value of the activation of output unit that corresponds to the effective upcoming duration or pitch. After each epoch, we evaluate the predictive performances – the percentage of correctly predicted upcoming pitch or duration – of the DAC on a random sample of 100 songs from the remaining 20% of the corpus (validation set), as well as 100 random songs from the training set to monitor the DAC performance. In contrast to [19], where training was stopped earlier, here we introduce a new stopping criterion: training is stopped when the average accuracy of the predictions over one epoch reaches a steady state on both the training and validation sets.

Composition of Melodies. At any given stage during (and after) learning the DAC can compose melodies. To do so, we iteratively add notes to a growing melody by sampling notes from the learned note transition distributions, while providing the previously sampled note as an input to the network. We first sample a duration from the duration network, which is then used as an additional input to the pitch network, from which finally sample a pitch:

$$d[n+1] \sim \text{Multinomial}\big(\Lambda(\mathbf{o}_D[n], T)\big), \tag{16}$$

$$p[n+1] \sim \text{Multinomial}\big(\Lambda(\mathbf{o}_P[n], T)\big), \tag{17}$$

where T is the sampling temperature and $\Lambda(\mathbf{x}, T) = \frac{\exp(\ln(\mathbf{x})/T)}{\|\exp(\ln(\mathbf{x})/T)\|_1}$. Note that a sampling temperature lower than 1.0 produces a more stereotypical distribution as the entropy is decreased, while a temperature higher than 1.0 will make the landscape of the predictions more flat as the temperature increases until reaching maximum entropy corresponding to the uniform distribution. A temperature of 1.0 corresponds to a sample from a multinomial distribution with the parameters given exactly by the outputs of the DAC.

3   Results

We trained and evaluated the Deep Artificial Composer (DAC, see Methods) network with a corpus comprising 2408 Irish folk melodies [27] and 585 Klezmer tunes [28]. The songs in the corpus contained on average 250 notes per melody for a total of 748'250 notes.

Training the DAC. The evolution of the accuracy of the DAC can be assessed by observing the shape of the output distributions of both networks at different training times (see Fig. 2). We observe an increase in the accuracy of the estimated distributions. Indeed, at the latest stages of training, both the duration and pitch network can relatively well predict the true upcoming pitches and durations even though the melody has never been by the DAC before. The predictive performance, or accuracy, of the DAC, corresponds to the percentage of transitions where the model gave the highest probability to the true upcoming event. On average, the accuracy of the trained DAC networks on melodies from the validation/training set is 50%/80%. The average predictive performance of the duration model is 80%/85%.

Rhythm is a rather stable feature of the musical corpora we trained on, i.e. typically several examples of the same rhythmical patterns are found in the corpus – in contrast to similar motifs of consecutive pitches which are much

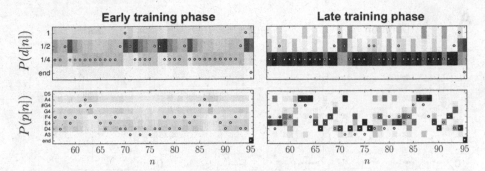

Fig. 2. DAC prediction performance during training: Evolution of the DAC predictions of duration (top) and pitch (bottom) during training from early (left) to late (right) training epochs. For each plot, the x-axis shows time steps, while the y-axis represents possible outputs of the networks. We show the last 40 predictions of the DAC on a randomly selected melody of the validation set. Small black and white circles represent the true upcoming notes, while shaded squares represent the probability distributions of the DAC's output units (white: $P = 0$, black: $P = 1$).

more variable. Thus, the duration network tends to show an overall higher accuracy. This observation further confirms our choice to condition the pitch transition distributions on the upcoming note duration, because by this the less accurate prediction is always conditioned by the more accurate one.

We invite the reader to convince himself of the increase in song structure as training progresses, by listening to mp3 renderings of DAC-generated songs during different training phases at https://goo.gl/eUO9BR.

Assessing the Novelty of Melodies Generated by the DAC. In order to assess the quality of the generated melodies, we designed an automated measure of novelty, as an indirect measure of one feature of musical creativity [29]. For this, we define $\mathcal{N}_m^C(s)$ as the novelty of a song s with respect to a corpus C and motif-size m. This consists in computing for the melody s which fraction of the song's transitions are *not found* in the musical corpus C, given only the last $m - 1$ notes. If all possible transitions of a melody with preceding motifs of size $m - 1$ can be found in the reference corpus, the novelty of the melody for the motif size m is 0.0. If all possible transitions with preceding motifs of size $m - 1$ are absent from the reference corpus, the novelty of the corresponding melody is 1.0.

We applied this measure on the pitch and duration features independently for motifs of size 2, 4, 8 and 16, comparing sequences of the training set to songs in the validation set, to data generated by our model, and to sequences of pitches and durations generated by a Markov chain model of order 4 and 8 trained on the duration and pitch sequences independently (see Fig. 3A). An ideal artificial composer would produce melodies that show a novelty computed against the training set of similar magnitude as the validation set ($\mathcal{N}_m^{\mathrm{Train}}(\mathrm{Valid})$ on Fig. 3A), across all motif sizes. Melodies produced by the DAC indeed show

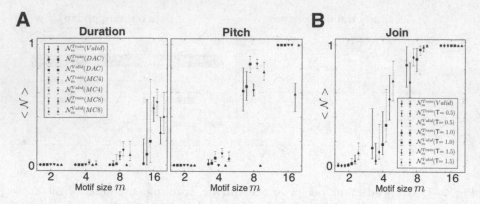

Fig. 3. Assessing DAC melodies by novelty measurement: The median, first and third quartiles of the novelty $\mathcal{N}$ of sequences (see text) with respect to the motif size m used for sequence comparison. (**A**) Novelty of sequences of pitches and durations separately (400 songs). Sampling temperature is 1.0. (**B**) Novelty of joint pitch and duration pairs for DAC songs generated at varying temperatures (400 songs each). The indicated corpora are: Valid = validation set, Train = training set, MC4 = Markov chain model of order 4 (400 songs), MC8 = Markov chain model of order 8 (400 songs).

a similar profile across motif sizes ($\mathcal{N}_m^{\mathrm{Train}}(\mathrm{DAC})$ on Fig. 3**A**), while Markov chain models tend to reproduce only motifs that they have already seen ("MC4" and "MC8" on Fig. 3**A**). This is especially true for motif sizes equivalent to the Markov chain order.

For both validation data and melodies generated by the DAC, the novelty profile gradually increases from 0.0 for motifs of size 2, up to completely novel choices of transitions for motifs of bigger sizes. For motifs of sizes bigger than 2, both validation melodies and sequences generated by the DAC are more novel in terms of their pitch sequence than their duration sequence. This further confirms the observation that rhythm is a stable feature of music that varies less across songs and motivate our choice to condition pitch predictions on duration (see Methods). In summary, the novelty profiles show that the DAC is able to create melodies that are *as novel with respect to the training data, as validation melodies from the same corpus*. However, because the DAC has to learn from examples, it generates melodies that are less novel with respect to songs in the training corpus than to songs from the validation set (compare $\mathcal{N}_m^{\mathrm{Train}}(\mathrm{DAC})$ to $\mathcal{N}_m^{\mathrm{Valid}}(\mathrm{DAC})$ in Fig. 3**A**).

The DAC not only generates songs that are closer to real songs (from the validation set) than Markov chain models in terms of novelty but by our design choice, it effectively models the relation between duration and pitch. To model the relation between duration and pitch with a Markov model, one would need to train it on the joint pitch and duration representation, which would yield to much sparser data and consequently even worse generative performance.

As a proxy for the final output of the model, we evaluate the novelty of joint sequences of notes (pitch-duration pairs) to study the effect of the sampling temperature on the novelty of DAC generated melodies (Fig. 3B). We observe that the sampling temperature parameter can be used to coax the novelty of the artificial composer networks. Higher temperatures result in a more exploratory composer, while lower temperatures yield a less *inspired* artificial composer (as seen by less novelty in the generated sequences). Songs produced by the DAC with a sampling temperature of 1.0 are the closest to real data (from the validation set) in terms of novelty, which is expected since training is performed at this temperature.

Analysis of Melodies Produced by the DAC. The novelty profile across motif sizes introduced in the last section (cf. Fig. 3A) nicely aggregates how similar melodies are with respect to the given corpus – we can convert the novelty $\mathcal{N}$ to a similarity by calculating $1 - \mathcal{N}$. Therefore, we computed the novelty (similarity) profiles of generated joint melodies (pitch and duration pairs, motif sizes $2, 4, 8$, and 16) with respect to *only Irish melodies* and *only Klezmer melodies* of the training set, and used this data to train a simple classifier (3-Nearest Neighbor [30]) of melodic style.

The trained classifier can well distinguish between Irish and Klezmer melodies, given the similarity $(1 - \mathcal{N})$ profile of melodies with respect to the Irish and Klezmer reference corpora (see Fig. 4). Classification is almost always correct (error rate $= 0.02$) on data from the validation set. We then used the

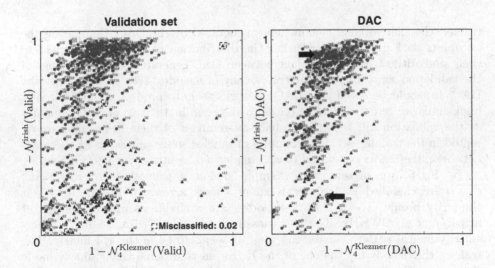

Fig. 4. Melody classification from the novelty profile: Similarities $(1 - \mathcal{N})$ of melodies from the validation set (left) and 400 DAC-generated songs (right) to either Klezmer melodies (x-axis) or Irish melodies (y-axis) of the training set. Similarity was calculated on motif-size $m = 4$. Each datapoint is labeled by the classification result (see text): circles for Irish and triangles for Klezmer. The few classification errors are highlighted with dashed squares. Sampling temperature for the generated melodies on the right figure is 1.0. Arrows point to the melodies displayed in Fig. 5.

Fig. 5. Examples of DAC-generated melodies: Two melodies generated at sampling temperature 1.0. One melody classified as Irish (**A**) and one classified as Klezmer (**B**). Bar plots (left) show the similarity $(1 - \mathcal{N})$ of the song with respect to songs of the training set of a given style, across motif-sizes. Notes highlighted in red shaded areas correspond to transitions of the last note that are present in the training data given the last 7 notes (corresponding a motif-size 8 based novelty of 0).

trained classifier to label new melodies generated by the DAC. Comparing the two plots on Fig. 4, we observe that the distribution of songs with respect to their similarities is very resemblant between DAC-generated songs and songs of the validation set, validating the consistency of melodic styles generated by the DAC. It should be noted that DAC networks seem to produce songs with very high similarity to either dataset (close to the coordinate axes) not as frequently as the validation set, which is probably a remnant of their being underrepresented in the training set. For examples of labeled songs generated by the DAC networks, the reader is referred to mp3 renderings at https://goo.gl/xTiYmg.

In Fig. 5, we present two example melodies generated by the DAC, that were classified as Irish or Klezmer (black arrows in Fig. 4) and their similarity profile $(1 - \mathcal{N})$. Both melodies are available as mp3 renderings at https://goo.gl/r6WKHv. The Irish-classified melody (Fig. 5A) exhibits coherent rhythmical patterns across the tune. It seems to be in the 4/4 metric. The scale of the melody fits the one of the E Dorian traditional Gaelic music mode throughout the entire melody, and it ends and starts on the fundamental. The similarity profiles (left) show that for motifs of a size smaller than 16, the melody shows larger similarity to Irish reference motifs than Klezmer reference motifs. Interestingly, the ascendant movement that appears at the end of the melody is coherent with the first note, and would permit a repetition of the whole melody – while we did not explicitly encode this notion in the model, this seems to be a staple of Irish music and is often found in the Irish corpus that the model was trained on. As such, it is also often present in DAC-generated melodies. Similarly, the melody labeled as Klezmer (Fig. 5B) shows coherent rhythmical patterns across the melody. The melody is in the scale of D Phrygian dominant traditional Klezmer mode and ends on the fundamental as well.

In summary, both example melodies imagined by the DAC present a very well-defined rhythmic and melodic structure that can be related to either Irish or Klezmer style. Notably, the modes of the two melodies fit either style. For both melodies, the artificial composer learned to end on the fundamental and display a well-defined structure, which can only be achieved if DAC networks were able to internally represent the mode through their hidden states.

4 Discussion

In this paper, we presented and evaluated the Deep Artificial Composer (DAC) – a model for the algorithmic composition of melodies. The use of a trainable architecture allows the model to automatically extract the invariant temporal dependencies specific to a given corpus of melodies. The DAC can, therefore, be trained on any monophonic music corpus in the MIDI format. In addition, we showed that DAC networks are able to learn similarities and divergences between different music styles, here on the example of Irish and Klezmer music. Indeed, most of the melodies generated by the trained artificial composer are consistent in style – when the model starts generating a melody with a structure close to an Irish folk song, the entire generated melody is close to the structure of an Irish tune. In similar ways, provided that it has been sufficiently trained, the artificial composer is also consistent in scale and rhythm.

Our model has to learn to correctly represent past events in order to be able to condition predictions on this internal representation of the history. By doing so, repetitions and more complex structures can be encoded and reused by the model during the generation of new melodies. However, probabilistic models are, by design, weak at repetition tasks: the only way for the model to exactly repeat the same sequence twice is to produce Dirac distributions. We have shown (see Fig. 2) that the DAC, as a probabilistic model, can come close to this. Additionally, because a sequence of durations is often redundant (durations are often repeated multiple times), the model has to learn an "internal clock" of the rhythm next to learning the transitions between different durations. We believe that a hybrid system with embedded grammars could improve the DAC, by externalizing some of the structure that is hard to produce with probabilistic networks. For example, learnable grammars such as the Sequitur algorithm [31] could be used to bias the network.

We demonstrated that the DAC as a model of melody composition produces melodies that are as similar to melodies in the training set as real tunes of the same style (from the validation set) are. To make this point, we introduced an indirect measure of creativity of generated sequences, by means of computing the novelty of produced sequences with respect to the trained corpus. Since the number of algorithms for automated music composition is increasing, there is currently a lack of tools and methods for the automated evaluation of creativity of these systems [32,33]. While care should be taken to define the notion of quantitative *creativity* more carefully, our quantitative assessment of creativity without the intervention of human judges by the measure of novelty is, to our knowledge, the first of its kind. We would like to note that this method could be applied to any system generating (musical or nonmusical) sequences.

Finally, we used the novelty measurement to classify the style of melodies produced by the DAC. A much more sophisticated alternative would be to include such a classifier in the DAC itself by adding labeled *style* input and output units. These could act as latent variables that, after training, could be used to coax the DAC toward the generation of melodies in the corresponding style. Since such a classifier would be able to work on the full sequence history, we believe that the small number of misclassified generated melodies could be further reduced.

In order to illustrate this work, `mp3` examples generated by `GarageBand` from DAC-generated `MIDI` files are available for download at https://goo.gl/CHhoAu. Note that a clarinet is used for generated songs classified as Klezmer and a harp for generated songs classified as Irish.

Acknowledgments. The authors thank Samuel P. Muscinelli and Johanni Brea for their guidance and helpful comments. This research was partially supported by the Swiss National Science Foundation (200020_147200) and the European Research Council grant no. 268689 (MultiRules).

References

1. Seeger, C.: On dissonant counterpoint. Mod. Music **7**(4), 25–31 (1930)
2. Xenakis, I.: Formalized Music: Thought and Mathematics in Composition, vol. 6. Pendragon Press, New York (1992)
3. Lovelace, A.: Notes on l. menabrea's "sketch of the analytical engine invented by charles babbage, esq.". Taylor's Scientific Memoirs 3 (1843)
4. Fernández, J.D., Vico, F.: Ai methods in algorithmic composition: a comprehensive survey. J. Artif. Intell. Res. **48**, 513–582 (2013)
5. Hochreiter, S.: The vanishing gradient problem during learning recurrent neural nets and problem solutions. Int. J. Uncertain. Fuzziness Knowl. Based Syst. **6**(2), 107–116 (1998)
6. Gers, F.A., Schmidhuber, J., Cummins, F.: Learning to forget: Continual prediction with LSTM. Neural Comput. **12**(10), 2451–2471 (2000)
7. Chung, J., Gulcehre, C., Cho, K., Bengio, Y.: Empirical evaluation of gated recurrent neural networks on sequence modeling. arXiv preprint (2014). arXiv:1412.3555
8. Graves, A.: Generating sequences with recurrent neural networks. arXiv preprint (2013). arxiv:1308.0850
9. Karpathy, A.: The unreasonable effectiveness of recurrent neural networks. (2015). http://karpathy.github.io/2015/05/21/rnn-effectiveness/. Accessed 1 Apr 2016
10. Cho, K., Van Merriënboer, B., Gulcehre, C., Bahdanau, D., Bougares, F., Schwenk, H., Bengio, Y.: Learning phrase representations using RNN encoder-decoder for statistical machine translation. arXiv preprint (2014). arXiv:1406.1078

11. Sutskever, I., Vinyals, O., Le, Q.V.: Sequence to sequence learning with neural networks. In: Advances in Neural Information Processing Systems, pp. 3104–3112 (2014)
12. Goodfellow, I., Pouget-Abadie, J., Mirza, M., Xu, B., Warde-Farley, D., Ozair, S., Courville, A., Bengio, Y.: Generative adversarial nets. In: Advances in Neural Information Processing Systems, pp. 2672–2680 (2014)
13. Gregor, K., Danihelka, I., Graves, A., Rezende, D.J., Wierstra, D.: Draw: a recurrent neural network for image generation. arXiv preprint (2015). arXiv:1502.04623
14. Gatys, L.A., Ecker, A.S., Bethge, M.: A neural algorithm of artistic style. arXiv preprint (2015). arXiv:1508.06576
15. Todd, P.M.: A connectionist approach to algorithmic composition. Comput. Music J. **13**(4), 27–43 (1989)
16. Boulanger-Lewandowski, N., Bengio, Y., Vincent, P.: Modeling temporal dependencies in high-dimensional sequences: application to polyphonic music generation and transcription. arXiv preprint (2012). arXiv:1206.6392
17. Franklin, J.A.: Computational models for learning pitch and duration using lstm recurrent neural networks. In: Proceedings of the Eighth International Conference on Music Perception and Cognition (ICMPC8), Adelaide, Australia, Causal Productions (2004)
18. Eck, D., Schmidhuber, J.: Finding temporal structure in music: blues improvisation with LSTM recurrent networks. In: Proceedings of the 2002 12th IEEE Workshop on Neural Networks for Signal Processing, pp. 747–756. IEEE (2002)
19. Colombo, F., Muscinelli, S.P., Seeholzer, A., Brea, J., Gerstner, W.: Algorithmic composition of melodies with deep recurrent neural networks. In: Proceedings of the First Conference on Computer Simulation of Musical Creativity (CSMC 2016), Huddersfield, UK (2016)
20. Walshaw, C.: abc notation. http://abcnotation.com/. Accessed 11 March 2016
21. Cuthbert, M., Ariza, C., Hogue, B., Oberholtzer, J.W.: music21, a toolkit for computer-aided musicology. http://web.mit.edu/music21/. Accessed 11 Nov 2016
22. Ames, C.: The markov process as a compositional model: a survey and tutorial. Leonardo **22**, 175–187 (1989)
23. Hébert, S., Peretz, I.: Recognition of music in long-term memory: are melodic and temporal patterns equal partners? Mem. Cogn. **25**(4), 518–533 (1997)
24. Bergstra, J., Breuleux, O., Bastien, F., Lamblin, P., Pascanu, R., Desjardins, G., Turian, J., Warde-Farley, D., Bengio, Y.: Theano: a CPU and GPU math expression compiler. In: Proceedings of the Python for Scientific Computing Conference (SciPy), vol. 4, p. 3. TX, Austin (2010)
25. Jozefowicz, R., Zaremba, W., Sutskever, I.: An empirical exploration of recurrent network architectures. In: Proceedings of the 32nd International Conference on Machine Learning (ICML 2015), pp. 2342–2350 (2015)
26. Kingma, D., Ba, J.: Adam: A method for stochastic optimization. arXiv preprint (2014). arXiv:1412.6980
27. Norbeck, H.: Henrik Norbecks's corpus of irish tunes. http://www.norbeck.nu/abc/. Accessed 03 Nov 2016
28. Chamber, J.: John chambers's corpus of klezmer tunes. http://trillian.mit.edu/jc/music/abc/Klezmer/. Accessed 03 Nov 2016
29. Gorney, E.: Dictionary of creativity: terms, concepts, theories and findings in creativity research (2007)
30. Fix, E., Hodges, J.L.: Discriminatory analysis-nonparametric discrimination: consistency properties. Technical report, DTIC Document (1951)

31. Nevill-Manning, C.G., Witten, I.H.: Identifying hierarchical strcture in sequences: a linear-time algorithm. J. Artif. Intell. Res. (JAIR) **7**, 67–82 (1997)
32. Jordanous, A.: A standardised procedure for evaluating creative systems: computational creativity evaluation based on what it is to be creative. Cogn. Comput. **4**(3), 246–279 (2012)
33. Loughran, R., ONeill, M.: Generative music evaluation: why do we limit to human? In: Proceedings of the first Conference on Computer Simulation of Musical Creativity (CSMC 2016), Huddersfield, UK (2016)

A Kind of Bio-inspired Learning of mUsic stylE

Roberto De Prisco, Delfina Malandrino, Gianluca Zaccagnino,
Rocco Zaccagnino[✉], and Rosalba Zizza

Dipartimento di Informatica, Università di Salerno, 84084 Fisciano, SA, Italy
zaccagnino@dia.unisa.it
http://music.dia.unisa.it

Abstract. In the field of Computer Music, computational intelligence approaches are very relevant for music information retrieval applications. A challenging task in this area is the automatic recognition of musical styles. The style of a music performer is the result of the combination of several factors such as experience, personality, preferences, especially in music genres where the improvisation plays an important role.

In this paper we propose a new approach for both recognition and automatic composition of music of a specific performer's style. Such a system exploits: *(1)* a one-class machine learning classifier to learn a specific music performer's style, *(2)* a music splicing system to compose melodic lines in the learned style, and *(3)* a LSTM network to predict patterns coherent with the learned style and used to guide the splicing system during the composition.

To assess the effectiveness of our system we performed several tests using transcriptions of solos of popular Jazz musicians. Specifically, with regard to the recognition process, tests were performed to analyze the capability of the system to recognize a style. Also, we show that performances of our classifier are comparable to that of traditional two-class SVM, and that it is able to achieve an accuracy of 97%. With regard to the composition process, tests were performed to verify whether the produced melodies were able to catch the most significant music aspects of the learned style.

1 Introduction

Music is one of the art forms that in many aspects has benefited of the use of computers: sound synthesis and design, digital signal processing, automatic composition, music information retrieval and so on. Computer Music is an emerging research area for the application of computational intelligence techniques, such as machine learning, pattern recognition, bio-inspired algorithms and so on. In this work we face interesting challenges in two different areas: the *music information retrieval* (MIR) and the *algorithmic music composition*. MIR is the interdisciplinary science of retrieving information from music, involving activities such as content-based organization, indexing, and exploration of digital music data sets, while algorithmic music composition addresses the problem of composing music by means of a computer program with no (or minimal) human intervention.

© Springer International Publishing AG 2017
J. Correia et al. (Eds.): EvoMUSART 2017, LNCS 10198, pp. 97–113, 2017.
DOI: 10.1007/978-3-319-55750-2_7

One of the problems to solve in MIR is how to model of *musical styles*. The style of a music performer is the result of his experience and mainly represents his personality or personal preferences. This is even more evident in musical genres where improvisation plays an important role. In this field computer systems could be trained to recognize the main stylistic features of performers. Many works explore the capabilities of machine learning methods to recognize musical style [1–6]. An application of such systems can be their use in cooperation with automatic composition algorithms to guide this process according to a given stylistic profile. In [7] several techniques to define algorithms for music composition have been discussed. In this work we investigate the use of a bio-inspired approach, specifically *music splicing systems* [8].

The main contributions of our paper can be summarized as follows.

1. To propose a machine learning approach for the *recognition* of a music performer's style. We describe a one-class support vector machine classifier to learn musical styles, trained on the significant features extracted from transcribed solos. We show that performances are comparable to traditional two-class SVM, and that the classifier achieves an average accuracy of 97%.
2. To propose a bio-inspired approach for automatic *composition* in a specific music style. We describe a music splicing composer of melodies in the style learned by the classifier. To build the composer we define a Long Short-Term Memory (LSTM) network to predict patterns in the learned style.
3. To assess our approach, we performed several tests by using transcriptions of solos from popular Jazz musicians. We analyzed the capability of our system in recognizing a style and we verified the stylistic coherence of the composed melodies. Moreover, we showed that the use of a LSTM network to predict pattern could involve better results respect to the traditional approaches.

Organization of the paper. In Sect. 2 we describe some relevant related works. In Sect. 3 we provide basic notions about music improvisation and the needed background to understand the paper. In Sects. 4, 5 and 6 we provide details about the machine learning recognizer, the LSTM predictor and the music splicing composer, respectively. Finally, in Sects. 7 and 8 we describe the experiment analysis and we conclude with final remarks and future directions.

2   Related Work

Several works focus on the musical styles recognition problem. Most of these explore the capabilities of machine learning methods. Pampalk et al. [1] use self-organizing maps (SOMs) to cluster music digital libraries according to sound features of musical themes. Whitman et al. [2] present a system based on neural networks and support vector machines able to classify an audio fragment into a given list of sources or artists. In [3], a neural system to recognize music types from sound inputs is described. An emergent approach to genre classification is used in [4], where a classification emerges from the data without any a priori given set of styles. The idea is to use co-occurrence techniques to extract musical similarity between titles or artists. Pitch histograms regarding the tonal pitch are used in [6] to describe blues fragments of Charlie Parker.

We remark that most of these works only take into account the problem of recognizing the styles in terms of music genre, without considering the style of a performer. Furthermore, they do not face the problem of the composition of music in a specific style. Several automatic composers, based on different approaches, have been already proposed in previous works: rules and expert-systems [9–11], systems based on a combination of formal grammars, analysis and pattern matching techniques [12], neural networks [13]. Several automatic composers are based on meta-heuristics, specifically on genetic algorithms [14–20]. Other works have been proposed for different musical genres or problems: thematic bridging [15], Jazz solos [16], harmonize chords progressions [17], monophonic Jazz composition given a chord progression [18], *figured* bass problem [19], and finally, *unfigured* bass problem [20]. In [8, 21] the authors define a bio-inspired approach for automatic composition, called music splicing systems. Starting from an initial set of precomposed music the system generates a language in which each word is the representation of music in the style of the chosen composer.

3 Background

We assume that the reader is familiar with the basics of music, machine learning and bio-inspired approaches. However, in this section we provide the basic notations used throughout the paper.

3.1 Improvisation in Music

Music improvisation refers to the ability of playing music extemporaneously, without planning or preparation, by inventing variations on a melody or creating new melodies. In this work we will use the term *solo* to indicate the transcription of an improvised melody. There are many musical genres in which the improvisation assumes a fundamental role during a performance. Usually the improviser's choices are conditioned by own personal experience or preferences, and also by the specific period in which music is collocated. It is interesting to notice that music performers of the same period used similar music choices during their improvisations. Furthermore, during the improvisation, musicians often borrow and customize existing musical pieces from previous performances (also for other musician's performance). Despite this, in each music performer it is possible to found music features that characterize his specific style, and an expert ear is able to perceive such aspects. Thus, given a corpus of solos, we can extract such significant features and using them to recognize his style. We remark that in our approach we are interested in modeling the style of a musician by looking at his music performances and not at the music genre or the music period to which the performer belongs to.

One of the key for extracting the significant features is to study the role that each music note assumes in the specific context (chord) in which it is played. Formally, let Ch be a music chord, $S(Ch)$ be a scale (sequence of notes) chosen by the musician and n be a note played on Ch. In this work we assume that each scale is composed of 7 notes. Then it is possible to associate a degree (position) to n respect to $S(Ch)$. We indicate with $Degree(n, S(Ch))$ such a position. The degrees are indicates with roman numerals, and obviously there 12 possible degrees, ($7°$ in $S(Ch)$ and 5 out of $S(Ch)$. For example let $Ch = Dm7$ and $S(Dm7) = (D, E, F, G, A, B, C)$ (dorian mode). Let $n = G$ then $Degree(G, S(Dm7)) = IV$. Let $n = Eb$ then $Degree(Eb, S(Dm7)) = bII$.

Notice that the musician's choice of $S(Ch)$ for a chord Ch is a crucial choice for the style of such a musician. Typically, it is possible to associate several scales to the same chord. For example, in Jazz music, a dominant chord is often substituted by the tritone chord. For example, let $Ch = G7$. Very traditional musicians could choose $S(G7) = (G, A, B, C, D, E, F)$ (mixolydian mode of Cmaj scale). An alternative choice is $S(Ch) = (Db, Eb, F, Gb, Ab, Bb, Cb)$ (mixolydian mode in Gbmaj scale). Obviously, the role of a note can be different depending on the chosen scale. In the previous example, let $n = F$. Then, if $S(G7) = (G, A, B, C, D, E, F)$ then $Degree(F, S(G7)) = VII$. Otherwise, if $S(Db7) = (Db, Eb, F, Gb, Ab, Bb, Cb)$ then $Degree(F, S(G7)) = III$. We can easily to say that the chord $G7$ has been substituted by the chord $Db7$. This process is known in Jazz as *substitution*. Further details can be found in [22].

3.2 Splicing Systems

Splicing systems are formal models for generating languages, i.e., sets of words [23]. We start with an initial set of words and we apply to these words the splicing operation by using rules in a given set. The set of generated words is joined to the initial set and the process is iterated on this new set until no new word is produced. The *language generated* is the collection of all these words.

Formally, a *splicing system* is a triple $S = (\mathcal{A}, \mathcal{I}, \mathcal{R})$, where $\mathcal{A}$ is a finite alphabet, $\mathcal{I} \subseteq \mathcal{A}^*$ is the initial language and $\mathcal{R} \subseteq \mathcal{A}^*|\mathcal{A}^*\$\mathcal{A}^*|\mathcal{A}^*$ is the set of rules, where $|, \$ \notin \mathcal{A}$. A splicing system S is finite when $\mathcal{I}$ and $\mathcal{R}$ are both finite sets. Let $L \subseteq \mathcal{A}^*$. We set $\gamma'(L) = \{w', w'' \in \mathcal{A}^* \mid (x, y) \vdash_r (w', w''), x, y \in L, r \in \mathcal{R}\}$. The definition of the splicing operation is extended to languages as follows: $\gamma^0(L) = L, \gamma^{i+1}(L) = \gamma^i(L) \cup \gamma'(\gamma^i(L)), i \geq 0$, and $\gamma^*(L) = \bigcup_{i \geq 0} \gamma^i(L)$.

Definition 1. *Let $S = (\mathcal{A}, \mathcal{I}, \mathcal{R})$ be a splicing system. We denote by $L(S) = \gamma^*(\mathcal{I})$ the splicing language generated by S. We say that L is a splicing language if there exists a splicing system S such that $L = L(S)$.*

The interested reader can refer to [24] for a detailed study of the theory of formal languages, and to [23,25] for further readings.

In [8] authors describe a music composer based on a splicing system for 4-voice chorale-like music. Similar to this earlier approach, the basic idea is to treat music compositions as words and to view the music compositional process as the result of operations on words.

3.3 One-Class Support Vector Machine (OCSVM)

A OCSVM algorithm maps input data into a high dimensional feature space (via a kernel) and iteratively finds the maximum margin hyperplane which best separates the training data from the origin. The OCSVM may be viewed as a regular two-class SVM where all the training data lies in the first class, and the origin is taken as the only member of the second class [26].

In terms of recognition, in this work we faced the following problem: given a melody, we want to verify whether it is coherent with a music performer's style. As we will see in Sect. 4, the idea is to associate a vector of significant features to the melody, and to use a One-class SVM for mapping the vector belonging to the training data (corpus of solos). Given the feature vector v_i corresponding to the improvised melody m_i, one-class SVM gives us the classifier (rule) $f(v_i) = \langle w, v_i \rangle + b$, where $\langle w, v_i \rangle + b$ is the equation of the hyperplane, with w being the vector normal to this hyperplane and b being the intercept. OCSVM solves an optimization problem of finding the rule f with the maximum geometric margin. We can use f to assign a label to a test example v_i. If $f(v_i) \geq 0$ then m_i is considered out of style, otherwise is considered in the style.

3.4 Music Composition Using LSTM Recurrent Neural Networks

We assume that the reader is familiar with the basics of machine learning more specifically artificial neural networks (ANN), but for further information see [27].

As well explained in [27], in the basic feedforward ANN there is a single direction in which the information flows: from input layer to output layer. But in a recurrent neural network (RNN), this direction constraint does not exist. The idea is to create an internal state (internal memory) of the network which allows it to exhibit dynamic temporal behavior.

The most straightforward way to compose music with an RNN is to use it as single-step predictor [28–30]. The network learns to predict notes at time $t + 1$ using notes at time t as inputs. A feedforward ANN would have no chance of composing music in this way, since it lacks the ability to store any information about the past. In "theory" an RNN does not suffer from this limitation. However, in "practice", RNNs do not perform very well this task given the problem of *vanishing gradients* which makes difficult for the networks to deal correctly with long-term dependencies. LSTM has succeed in similar domains where other RNNs have failed, and several works [31] showed that it is also a good mechanism for learning to compose music. For further information about LSTM see [32].

4 The Recognizer

Our intuition is that in each music perfomer it is possible to found music features that characterize his specific style. Thus, given a corpus of solos by several performers, to recognize a specific style we can extract such significant features and using them to make recognition. We remark that in our approach such features are *automatically* extracted from the corpus. In the following, we first introduce the model to extract the features and then present our machine learning approach. We will indicate with $\mathcal{M}$ the corpus solos by several musicians used for the definition of the recognizer. We assume to have a small subset $\mathcal{M}'_X$ that are melodies improvised by a specific musician X. So we have a classification problem in which the melodies in $\mathcal{M}$ can belong to two possible classes (the class of melodies coherent with the X's style and the class of melodies not coherent with the X's style. We remark that all the melodies in $\mathcal{M}'_X$ was improvised by X. Thus, the melodies in $\mathcal{M} - \mathcal{M}'_X$ can be coherent with the X's style or not. The goal of the classifier is to classify correctly the melodies in $\mathcal{M}$ in these two classes. Let v_j be the feature vector of some melody m_j, obtained through the feature extraction model to be discussed shortly. The recognizer will be defined to classify m_j by using v_j as input. In the following sections we indicate the recognizer with $\mathcal{R}$.

4.1 Feature Model

We are interested in a model that captures the most significant features of a melody. We use the n-gram model introduced in [33], which identifies *tokens* in melodies whose importance can than be determined through some *statistical measure*. An n-gram refers to n tokens which are dependent on each other.

Once n-grams have been constructed by the training set of melodies, we use a statistical measure to calculate their relative importance. The result is list of n-grams, ordered according to their importance. Depending on the value of n we can have different models, and different size of such a list, so for practical reasons, we fix a maximum size of the list. Notice that the n-grams in such a list have been selected according to their importance, but there is no theoretical support to their style classification capability. Thus, we apply a Random Forest selection procedure to keep those that better contribute to the classification (details in Sect. 7). In Fig. 1 we provide an overall view of the features extraction process.

Fig. 1. Feature extraction process.

As we will see in Sect. 7, for each experiment we have tried several values of n but found no significant improvement beyond $n = 24$. In the following paragraphs, we will describe the tokens and statistical measure used in our approach.

The token. In our approach a token is the set of relevant information about a music note. Thus a n-gram represents a sequence of n notes. In order to describe a music note we decided to use three parameters (k_1, k_2, and k_3) that better explain the relation between melody and harmony, and specifically the role of the note respect to the chord in which is played. We refer to such parameters as *token's parameters*. The parameter k_1 indicates the *name chord*. In our approach we consider 12 possible music name chords, i.e., C, $C\#$, D, $D\#$, E, F, $F\#$, G, $G\#$, A, $A\#$, B. Thus we have 12 possible values for k_1. Specifically 0 for C, 1 for $C\#$, 2 for D and so on. The parameter k_2 refers to the *type chord*. In order to consider the most used type of chords, in this work we consider the chord types derived from the modes of the major, melodic minor and harmonic minor scales (see [22] for further details). In Table 1 we describe such type chords.

Table 1. Modes of the major, melodic minor and harmonic minor scales.

Description $S(Ch)$	Mode	Chord type	k_2 value
Major scale			
ionic	I	maj7	0
dorian	ii	m7	1
phrygian	iii	m7b9	2
lydian	IV	maj7#11	3
mixolydian	V	7	4
aeolian	vi	m7b6	5
locrian	vii	m7b5	6
Melodic minor scale			
ipoionic	i	m(maj7)	7
dorian b2	ii	m7b9	8
augmented lydian	III	maj7#5	9
dominant lydian	IV	7#11	10
mixolydian b6	V	7b6	11
locrian #2	vi	m7b5#2	12
superlocrian	vii	7alt	13
Harmonic minor scale			
ipoionic b6	I	m(maj7)	14
locrian #6	ii	m7b5b9b13	15
augmented lydian	iii	maj7#5	16
minor lydian	iv	m7#11	17
mixolydian b2b6	V	7b9b13	18
lydian #2	VI	maj7#9#11	19
diminished superlocrian	vii	○	20

The parameter k_3 indicates the *role* of the note respect to the chord. As explained in Sect. 1 for each scale there are 12 possible positions and so k_3 is an integer in the range from 0 to 11.

To summarize, the *3-tuple* $K^i = [k_1^i, k_2^i, k_3^i]$ is the token that describes the music note played at i^{th} time interval. For example $K^6 = [9, 4, 7]$ says that at the 6th time interval the music note played has degree V ($k_3 = 7$ means degree V) in the scale corresponding to the chord $Bb7$ ($k_1 = 9$ means name chord Bb and $k_2 = 4$ means type chord 7). Thus the note played is D.

We remark that in our approach, the duration of the note is not considered as parameter of the token. Consequently, the rhythmic structure of melodies is not considered in this phase. This because in our opinion, the duration information is not important for the role respect to the chord, but it is important for the performance and the execution of the melody. Thus, as we will see in Sect. 6.5 the duration of the notes will be established by the composer at the end of the composition process, by using a specific operator.

The statistical measure. In our approach the statistical measure used to evaluate the relative importance of n-grams is the *term frequency with inverse document frequency* ($tfidf$). Central to this measure is the term t. In our approach a term t is a n-gram, i.e., a sequence of n tokens, which corresponds to a sequence of n music notes. We use the boolean term frequency (tf) measure such that $tf(t, m_j) = 1$ if $t \in m_j$ (the sequence of notes corresponding to t occurs in the melody m_j), 0 otherwise. The inverse document frequency measure idf is defined as: $idf(t, m_j, \mathcal{M}) = \log(|\mathcal{M}|/|\{\ m_j \in \mathcal{M} : t \in m_j\ \}|)$.

Finally we obtain tf with idf for each $t \in m_j$ over the corpus $\mathcal{M}$, as: $tfidf(t, m_j, \mathcal{M}) = tf(t, m_j) \cdot idf(t, m_j, \mathcal{M})$.

Intuitively the $tfidf$ measure gives more weight to terms that are less common in $\mathcal{M}$, since such terms are more likely to make the corresponding melody stand out. The $tfidf$ measure thus transform our corpus $\mathcal{M}$ to the feature vector space. The feature vector corresponding to a melody m_j is denoted by v_j. The ith component of v_j, denoted with $v_j[i]$ is equal to $tfidf(t_i, m_j, \mathcal{M})$.

4.2   The OCSVM classifier

As explained before, let m_j be a melody we define a feature vector corresponding to m_j denoted by v_j. Such a vector contains an element for each significant feature of the melody, and the value of this element is the $tfidf$ explained in Sect. 4.1. In our approach we use a OCSVM for mapping v_j. The OCSVM gives us the classifier (rule) $f(v_j) = \langle w, v_j \rangle + b$, where $\langle w, v_j \rangle + b$ is the equation of the hyperplane, with w being the vector normal to this hyperplane and b being the intercept. If $f(v_j) \geq 0$ then m_j is considered as melody not coherent with the X's style, otherwise m_j is coherent with the X's style. As we will see in Sect. 7 we use the traditional supervised two-class support vector machine (SSVM) [34] as benchmark for the performance of our OCSVM classifiers. In Fig. 2 we provide an overall view of the recognition process.

Fig. 2. Recognition process.

5   The Predictor

In this section we describe a LSTM network to predict musical patterns that follows the style learned by the recognizer. In our framework, we first build the recognizer $\mathcal{R}$ as described in Sect. 4, and then we build the predictor by using information and data from $\mathcal{R}$.

Thus, let $\mathcal{M}$ be a corpus of solos and $\mathcal{M}'_X \subseteq \mathcal{M}$ be a set of solos performed by a musician X. Let $\mathcal{R}$ be the recognizer and n the value used for the construction of the n-grams as described in Sect. 4.1. Then, the idea is to define a machine learning predictor $\mathcal{P}$ that given a n-gram at time i have to predict the n-gram at time $t + 1$. This equivalent to saying that given the sequence of n music notes at time i, $\mathcal{P}$ has to predict the sequence of n music notes at time $i + 1$.

The training set. We define a data set of n-grams as follows. Let $calT \subseteq \mathcal{M}$ be the training set used for the training of $\mathcal{R}$. For each $m_j \in \mathcal{T}$ such that $\mathcal{R}$ says that m_j is coherent with the X's style ($f(m_j) < 0$), we consider the sequence of n-grams extract by m_j. So, let $Ngrams(m_j) = (ng_1, \ldots, ng_{k_j})$ be the sequence of n-grams extract from m_j, we insert the pair (n_i, n_{i+1}) in the training set for $\mathcal{P}$, for each $1 \leq i \leq k_j - 1$.

The architecture. In our approach the predictor $\mathcal{P}$ is a LSTM network and we now describe the steps and reasons for its definition.

1. First, we define a ANN that must predict the music note at time $i + 1$ given music note at time i. As described in Sect. 4, each note is a token K^i which is a triple $K^i = [k_1^i, k_2^i, k_3^i]$. Thus, input and output layers have size 3. In our approach we use four hidden layers having size 3.
2. To add "recurrency", we take the output of each hidden layer, and feed it back to itself as an additional input. Each node of the hidden layer receives both the list of inputs from the previous layer and the list of outputs of the current layer in the last time.
3. To solve the problem of shot-term memory we use LSTM nodes.
4. We need that the network has to be *(1) time-invariant*, i.e., identical for each step, and *(2) note-invariant*, i.e., identical for each note. To this, we build a stack of n identical RNN networks, one for each token. The overall network can be viewed as a predictor of the n-gram at time $i + 1$ given n-gram at time i. We use a "biaxial RNN" approach: there are two axis, i.e., time-axis and note-axis; each recurrent layer transforms input to outputs, and also sends recurrent connections along one of these axis. The first two layers

have connections across time steps, but are independent across notes. The last two layers have connections between notes, but are independent between time steps.

In Fig. 3 we provide an overall view of the construction of $\mathcal{P}$. As we will see in Sect. 6, it has a fundamental role for the construction of the composer.

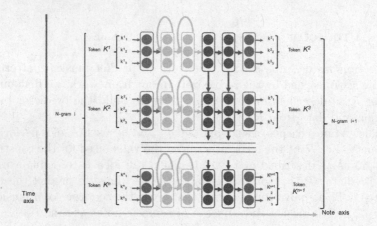

Fig. 3. LSTM recurrent neural network architecture.

6   The Composer

In this section we describe a music splicing system to compose melodies coherent with the style learned by the recognizer. In our framework, we first build the recognizer $\mathcal{R}$ as described in Sect. 4, then we build the predictor $\mathcal{P}$ as described in Sect. 5 and finally we build the composer.

Thus, let $\mathcal{M}$ a corpus of transcribed solos and $\mathcal{M}'_X \subseteq \mathcal{M}$ a corpus of solos performed by a specific musician X. The idea is to build a music splicing composer which produces melodies coherent with the X's style.

As explained in [8] in order to define a music splicing composer we need to define an alphabet, an initial set and a set of rules. Such a system is defined by an initial set of melodies coherent with the X's style, and a set of rules built by using $\mathcal{P}$. The language generated contains words that represent pieces of "new" melodies coherent the X's style. Formally, a music splicing system is a triple $\mathcal{S}_{\text{MSS}} = (\mathcal{A}_{\text{MSS}}, \mathcal{I}_{\text{MSS}}, \mathcal{R}_{\text{MSS}})$. Among the generated words we choose the best solution according to the evaluation function that we will define in Sect. 6.4.

6.1   The Alphabet $\mathcal{A}_{\text{mss}}$

We set $\mathcal{A}_{\text{MSS}} = \mathcal{A}_N \cup \mathcal{A}_T \cup \mathcal{A}_D \cup \mathcal{A}_S$ where $\mathcal{A}_N$ is the *chord name alphabet*, $\mathcal{A}_T$ is the *chord type alphabet*, $\mathcal{A}_D$ is *note degree alphabet* and $\mathcal{A}_S$ is the *separator alphabet*. Specifically, $\mathcal{A}_N = \mathcal{A}_D = \{0, \ldots, 11\}$, $\mathcal{A}_T = \{0, \ldots, 20\}$ and $\mathcal{A}_S = \{\tau, \mu_N, \mu_T, \mu_D\}$. We use $\mathcal{A}_{\text{MSS}}$ to represent melodies as words. As explained in Sect. 4 each token represents a music note. Thus, given a melody $m = (n_1, \ldots, n_l)$, for each note n_i, the token K^i is represented as a word w_i over $\mathcal{A}_{\text{MSS}}$. Specifically $w_i = \mu_N x_i \mu_T y_i \mu_D v_i$, where $x_i \in \mathcal{A}_N$, $y_i \in \mathcal{A}_T$ and $y_i \in \mathcal{A}_D$, for each $1 \leq i \leq l$. Thus, m is represented by the word $\mathcal{W}(m) = \tau w_1 \tau w_2 \tau \cdots \tau w_n \tau$.

Example 1. Let us consider the melody m shown in Fig. 4 (first 3 measures of the standard Jazz "All blues" of Miles Davis). As explained before, our representation does not take into account the duration of the notes (and tied notes are considered as a single note). We have $m = (n_1, \ldots, n_{10})$. Each n_i is played in G7 chord. The chord name G has value 7. Chord G7 is the associated scale is the mixolydian mode of the major scale (see Table 1), thus the chord type has value 4. Now, let us analyze the role of notes respect the G7. We have $n_1 = n_3 = n_5 = n_7 = n_9 = B$ which has degree *iii* (position value 4), $n_2 = n_6 = n_{10} = D$ which has degree V (position value 7) and $n_4 = n_8 = C$ has degree IV (position value 5). Thus, we have 10 tokens: $K^1 = K^3 = K^5 = K^7 = K^9 = (7, 4, 4)$, $K^2 = K^6 = K^{10} = (7, 4, 7)$ and $K^4 = K^8 = (7, 4, 5)$. So, $\mathcal{W}(m) = \tau w_1 \tau \cdots \tau w_{10} \tau$ and $w_1 = \mu_N 7 \mu_T 4 \mu_D 4$, $w_2 = \mu_N 7 \mu_T 4 \mu_D 7$, $w_3 = \mu_N 7 \mu_T 4 \mu_D 4$, $w_4 = \mu_N 7 \mu_T 4 \mu_D 5$, $w_5 = \mu_N 7 \mu_T 4 \mu_D 4$, $w_6 = \mu_N 7 \mu_T 4 \mu_D 7$, $w_7 = \mu_N 7 \mu_T 4 \mu_D 4$, $w_8 = \mu_N 7 \mu_T 4 \mu_D 5$, $w_9 = \mu_N 7 \mu_T 4 \mu_D 4$ and $w_{10} = \mu_N 7 \mu_T 4 \mu_D 7$.

Fig. 4. First 3 measures of Miles Davis's All Blues.

6.2 Initial Set and Rules Definition

The idea is to start from an initial set of melodies that we known to be coherent with the X's style. Thus, we consider the following set of n-grams (the n used by $\mathcal{R}$): let $\mathcal{T}$ be the training set used for the training of $\mathcal{R}$. For each $m_j \in \mathcal{T}$ such that $\mathcal{R}$ says that m_j is coherent with the X's style ($f(m_j) < 0$) we consider the list $Ngrams(m_j)$ of n-grams extract by m_j. So, let $Ngrams(m_j) = (ng_1, \ldots, ng_{k_j})$, we insert $\mathcal{W}(ng_i)$ in the $\mathcal{I}_{\text{MSS}}$, for each $1 \leq i \leq k_j - 1$.

We remark that the definition of the rules is crucial because they determine the language being generated. Unlike the problem described in [8], in which it was possible to model rules by using the theory of music harmony, in our case we have not a method to model rules to compose melodies coherent with the X's style. Thus, we can use $\mathcal{P}$ as follows.

1. *Group 1 (forcing sequence n-grams).* Let $w_i \in \mathcal{I}_{MSS}$ and n_i such that $w_i = \mathcal{W}(n_i)$, these rules force the composer to paste words corresponding to n-grams that really follows n_i in the training set $\mathcal{T}$. Formally, let $w_i \in \mathcal{I}_{MSS}$, and n_i such that $w_i = \mathcal{W}(n_i)$, we define the rule $r = w_i|\epsilon\$\epsilon|\mathcal{W}(\mathcal{P}(n_i))$ where, $\mathcal{P}(n_i)$ is the n-gram predicted by $\mathcal{P}$ with n_i as input.

2. *Group 2 (forcing sequence tokens).* Given two words $w_i, w_j \in \mathcal{I}_{MSS}$ and n_i, n_j such that $w_i = \mathcal{W}(n_i)$, and $w_j = \mathcal{W}(n_j)$. If n_i and n_j share a token, then these rules force the composer to cut w_i and w_j in correspondence of such a token, and paste the remaining words. Formally, let $w_i, w_j \in \mathcal{I}_{MSS}$, n_i such that $w_i = \mathcal{W}(n_i)$, n_j such that $w_j = \mathcal{W}(n_j)$, where $n_i = (k_{i,1}, \ldots, k_{i,n})$, $n_j = (k_{j,1}, \ldots, k_{j,n})$ and there exist l_1, l_2 such that $k_{i,l_1} = k_{j,l_2}$. Then, we define the rule $r = \mathcal{W}(k_{i,1}, \ldots, k_{i,l_1-1})\epsilon\$\epsilon|\mathcal{W}(k_{j,l_2}, \ldots, k_{j,n_j})$.

6.3   Implementation Details

In general, given a splicing system $\mathcal{S} = (\mathcal{A}, \mathcal{I}, \mathcal{R})$, the generated language $L(\mathcal{S})$ is an infinite set of words, and the number of iterations of the splicing operation to generate it is unbounded. Of course, for practical reasons, we need to fix bounds for both these parameters (cardinality of the language and number of iterations). Thus, we fix a number k of iterations and a maximal cardinality p_{max}. We also define k languages as follows. We set $L_0 = \mathcal{I}_{MSS} = \gamma^0(\mathcal{I}_{MSS})$. For any i, $1 \leq i \leq k$, we consider $L'_i = L_{i-1} \cup \gamma'(L_{i-1})$, which corresponds to enlarge L_{i-1} by an application of all the rules in $\mathcal{R}_{MSS}$ to all possible pairs of words in L_{i-1}. If $\mathrm{Card}(L'_i) \leq p_{max}$, then $L_i = L'_i$. Otherwise, L_i is obtained from L'_i by erasing the $\mathrm{Card}(L'_i) - p_{max}$ words in L'_i that are the worst with respect to an evaluation function, i.e., the quality of the melody in terms of stylistic coherence. Therefore, to measure the quality of the compositions and to choose the better solutions we consider such a function. Finally, we define $L(k, p_{max}) = \cup_{1 \leq i \leq k} L_i$ as the (k, p_{max})-language generated by $\mathcal{S}_{MSS}$. We remark that $L(k, p_{max})$ is the language considered during the experiments described in Sect. 7.

6.4   The Evaluation Function

As we will see in Sect. 7, in order to use the composer, from a practical point of view we need to set some parameters, such as, maximum size of the generated language and so a function f_e to evaluate the melodies produced, used to select at each splicing step the best compositions.

As for the definition of the rules described before, also in this case we have not music rules to use for defining a function that evaluate a composition in terms of "stylistic" goodness. So the idea is to use the predictor $\mathcal{P}$ as follows. Let w be a word generated by the composer $\mathcal{S}_{MSS}$. Let $m = (n_1, \ldots, n_l)$ the melody such that $w = \mathcal{W}(m)$. Now, let $Ngrams(m) = (ng_1, \ldots, ng_{l-1})$ be the sequence of n-grams extract from m. We define f_e as:

$$f_e(m) = \sum_{1 \leq i \leq l-1} (tfidf(ng_i, m, \mathcal{M})) + Diff(ng_{i+1}, \mathcal{P}(ng_i))$$

where $Diff(ng_i, \mathcal{P}(ng_i))$ is the *difference* between ng_{i+1} and $\mathcal{P}(ng_i)$ that is the n-gram predicted by $\mathcal{P}$ with ng_i as input. Such a difference is defined as follows: let $K^{i+1} = [k_1^{i+1}, k_2^{i+1}, k_3^{i+1}]$ be the token for ng_{i+1} and $K'^{i+1} = [k_1'^{i+1}, k_2'^{i+1}, k_3'^{i+1}]$ be the token for $\mathcal{P}(ng_i)$. Then $Diff(ng_{i+1}, \mathcal{P}(ng_i)) = |k_1^{i+1} - k_1'^{i+1}| + |k_2^{i+1} - k_2'^{i+1}| + |k_3^{i+1} - k_3'^{i+1}|$.

6.5 The Rhythmic Transformation of the Melodies

We remark that when a word w is generated, it only represents a sequence of music notes $m = (n_1, \ldots, n_k)$. In order to add a rhythmic structure to m we use an operator that applies the following operations (each operation is performed with a uniform distribution of probabilities):

1. *Duration:* initially the duration of each note is assumed to be equal a beat duration. For each note n_i, the operator changes the duration of $n_i \in m$. If the duration of n_i is increased of a value inc then the duration of the remaining notes is decreased of $inc/(k-1)$; otherwise, if the duration of n_i is decreased of a value dec then the duration of the remaining notes is increased of $dec/(k-1)$.
2. *Rest notes:* $(n_1, \ldots, n_k)$ for each note $n_i \in m$, the operator inserts a rest note having the duration dur_i of n_i. So the duration of the remaining notes is decreased of dur_i/k.
3. *Tied notes:* For each note $n_i, n_{i+1} \in m$ such that $n_i = n_{i+1}$, the operator inserts a ligature between n_i and n_{i+1}.
4. *Triplet notes:* Given the list of notes $(n_1, \ldots, n_k)$, for each note n_i, n_{i+1}, n_{i+2}, the operator creates a triplet by using these notes.

7 Experimental Analysis

In this section we report the results of tests that we carried out to assess the validity of our approach. We implemented the system in Python by using the Anaconda and Scikit-learn libraries. We focused on Jazz music and thus we create a data set $\mathcal{M}$ of transcribed solos in MusicXML format, from three very popular Jazz musicians: Louis Armstrong (A), Charlie Parker (P), and Miles Davis (D). Formally, $\mathcal{M} = \mathcal{M}_\mathcal{A} \cup \mathcal{M}_\mathcal{P} \cup \mathcal{M}_\mathcal{D}$ where $\mathcal{M}_\mathcal{A}$, $\mathcal{M}_\mathcal{P}$ and $\mathcal{M}_\mathcal{D}$ are the sets of Armstrong's, Parkers's and Davis's solos respectively. In details, $|\mathcal{M}_\mathcal{A}| = 50$, $|\mathcal{M}_\mathcal{P}| = 50$ and $|\mathcal{M}_\mathcal{D}| = 50$, so $|\mathcal{M}| = |\mathcal{M}_\mathcal{A}| + |\mathcal{M}_\mathcal{P}| + |\mathcal{M}_\mathcal{D}| = 150$. The average number of notes for each melody in $\mathcal{M}$ is 318.

To validate our approach we performed 3 experiments, one for each chosen musician. For each experiment we fixed $X \in \{A, P, D\}$ and we selected the reference set $\mathcal{M}' \subseteq \mathcal{M}_\mathcal{X}$. Then, by using $\mathcal{M}_\mathcal{X}$ and $\mathcal{M}'$ we build a OCSVM recognizer $\mathcal{R}_X$ (Sect. 4), a predictor $\mathcal{P}_X$ (Sect. 5) and a composer $\mathcal{C}_X$ (Sect. 6). The aim of each experiment is to validate and to verify the capability of: *(1)* $\mathcal{R}_X$ in recognizing the X's style, *(2)* $\mathcal{P}_X$ in predicting patterns that follows the X's style and *(3)* $\mathcal{C}_X$ in composing melodies that follows the X's style.

Recognition performances. First we use a traditional supervised two-class support vector (SSVM) machine as a benchmark for the performance of $\mathcal{R}_X$. For $\mathcal{R}_X$ we build a training set by considering the 80% of melodies in $\mathcal{M}'$ (randomly chosen). For the SSVM, we build a training set by considering the 80% of melodies in $\mathcal{M}_X$ (randomly chosen) and a testing set by considering the 20% of melodies in $\mathcal{M}_X$ (randomly chosen). As we can see in Table 2 (left) best results have been obtained with $n = 24$, and in this case two classifiers achieve very similar rates, with true positive and negative rates of up to 0.97 and false positive and negative of only 0.03. This prove that the performance of $\mathcal{R}_X$ are comparable to the traditional SSVM. $\mathcal{R}_X$, where training set of only a subset of melodies in $\mathcal{M}_X$ is used for the learning, has the advantage that it requires fewer examples of melodies that follows the X's style. As a result we obtain that the one-class support vector machines were proven to roughly as good as two-class support vector machines for this problem.

Prediction efficacy. In this section we describe the results of a study aiming at assess the efficacy of $\mathcal{P}$ in terms of abilities in predicting patterns that follows the X's style. As explained in Sect. 5 the accuracy of $\mathcal{P}$ is also validated and compared against a feedforward ANN and a recurrent ANN trained on the same training set. Table 2 (right) summarizes the (most significant) best average results (over the three experiments) about the prediction rate. As we can see, in a range of 22000 epochs, we have obtained the highest prediction rate using the LSTM recurrent neural network, with Back propagation training, $M = 0.6$ (momentum) and $L = 0.7$ (learning rate).

Table 2. Classifiers performances (left) and Test with average prediction rate (right)

Feature model	Classifier	$\mathcal{M}_X$ Yes	Not	$\mathcal{M} - \mathcal{M}_X$ Yes	Not
$n = 12$	$\mathcal{R}_X$	0.91	0.09	0.06	0.94
	SSVM	0.90	0.1	0.04	0.96
$n = 16$	$\mathcal{R}_X$	0.94	0.06	0.05	0.95
	SSVM	0.94	0.06	0.04	0.96
$n = 24$	$\mathcal{R}_X$	0.3	0.97	0.04	0.96
	SSVM	0.3	0.97	0.03	0.97

Representation	M	L	Prediction rate [%]
LSTM	0.6	0.7	95.1
LSTM	0.5	0.4	93.9
Recurrent NN	0.5	0.5	89.1
ANN	0.4	0.6	85.4
LSTM	0.7	0.3	85.1

Music quality. The quality and the stylistic coherence of music produced by $\mathcal{C}_X$ has been evaluated by music experts among conservatory's teachers and professional musicians. All participants had more than 10 years of experience in the music field. For each experiment we select the best 4 solutions (available online[1]) produced by $\mathcal{C}_X$ according to the evaluation function f_e (see Sect. 6.4). We asked participants to listen to such a list and respond to the following questions: *(1) "How do you rate the quality of music?"*, and *(2) "How do you rate the*

---

[1] http://goo.gl/FWn2EX.

coherence of the music with the X's style?" (rating on a 7-point Likert scale).
We also asked participants to provide a motivation about their judgments.

Our experts rated very positively both quality and stylistic coherence
($M = 6.2$ and $M = 6.5$, respectively). They also were impressed about the sound-
ness of the music produced by $\mathcal{C}_X$. Nevertheless, some participants felt that in
some case the rhythm is too elaborate. The motivation is that the automatic
rhythmic operator defined in Sect. 6.5 sometimes is not able to create appropriate rhythms.

8   Conclusion

Several works present different systems for recognizing a music style, mostly
based on machine learning approach. Conversely, relatively few works describe
automatic systems composer of music coherent with a specific style.

In this paper we propose a new approach for the recognition of a music performer's style and automatic composition of melodies coherent with such a style.
Specifically, we describe a machine learning approach for recognition based on a
one-class machine learning classifier, and a bio-inspired approach for automatic
composition based on music splicing systems. We performed several tests to analyze the efficacy of our system in terms of a style recognition and coherence of
the produced melodies with such a style. Performances of our classifier are comparable with the traditional two-class SVM and it is able to achieve an accuracy
of 97%. Some future works will include new experiments to widen the corpus of
solos to consider other Jazz musicians and other musical genres. Furthermore, to
address comments obtained during the music quality evaluation we will define
a stylistic rhythmic operator, by using the same approach used for the melodic
features. The idea is to individuate the more significant rhythmic features of a
music style, and using such features during the composition process.

References

1. Pampalk, E., Dixon, S., Widmer, G.: Exploring music collections by browsing different views. In: Proceedings of the Fourth International Conference on Music
   Information Retrieval, Baltimore (2003)
2. Whitman, B., Flake, G., Lawrence, S.: Artist detection in music with minnow-
   match. In: Proceedings of the 2001 IEEE Signal Processing Society Workshop
   Neural Networks for Signal Processing XI, pp. 559–568. IEEE (2001)
3. Soltau, H., Schultz, T., Westphal, M., Waibel, A.: Recognition of music types.
   In: Proceedings of the IEEE International Conference on Acoustics, Speech, and
   Signal Processing (1998)
4. Pachet, F., Westermann, G., Laigre, D.: Musical data mining for electronic music
   distribution. In: First International Conference on WEB Delivering of Music
   (WEDELMUSIC 2001), Florence, Italy, November 23–24, 2001, pp. 101–106 (2001)
5. Dannenberg, R.B., Thom, B., Watson, D.: A machine learning approach to musical style recognition. In: International Computer Music Conference, pp. 344–347
   (1997)

6. Tzanetakis, G., Ermolinskyi, A., Cook, P.: Pitch histograms in audio and symbolic music information retrieval. In: Fingerhut, M. (ed.) Third International Conference on Music Information Retrieval: ISMIR 2002, pp. 31–38 (2002)

7. Miranda, E.: Composing Music with Computers. Focal Press, Oxford (2001)

8. Felice, C., Prisco, R., Malandrino, D., Zaccagnino, G., Zaccagnino, R., Zizza, R.: Chorale music splicing system: an algorithmic music composer inspired by molecular splicing. In: Johnson, C., Carballal, A., Correia, J. (eds.) Evo-MUSART 2015. LNCS, vol. 9027, pp. 50–61. Springer, Cham (2015). doi:10.1007/978-3-319-16498-4_5

9. Ebcioglu, K.: An expert system for harmonizing four-part chorales. In: Machine Models of Music, pp. 385–401 (1992)

10. Sundberg, J., Askenfelt, A., Frydén, L.: Musical performance: a synthesis-by-rule approach. Comput. Music J. **7**(1), 37–43 (1983)

11. Friberg, A.: Generative rules for music performance: a formal description of a rule system. Comput. Music J. **15**(2), 56–71 (1991)

12. Cope, D.: Experiments in Musical Intelligence. Computer Music and Digital Audio Series, A-R Editions (1996)

13. Lehmann, D.: Harmonizing melodies in real-time: the connectionist approach. In: Proceedings of the International Computer Music Association, pp. 27–31 (1997)

14. Wiggins, G., Papadopoulos, G., Amnuaisuk, S., Tuson, A.: Evolutionary methods for musical composition. In: CASYS 1998 (1998)

15. Horner, A., Goldberg, D.: Genetic algorithms and computer assisted music composition. Technical report, University of Illinois (1991)

16. Biles, J.A.: GenJam: a genetic algorithm for generating jazz solos. In: International Computer Music Conference, pp. 131–137 (1994)

17. Horner, A., Ayers, L.: Harmonization of musical progression with genetic algorithms. In: International Computer Music Conference, pp. 483–484 (1995)

18. Biles, J.A.: GenJam in perspective: a tentative taxonomy for GA music and art systems. Leonardo **36**(1), 43–45 (2003)

19. Prisco, R., Zaccagnino, R.: An evolutionary music composer algorithm for bass harmonization. In: Giacobini, M., Brabazon, A., Cagnoni, S., Caro, G.A., Ekárt, A., Esparcia-Alcázar, A.I., Farooq, M., Fink, A., Machado, P. (eds.) EvoWorkshops 2009. LNCS, vol. 5484, pp. 567–572. Springer, Heidelberg (2009). doi:10.1007/978-3-642-01129-0_63

20. De Prisco, R., Zaccagnino, G., Zaccagnino, R.: Evobasscomposer: a multi-objective genetic algorithm for 4-voice compositions. In: Genetic and Evolutionary Computation Conference, GECCO 2010, Portland, Oregon, USA, July 7–11, pp. 817–818 (2010)

21. De Felice, C., De Prisco, R., Malandrino, D., Zaccagnino, G., Zaccagnino, R., Zizza, R.: Splicing music composition. Inf. Sci. **385**—-**386**, 196–212 (2017)

22. Levine, M.: The jazz theory book. Curci (2009)

23. Head, T.: Formal language theory and DNA: an analysis of the generative capacity of specific recombinant behaviours. Bull. Math. Biol. **49**, 737–759 (1987)

24. Hopcroft, J.E., Motwani, R., Ullman, J.D.: Introduction to Automata Theory, Languages, and Computation, 3rd edn. Addison-Wesley, Reading (2006)

25. Păun, G.: On the splicing operation. Discrete Appl. Math. **70**, 57–79 (1996)

26. Schölkopf, B., Platt, J.C., Shawe-Taylor, J.C., Smola, A.J., Williamson, R.C.: Estimating the support of a high-dimensional distribution. Neural Comput. **13**(7), 1443–1471 (2001)

27. Bishop, C.M.: Pattern Recognition and Machine Learning (Information Science and Statistics). Springer-Verlag New York, Inc., Secaucus (2006)

28. Todd, P.M.: A connectionist approach to algorithmic composition. In: Todd, P.M., Loy, D.G. (eds.) Music and Connectionism, pp. 173–194. MIT Press/Bradford Books, Cambridge (1991)
29. Bharucha, J.J., Todd, P.M.: Modeling the perception of tonal structure with neural nets. Comput. Music J. **13**(4), 44–53 (1989)
30. Mozer, M.: Neural network music composition by prediction: exploring the benefits of psychoacoustic constraints and multi-scale processing. In: Connection Science, pp. 247–280 (1994)
31. Eck, D., Schmidhuber, J.: A First Look at Music Composition Using LSTM Recurrent Neural Networks. Technical report (2002)
32. Gers, F.A., Schmidhuber, J.: Recurrent nets that time and count. In: Proceedings of the IJCNN 2000, International Joint Conference on Neural Networks, Como, Italy (2000)
33. Hsiao, C.H., Cafarella, M., Narayanasamy, S.: Using web corpus statistics for program analysis. In: OOPSLA. ACM (2014)
34. Muller, K.R., Mika, S., Rt, G., Tsuda, K., Schlkopf, B.: An introduction to kernel-based learning algorithms. IEEE Trans. Neural Networks **12**(2), 181–201 (2001)

Using Autonomous Agents to Improvise Music Compositions in Real-Time

Patrick Hutchings[✉] and Jon McCormack

sensiLab, Faculty of Information Technology,
Monash University, Caulfield East, Australia
{Patrick.Hutchings,Jon.McCormack}@monash.edu

Abstract. This paper outlines an approach to real-time music generation using melody and harmony focused agents in a process inspired by
jazz improvisation. A harmony agent employs a Long Short-Term Memory (LSTM) artificial neural network trained on the chord progressions
of 2986 jazz 'standard' compositions using a network structure novel to
chord sequence analysis. The melody agent uses a rule-based system of
manipulating provided, pre-composed melodies to improvise new themes
and variations. The agents take turns in leading the direction of the
composition based on a rating system that rewards harmonic consistency
and melodic flow. In developing the multi-agent system it was found that
implementing embedded spaces in the LSTM encoding process resulted
in significant improvements to chord sequence learning.

Keywords: Multi-agent systems · Music composition · Artificial neural
networks

1 Introduction

1.1 Virtual Improvisers

Generating original music in real-time for live performance or scoring of dynamic
media such as games presents many unique challenges. Music should adapt to the
changing mood while ensuring musically consistent results. This paper presents
an approach to real-time, 'vertical' [10] composition that uses two collaborative
virtual agents: a harmony improviser and melody improviser. The approach was
developed to create an original performance system that demonstrates common
harmonic structures and melodic consistency with real-time adaptability. The
specific technique of using improvising role-focused agents comes from observations of co-operative composition in small jazz ensembles where players face the
challenges of real-time composition through co-agency [3].

   The system uses short melodic themes provided by a human composer to
seed the improvisational process. The agents explore variations of the themes
in parallel with chord progressions based on patterns found in well-known jazz
compositions. Agents self-rate their proposed melodies or chord sequences based
on harmonic suitability to select a melody or harmony lead state (see Fig. 1).

© Springer International Publishing AG 2017
J. Correia et al. (Eds.): EvoMUSART 2017, LNCS 10198, pp. 114–127, 2017.
DOI: 10.1007/978-3-319-55750-2_8

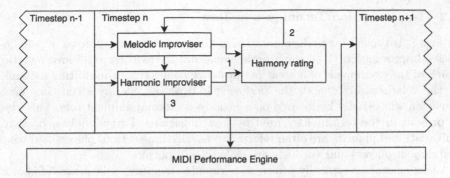

Fig. 1. Overview of composition system by time-step with data flow following the numbered pathways. Improvisation agents propose sequences that are then rated. The agent with the lower rated proposal then moves to an accompanying role and generates a suitable accompaniment for the higher rated proposal. The music is performed and the harmonised sequence is used as input for the next time-step.

The two agents have distinct roles and therefore utilise different techniques. The harmony agent is built on an LSTM neural network trained on the chord progressions of 2986 jazz standards. The melody agent uses pre-seeded melodies and permutations to generate phrases using a knowledge-based model of melodic development.

The basis for the use of these specific techniques comes from an evaluation of the relation between harmony and melody in melodic composition and co-operative composition techniques used by jazz ensembles. These background concepts are presented along with related works and followed by a rationale for specific techniques explored.

2 Background

2.1 Melody and Harmony

In melodic music, melody and harmony typically function in a codependent relationship. Harmony, whether explicit or implied, provides context for melodic phrases and can have a dramatic effect on their affective qualities. Conversely the melody's shape can suggest a harmonic direction or imply a harmony that isn't performed. A composers inspiration may come in the form of a melody, for which they then searches for an effective harmonisation, or a pleasing harmonic progression may inspire new melodic phrases.

The relationship between melody and harmony has proved to be a significant challenge to manage algorithmically and one of the many hurdles on the path to emulating human-like composition skills with computers. The challenge is compounded when generating music in real-time as it involves structures of various length that require consideration beyond the next few moments of music. It is useful to look at music traditions that involve real-time composition to see working systems of dynamically shaping harmony and melody into an unknown future.

2.2   Improvisation Techniques in Jazz

In jazz it is typical for performers to take different roles in a process of collaborative improvisation [3]. Among professional musicians these roles aren't static; much of the excitement of a great performance comes from the shifting dynamic of the relationship *between* the performers. However, at any given time each musician will usually be focused on a melody or accompaniment role. This idea is present in the terminology used by jazz musicians. Percussionists, bassists, guitarists and pianists are often referred to as 'rhythm-section' players and wind instrument players and vocalists are called 'lead' players.

Jazz ensembles typically prioritise harmonic structure. A sequence of chords are repeated and each musician improvises melodic lines or accompaniments to match the sequence [12]. Rhythm-section players often change the tempo, key and rhythmic feel using their ears and intuition. Lead players reshape, or replace melodies, but the fundamental shape of the harmonic structure is usually maintained. By focusing on one role, musicians have the artistic freedom to contribute to the direction of the overall composition structure while avoiding unwanted chaos that can come from overlapping parts.

With professional performers the concept of accompaniment ('comping') goes beyond a simple dynamic of leader and supporter to a relationship of co-agency. The accompanist and soloist work towards a common goal of improvising a creative and original composition while maintaining autonomy to contribute their own artistic signature [2].

3   Related Works

3.1   Computer Improvisation

Various approaches have been used to generate music in real-time both as a stand-alone composition and for interactive use with human performers. Impro-visor [13] is a software package that can be used to generate jazz melodies using probabilistic grammars. Different grammars trained on solos of well known jazz musicians generate melodies in real-time and demonstrate idiomatic qualities of jazz. Biles' GenJammer [5] uses a genetic algorithm to produce jazz styled melodic phrases.

Pachet's Continuator [17] uses a method of detecting phrase endings, sequence patterns and global properties to trade lines with human improvisers. The sequence pattern analysis uses a Markov model which allows for training to occur during a performance.

Pachet and Roy presented a model of generating lead sheets, including chord sequences, trained on a specific composer to produce new chord sequences in a similar style [18]. Constraints were used to filter procedures to control the bar-relative timing of chords and avoid producing long sequences appearing in the training material. Papadopoulos, Roy and Pachet [19] used Markov constraints for an online lead sheet generation tool that produces music of various styles. The authors pointed out that statistical models, such as Markov models, are more effective at developing short passages than larger, global structures of a composition.

Eigenfeldt and Pasquier's Kinetic Engine [8,9], uses a multi-agent model to generate music using melody, harmony and rhythm agents. Different techniques have been used in different iterations of the software including genetic algorithms and Markov chains.

3.2   LSTM Neural Networks

LSTM neural networks are a class of recurrent neural networks that use gate layers to simulate memory. The memory mechanism in LSTM networks are well suited to learning temporal sequences such as in music. Eck and Schmidhuber [7] used LSTM neural networks to generate chord sequences and melody/chord combinations trained on a jazz-version of the 12 bar blues and simple provided melodies using a single pentatonic scale.

Recently larger datasets have been explored for training LSTM networks. The use of graphics processing units (GPUs) has significantly decreased the training and running time of deep neural networks by running calculations for thousands of artificial neurons in parallel and current hardware is capable of training networks on datasets with millions of tokens in hours or days. Sturm et al. [22] trained an LSTM neural network on 23,958 folk compositions using a large 'deep' network structure. The network has produced thousands of folk compositions published on a regular basis online. The demonstrated effectiveness of LSTM networks for composition in a specific musical style inspired the exploration of LSTM networks in the research presented here.

A study by Choi, Fazekas and Sandler [6] used chord sequences from the jazz Real Books to train a deep LSTM networks to produce new chord sequences. They demonstrated the potential of the approach by generating chord sequences that exhibited structures commonly seen in jazz. Empirical measurements of validation perplexity and details of the model design including method of encoding tokens and justification of network size were not published. These findings invited an exploration of the role of hyper-parameters such as the number of LSTM units per layer, number of hidden layers, learning rate and encoding techniques on perplexity on the same dataset in this paper.

4   Rationale

There are many different possible approaches to real-time music generation. The rationale for the system design presented in this paper is based on the background concepts and analysis of related works.

4.1   Adapting Human-Made Content

The ability to combine the skills of professional composers with the dynamic variability of real-time algorithmically controlled arranging has direct applications in scoring interactive media. A professional composer can develop musical themes based on knowledge of the broader context of the listener, listening environment and associated visual content or narrative, which is difficult to simulate algorithmically.

Augmenting a professionally composed piece of music with an intelligent means of real-time adaptation is desirable if the adaptation does not come at significant cost to the overall quality of the piece. Using common melody manipulations on pre-composed melodies provides a means of adaptation with a high confidence of musical consistency in style and quality. It is also an improvisation device observed in jazz [4].

4.2   Two Agent Approach

Three commonly appearing role types were identified in small jazz ensembles: harmony, melody and percussive rhythm. Harmony and melody were of particular interest in this research as their co-dependent relationship is a challenge to manage in real-time. Rhythmic relationships can also still be explored without percussive rhythm and many ensembles do not contain percussive instruments. For these reasons a system with a harmony and melody agent was developed first and proposals for the development of larger systems with expanded roles are discussed in future works.

For jazz musicians, knowledge of the 'standard' repertoire allows for a shared starting point for a performance and establishes the clichés of the idiom. Deep learning provided a method of training the harmony agent on thousands of jazz standards to gain this familiarization with the harmonic language of jazz. For the melody agent priority was given to adaptation of provided melodies to dynamic harmonic context, so a system based on modifying existing themes through commonly used melody manipulation techniques was tested.

4.3   Long Short-Term Memory Neural Networks

An advantage of using an LSTM model with jazz compositions is that jazz standards typically use a repeated chord sequences of several hundred beats, a length suited to LSTM models that can be trained and run on current hardware. Jazz chord sequences provide new challenges as a training corpus compared to other genres of music such as folk and pop because of the range of chord types and flexible approach to tonality. Jazz also proves an interesting genre for training a system designed for real-time music generation as it is a genre built on improvisation, as already described.

While whole compositions can be created by sampling from a single network, such systems have limited real-time controls and continue to lack the melodic and rhythmic consistency that human composers can achieve with relative ease. Generating chord sequences is a problem where an expansive knowledge of the idiom is of use but intricate temporal detail is not as important as it is in melody generation. An LSTM network was chosen for the harmonic agent to use these strengths and alternate methods were selected for melodic generation where the network would not perform as effectively or flexibly.

Deep artificial neural networks should capture greater stylistic range within a single network due to their larger size and memory mechanisms. The memory mechanism in LSTM networks also increases the potential for learning more distant, timing dependent relationships in composition structures.

Recent advancements in artificial neural network techniques have shown impressive results in language processing [21] that have not yet been applied to music composition or symbolic music analysis. The use of embedded spaces in place of one-hot encoding and peephole mechanisms have been shown to result in significant performance improvements in some cases [11].

5   Harmonic Improviser

5.1   Method

The chord progressions of jazz standards from the Real Book series of jazz books were used to train an LSTM neural network. A script was developed to parse the chord symbols from 3020 Band-in-a-box files of Real Book standards used by musicians as practice accompaniments. As a range of time-signatures are used in jazz and chord changes often happen mid-bar, chords were entered into a database on a per-beat basis. Arrangements with more than 400 beats were removed to focus on simple arrangements leaving 2986 compositions in the database used for training the neural network. Chords were simplified to four note spellings and represented by word tokens. For example C9 and C7b13 would both be simplified to C7.

Analysing tonality in jazz chord progressions is complex as it is common for compositions to move between multiple keys even within a 32 bar melody and atonality was widely embraced in modern jazz. For this reason it was not possible to normalise the scores to a common key. Each composition was transposed in to all 12 keys in the modern western equal temperament tuning system resulting in a corpus of 35832 sequences with a vocabulary of 1668 chord types. The data set covers a wide selection of music styles within the loose 'jazz' label as stylistic flexibility was desirable and 'deep learning' neural networks perform well with large, diverse datasets.

A recurrent neural network (RNN) with LSTM was built using Tensorflow [1]. Initial experiments began with three hidden layers and 512 LSTM units per layer,

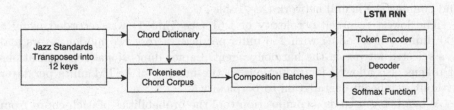

Fig. 2. Training the LSTM neural network on jazz standards

based on the model effectively used by Sturm, Santos and Korshunova [22] to generate folk compositions. A validation set was made from randomly selecting 10% of the tokenised corpus. All of the sequences were padded to have a consistent length of 400 tokens represented with one-hot encoding.

A softmax function was used to output the predicted probability of each token in the vocabulary being the next chord in an input sequence. This was achieved by utilising a sequence-to-sequence LSTM neural network structure. By using a common dictionary for the encoder and decoder and feeding the decoder with the token succeeding the token in the training batch fed to the encoder, the network can be trained to predict the next token in the batch (see Fig. 2). An ADAM [14] optimiser was used to optimise the adaptation of learning rates during training. Experimentation with learning rate and layer size resulted in some reductions to perplexity.

A multiple dimension encoding method was implemented as an alternative to one-hot encoding following the success of Rendel [21] using high dimension encoding in LSTM networks for natural language processing applications. Encoders using 10, 100, 500 and 1000 dimensions were tested.

In chord sequences the number of chords used can be small but establishing a time signature and cadence is important. In highly repetitive sections it is still important to maintain a consistent number of beats per bar, which can be a challenge in beat-by-beat based systems. Improvements were made in this area by implementing a peephole mechanism in the LSTM network where the gate layers depend on the internal state as well as the hidden state of the previous values of the cell. Peephole mechanisms have been shown to be beneficial in learning sequences where precision in timing is important [11], making them particularly relevant to modelling music structures.

Networks with 192, 256 and 512 units per layer and 2 to 3 layers were tested. Networks larger than this were deemed to be unsuited for the task of generating chord sequences in real-time due to the computational cost of running the network for each beat of music and a trend towards better training validation was not observed with increased network size. In larger networks it was observed that validation loss measures stayed above training loss which is a sign of over-fitting.

5.2   Results

Variations to the size and number of hidden layers and the number of dimensions used for token encoding affected the prediction perplexity and speed of training and running the neural network (see Table 1)

The lowest recorded perplexity of 1.61 (see Table 1) was recorded using a 100 dimension encoding with 256 units per layer across two hidden layers and was selected for use in the harmony agent. Larger dimensional representations of tokens also allowed for reductions in the number of LSTM units per layer without significant degradation to perplexity.

When being used as a composition tool the probabilities of each chord from the dictionary coming next in the sequence (C_p) can be sampled using different algorithms. A single chord token can be used to seed a new composition.

Table 1. Best prediction perplexity for different network sizes and token encodings after 20 epochs.

Encoding	One-hot	10 dim	100 dim	500 dim	1000 dim
512 units, 3 layers	2.86	2.30	2.58	1.71	1.77
512 units, 2 layers	2.75	2.05	2.59	1.93	1.74
256 units, 3 layers	3.12	2.07	1.70	1.70	1.71
256 units, 2 layers	3.10	2.59	**1.61**	1.63	1.72
192 units, 3 layers	3.42	2.10	1.83	1.73	1.74
192 units, 2 layers	4.20	1.86	1.80	1.78	1.74

After the seeding token has been used to feed the encoder the most recently selected output chord is fed into the encoder for each beat.

To operate in co-operation with other improvising agents the sampling algorithm should be dependent on the actions of other agents. This is analogous to a performer who has knowledge of the standard practices of the jazz idiom but has the freedom to take detours based on the broader musical context of the performance.

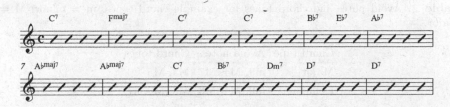

Fig. 3. A chord sequence generated by seeding the network with a single 'C7' token and a roulette wheel method of sampling the softmax output.

In the example presented in Fig. 3 the chord changes typically occur once every four beats implying a 4/4 time signature. The entire 12 bar sequence can be analysed in one key and typical jazz harmony techniques are evident. Bar 7 contains a flat-direction major chord, bars 11 and 12 contain secondary dominant chords. Root movements in fourths appear regularly throughout the composition. A selection of other chord progressions are presented in the evaluation.

6 Integrating Melody

6.1 Co-Agency

A melody role agent was introduced to function in co-operation with the harmony agent. The two agents work in a process of co-agency and alternate between leading and accompanying modes. The melody improviser uses a database of phrases based on common manipulations and permutations of human composed melodies and a knowledge-based model of harmony. The two agents use a rating system to find viable candidates for harmonic and melodic direction. To develop co-agency the ratings are compared to select which agent takes on a temporary leading role.

6.2 Phrase Database

The melody agent builds a database of melodic phrases based on seeding melodies of one to four bars length provided as monophonic MusicXML files. Each melody is reversed, inverted, augmented and diminished, separately and in combination. The produced phrases are then sliced, by barline as present in the MusicXML file and by equal divisions of the whole phrase down to four note batches. Each new phrase is then transposed in 12 keys and stored in the database with duplicate entries removed.

For a given phrase (P) each note (P_i) is given a rating for a provided chord sequence. Three categories were established for rating notes. Chord tones, usable tones and avoid notes. Avoid notes were based on Levine [15], extended to include dissonant notes outside of the diatonic scale (see Table 2).

Table 2. Avoid notes and chord tones for example chord types. m = minor, M = major.

Chord type	Avoid notes	Chord tones
Major	m9, M4, M7	M3, M5
Dominant	m9, M4, M8	M3, M5, n7
Minor	m9	m3, M5

A consonance function f(x) rates avoid notes, usable tones and chord tones with consonance ratings of 0, 0.5 and 1 respectively. Variations of these values could be made to adjust tension levels. A rating modifier (b) was introduced to accommodate passing tones by increasing the rating of avoid notes and usable tones by 0.25 each. The total melody rating (M_r) for each melodic phrase of length L is the average of the ratings of each note in the phrase at the time they would be played.

6.3 Lead States

The system has two distinct states. A melody lead state and harmony lead state. In the harmony lead state chord progressions are generated by the harmony agent by selecting the chord symbol with the highest C_p for each beat in the next two bars. The average C_p is recorded and used as the chord rating (C_r) The melody agent then performs a beam search on the melody database to find a melodic line with M_r (see Eq. 1) above a desired threshold to play over the chord progression. If no phrase can be found with a rating above a desired threshold then the melody improviser rests until a phrase is accepted. In the melody lead state the melody agent selects a melody to play and the harmony agent searches the output of the neural network for each upcoming beat from highest to lowest C_p to find a harmonic backing for the melody with Mr above the desired threshold.

$$M_r = \frac{1}{L} \sum_{i=1}^{L} f(P_i) + b \tag{1}$$

When in the harmony lead state, a change of state is triggered when a melodic line is found with Mr greater than C_r for the upcoming sequence or a melodic line is found with M_r above the set threshold that goes longer than two bars. In the melody lead state, a change of state is triggered when the harmonic improviser proposes a chord sequence with C_r greater than M_r for the selected melodic line (see Fig. 4).

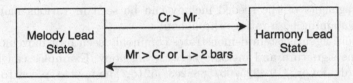

Fig. 4. State changes are triggered by differences in the rating for each agent's proposed improvisation. Melody lead states are also triggered by proposing melodies that go beyond the 2 bar sequences proposed by the harmony agent.

Joining melodic lines end-to-end allows longer melodies to be developed. Restricting the database search to lines with a first note within a specific pitch range from the last performed note reduces computational cost and encourages melodic flow. Results presented in this paper used a maximum pitch leap of a perfect fourth as it produced pleasing results.

Random rests were introduced in between melodic fragments to introduce rhythmic variation and to prevent the agent playing non-stop. With experimentation a probability of 0.1 for each of a crotchet or quaver rest being inserted was deemed pleasant and set as a default value.

By basing the change of state on the rating of the proposed harmonic and melodic direction of the music, both agents are contributing to the structural development of the performance.

6.4 Results

Static chord sequences from the Real Book dataset and the harmony agent were used to test the generation of melodies.

Fig. 5. Melody improvised for a static chord progression. Solid brackets show sources of fragments used, identified by colour. Note consonance ratings appear beneath each note. (Color figure online)

In Fig. 5 three phrases developed over six bars are shown. The melody played in the first two bars was provided by the first author as a seeding melody. In the melody-lead state the chord progression from C7 to Fmaj7 was selected by the harmony agent. The harmony agent proposed a continuation with a C7 chord ($C_p = 0.92$) and triggered a harmony-lead state. Throughout the three phrases fragments of the seeded melody can be seen in various manipulated forms, maintaining high ratings throughout.

The multi-agent system demonstrates commonly seen root movements, consistent time signature and melodic fit with harmony. Examples can be heard at https://dx.doi.org/10.4225/03/58882e3f5521d. The next step was to evaluate the overall quality of the generated music alongside those from existing systems and professional jazz musicians.

6.5   Evaluation

To evaluate the generated music an online survey was developed. The survey was open to participants that self-identified as professional jazz musicians with a bachelor degree in music or at least three years of professional performance experience. Advertising for the survey was done on social media sites groups for jazz musicians and jazz organisations.

In the survey 10 short scores and accompanying music files were presented to participants with chords and melodies pre-generated using different techniques and sources, one at a time (see Table 3). Participants were asked to rate each score using a 5-point Likert scale in categories of: Melodic creativity, thematic

Table 3. Sources of chords and melodic lines for scores presented in the survey. S = jazz standards, I = Impro-visor software package, A = multi-agent system, M = manually composed

Score no:	1	2	3	4	5	6	7	8	9	10
Melody	M	M	I	I	A	A	A	A	I	I
Chords	S	S	S	S	S	S	A	A	A	A

development, harmonic creativity, harmonic consistency, overall musicality and demonstrated knowledge of the jazz idiom.

Music files were produced using a high sample-rate virtual piano to maintain consistent performance characteristics of phrasing and tone for each. Manually composed melodies were produced by the lead author, who has over 10 years experience as a professional jazz musician. The 10 scores used in the survey can be heard at https://dx.doi.org/10.4225/03/58882e3f5521d.

At the time of writing 33 participants have completed the online survey. Calculating the median response for each score in each assessment category shows that improvisations produced with the multi-agent system were assessed at similar levels to those from Impro-visor and jazz standards (see Table 4).

Table 4. Median Likert scale values for each score presented in the online survey. 1 = very poor, 2 = poor, 3 = fair, 4 = good, 5 = very good. Highlighted columns represent scores with both harmony and melody generated by the multi-agent system.

Score no:	1	2	3	4	5	6	7	8	9	10
Harmonic Consistency	4	4	3	4	4	4	3	3	3	2
Chord Sequence Creativity	2	2	2	1	2	2	3	3	3	3
Thematic Development	3	3	3	4	3	3	4	3	3	2
Melodic Creativity	2	3	2	2	3	4	3	3	2	3
Overall Musicality	4	3	4	3	3	3	3	4	3	3
Demonstrated Knowledge of Jazz Idiom	4	4	3	3	3	3	3	3	3	3

7 Discussion

Harmonic consistency for tracks with chords generated by the system was on average rated lower than for those using standard jazz progressions but creativity was rated higher for the same tracks. A higher average rating in chord sequence creativity was likely due to professional musicians being familiar with the standard progressions and therefore deeming them uncreative. The novelty of the sequences produced by the system could be seen as both a sign of creativity and a lack of consistency.

The combination of agents with fundamentally different mechanics results in interesting dynamics and allows the strengths of those mechanics to be focused in an area of most effect. In this system the melodic agent has no training data that would allow for deep explorations of the idiom, but has a high level of musical consistency. The harmonic agent was trained on a large corpus and has a greater knowledge representation of the broader harmonic language of jazz but can be less consistent due to the wider range of possible outputs. Their combination was intended to produce a system that generates music that is identified as being creative with melodies that display a high level of musicality. The results of the survey support the general design of the system as musicality and creativity scores for generated compositions were equal or higher to music from other compared systems.

7.1 Future Work

The multi-agent system presented in this paper utilises only two agents and the database method of melodic development contains only very simple rules of manipulation. Further explorations of more sophisticated melody rules could result in improved results within the multi-agent framework. Other roles such as percussive rhythm and counter-point melody are intended in future iterations. Systems using probabilistic grammars and Markov chains have been effective for learning small time-scale structures in music which could be suited to rhythmic agents.

Jazz standards were used as training data for reasons outline in Sect. 4.3 but data sets from other music genres could also be used. Data collection for other genres such as pop and rock music is currently the subject of investigation.

In recent years, dynamic layering of instrument parts has become common in commercial game releases, where different environment properties, such as avatar movement speed, number of competing agents and player actions, add or remove layers of the score [20]. A pre-recorded instrument layering approach provides a guarantee that adding or removing layers will always work well harmonically, but more flexible, note-level arranging techniques require careful consideration of the harmonic relationships between different instrumental parts. With more stylistic range and perceptual agency [16] of the improvisers spread to non-music content for the purpose of adaptation, the model on which this system is built could have direct applications in this area.

7.2 Conclusion

The results and evaluation support the use of a multi-agent system for real-time music generation. In developing the system a significant finding was made in the application of neural network designs used in language processing for learning chord sequences. Specifically the use of embedded spaces and peephole mechanisms were found to have demonstrable benefits in this area.

References

1. Abadi, M., Agarwal, A., Barham, P., Brevdo, E., Chen, Z., Citro, C., Corrado, G.S., Davis, A., Dean, J., Devin, M., Ghemawat, S., Goodfellow, I., Harp, A., Irving, G., Isard, M., Jia, Y., Jozefowicz, R., Kaiser, L., Kudlur, M., Levenberg, J., Mané, D., Monga, R., Moore, S., Murray, D., Olah, C., Schuster, M., Shlens, J., Steiner, B., Sutskever, I., Talwar, K., Tucker, P., Vanhoucke, V., Vasudevan, V., Viégas, F., Vinyals, O., Warden, P., Wattenberg, M., Wicke, M., Yu, Y., Zheng, X.: TensorFlow: large-scale machine learning on heterogeneous systems (2015). http://tensorflow.org/, software available from tensorflow.org
2. Barrett, F.J.: Coda–creativity and improvisation in jazz and organizations: implications for organizational learning. Organ. Sci. **9**(5), 605–622 (1998)
3. Bastien, D.T., Hostager, T.J.: Jazz as a process of organizational innovation. Commun. Res. **15**(5), 582–602 (1988)

4. Berliner, P.: Thinking in jazz: composing in the moment. Jazz Educ. J. **26**, 241 (1994)
5. Biles, J.A.: Genjam in transition: from genetic jammer to generative jammer. In: Generative Art, vol. 2002 (2002)
6. Choi, K., Fazekas, G., Sandler, M.: Text-based LSTM networks for automatic music composition. arXiv preprint arXiv:1604.05358 (2016)
7. Eck, D., Schmidhuber, J.: A first look at music composition using LSTM recurrent neural networks. Istituto Dalle Molle Di Studi Sull Intelligenza Artificiale 103 (2002)
8. Eigenfeldt, A., Pasquier, P.: A realtime generative music system using autonomous melody, harmony, and rhythm agents. In: XIII Internationale Conference on Generative Arts, Milan, Italy (2009)
9. Eigenfeldt, A., Pasquier, P.: Realtime generation of harmonic progressions using controlled Markov selection. In: Proceedings of ICCC-X-Computational Creativity Conference, pp. 16–25 (2010)
10. Folkestad, G., Hargreaves, D.J., Lindström, B.: Compositional strategies in computer-based music-making. Br. J. Music Educ. **15**(01), 83–97 (1998)
11. Gers, F.A., Schraudolph, N.N., Schmidhuber, J.: Learning precise timing with LSTM recurrent networks. J. Mach. Learn. Res. **3**, 115–143 (2002)
12. Johnson-Laird, P.N.: How jazz musicians improvise. Music Percept. Interdisc. J. **19**(3), 415–442 (2002)
13. Keller, R.M., Morrison, D.R.: A grammatical approach to automatic improvisation. In: Proceedings, Fourth Sound and Music Conference, Lefkada, Greece, July. Most of the soloists at Birdland had to wait for Parker's next record in order to find out what to play next. What will they do now (2007)
14. Kingma, D., Ba, J.: Adam: a method for stochastic optimization. arXiv preprint arXiv:1412.6980 (2014)
15. Levine, M.: The Jazz Theory Book. O'Reilly Media Inc., Sebastopol (2011)
16. Monson, I.: Jazz as political and musical practice. In: Musical Improvisation: Art, Education, and Society, pp. 21–37 (2009)
17. Pachet, F.: Enhancing individual creativity with interactive musical reflexive systems. In: Musical Creativity, pp. 359–375 (2006)
18. Pachet, F., Roy, P.: Imitative leadsheet generation with user constraints. In: ECAI, pp. 1077–1078 (2014)
19. Papadopoulos, A., Roy, P., Pachet, F.: Assisted lead sheet composition using Flow-Composer. In: Rueher, M. (ed.) CP 2016. LNCS, vol. 9892, pp. 769–785. Springer, Cham (2016). doi:10.1007/978-3-319-44953-1_48
20. Plans, D., Morelli, D.: Experience-driven procedural music generation for games. IEEE Trans. Comput. Intell. AI Games **4**(3), 192–198 (2012)
21. Rendel, A., Fernandez, R., Hoory, R., Ramabhadran, B.: Using continuous lexical embeddings to improve symbolic-prosody prediction in a text-to-speech front-end. In: 2016 IEEE International Conference on Acoustics, Speech and Signal Processing (ICASSP), pp. 5655–5659. IEEE (2016)
22. Sturm, B.L., Santos, J.F., Ben-Tal, O., Korshunova, I.: Music transcription modelling and composition using deep learning. arXiv preprint arXiv:1604.08723 (2016)

Generating Polyphonic Music Using Tied Parallel Networks

Daniel D. Johnson[(✉)]

Harvey Mudd College, Claremont, CA 91711, USA
ddjohnson@hmc.edu

Abstract. We describe a neural network architecture which enables prediction and composition of polyphonic music in a manner that preserves translation-invariance of the dataset. Specifically, we demonstrate training a probabilistic model of polyphonic music using a set of parallel, tied-weight recurrent networks, inspired by the structure of convolutional neural networks. This model is designed to be invariant to transpositions, but otherwise is intentionally given minimal information about the musical domain, and tasked with discovering patterns present in the source dataset. We present two versions of the model, denoted TP-LSTM-NADE and BALSTM, and also give methods for training the network and for generating novel music. This approach attains high performance at a musical prediction task and successfully creates note sequences which possess measure-level musical structure.

1 Introduction

There have been many attempts to generate music algorithmically, including Markov models, generative grammars, genetic algorithms, and neural networks; for a survey of these approaches, see Nierhaus [17] and Fernández and Vico [5]. Neural network models are particularly flexible because they can be trained based on the complex patterns in an existing musical dataset, and a wide variety of neural-network-based music composition models have been proposed [1,7,14,16,22].

One particularly interesting approach to music composition is training a probabilistic model of polyphonic music. Such an approach attempts to model music as a probability distribution, where individual sequences are assigned probabilities based on how likely they are to occur in a musical piece. Importantly, instead of specifying particular composition rules, we can train such a model based on a large corpus of music, and allow it to discover patterns from that dataset, similarly to someone learning to compose music by studying existing pieces. Once trained, the model can be used to generate new music based on the training dataset by sampling from the resulting probability distribution.

Training this type of model is complicated by the fact that polyphonic music has complex patterns along multiple axes: there are both sequential patterns between timesteps and harmonic intervals between simultaneous notes. Furthermore, almost all musical structures exhibit transposition invariance. When music is written in a particular key, the notes are interpreted not based on their absolute

© Springer International Publishing AG 2017
J. Correia et al. (Eds.): EvoMUSART 2017, LNCS 10198, pp. 128–143, 2017.
DOI: 10.1007/978-3-319-55750-2_9

position but instead relative to that particular key, and chords are also often classified based on their position in the key (e.g. using Roman numeral notation). Transposition, in which all notes and chords are shifted into a different key, changes the absolute position of the notes but does not change any of these musical relationships. As such, it is important for a musical model to be able to generalize to different transpositions.

Recurrent neural networks (RNN), especially long short-term memory networks (LSTM) [8], have been shown to be extremely effective at modeling single-dimensional temporal patterns. It is thus reasonable to consider using them to model polyphonic music. One simple approach is to treat all of the notes played at any given timestep as a single input vector, and train an LSTM network to output a vector of probabilities of playing each note in the next timestep [4]. This essentially models each note as an independent event. While this may be appropriate for simple inputs, real-world polyphonic music contains complex harmonic relationships that would be better described using a joint probability distribution. To this end, a more effective approach combines RNNs and restricted Boltzmann machines to model the joint probability distribution of notes at each timestep [3].

Although both of the above approaches enable a network to generate music in a sequential manner, neither are transposition-invariant. In both, each note is represented as a separate element in a vector, and thus there is no way for the network to generalize intervals and chords: any relationship between, say, a G and a B, must be learned independently from the relationship between a Gb and a Bb. To capture the structure of chords and intervals in a transposition-invariant way, a neural network architecture would ideally consider *relative* positions of notes, as opposed to absolute positions.

Convolutional neural networks, another type of network architecture, have proven to be very adept at feature detection in image recognition; see Krizhevsky et al. [12] for one example. Importantly, image features are also multidimensional patterns which are invariant over shifts along multiple axes, the x and y axes of the image. Convolutional networks enable invariant feature detection by training the weights of a convolution kernel, and then convolving the image with the kernel.

Combining recurrent neural networks with convolutional structure has shown promise in other multidimensional tasks. For instance, Kalchbrenner et al. [10] describe an architecture involving LSTMs with simultaneous recurrent connections along multiple dimensions, some of which may have tied weights. Additionally, Kaiser and Sutskever [9] present a multi-layer architecture using a series of "convolutional gated recurrent units". Both of these architectures have had success in tasks such as digit-by-digit multiplication and language modeling.

In the current work, we describe two variants of a recurrent network architecture inspired by convolution that attain transposition-invariance and produce joint probability distributions over a musical sequence. These variations are referred to as Tied Parallel LSTM-NADE (TP-LSTM-NADE) and Biaxial LSTM (BALSTM). We demonstrate that these models enable efficient encoding of both temporal and pitch patterns by using them to predict and generate musical compositions.

1.1   LSTM

Long Short-Term Memory (LSTM) is a sophisticated architecture that has been shown to be able to learn long-term temporal sequences [8]. LSTM is designed to obtain constant error flow over long time periods by using *Constant Error Carousels* (CECs), which have fixed-weight recurrent connections to prevent exploding or vanishing gradients. These CECs are connected to a set of nonlinear units that allow them to interface with the rest of the network: an *input gate* determines how to change the memory cells, an *output gate* determines how strongly the memory cells are expressed, and a *forget gate* allows the memory cells to forget irrelevant values. The formulas for the activation of a single LSTM block with inputs $\mathbf{x}_t$ and hidden recurrent activations $\mathbf{h}_t$ at timestep t are given below:

$$\mathbf{z}_t = \tanh(W_{xz}\mathbf{x}_t + W_{hz}\mathbf{h}_{t-1} + \mathbf{b}_z) \qquad \text{block input}$$
$$\mathbf{i}_t = \sigma(W_{xi}\mathbf{x}_t + W_{hi}\mathbf{h}_{t-1} + \mathbf{b}_i) \qquad \text{input gate}$$
$$\mathbf{f}_t = \sigma(W_{xf}\mathbf{x}_t + W_{hf}\mathbf{h}_{t-1} + \mathbf{b}_f) \qquad \text{forget gate}$$
$$\mathbf{c}_t = \mathbf{i}_t \odot \mathbf{z}_t + \mathbf{f}_t \odot \mathbf{c}_{t-1} \qquad \text{cell state}$$
$$\mathbf{o}_t = \sigma(W_{xo}\mathbf{x}_t + W_{ho}\mathbf{h}_{t-1} + \mathbf{b}_o) \qquad \text{output gate}$$
$$\mathbf{h}_t = \mathbf{o}_t \odot \tanh(\mathbf{c}_t) \qquad \text{output}$$

where W denotes a weight matrix, b denotes a bias vector, $\odot$ denotes elementwise vector multiplication, and σ and tanh represent the logistic sigmoid and hyperbolic tangent elementwise activation functions, respectively. We are omitting so-called "peephole" connections, which use the contents of the memory cells as inputs to the gates. These connections have been shown not to have a significant impact on the network performance [6]. Like traditional RNN, LSTM networks can be trained by backpropagation through time (BPTT), or by truncated BPTT. Figure 1 gives a schematic of an LSTM block.

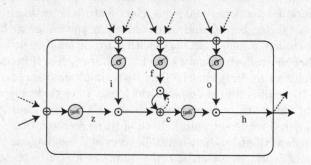

Fig. 1. Schematic of a LSTM block. Dashed lines represent a one-timestep delay. Solid arrow inputs represent $\mathbf{x}_t$, and dashed arrow inputs represent $\mathbf{h}_{t-1}$, each of which are scaled by learned weights (not shown). $\oplus$ indicates a sum, and $\odot$ indicates elementwise multiplication.

1.2   RNN-NADE

The RNN-RBM architecture, as well as the closely related RNN-NADE architecture, are attempts to model the joint distribution of a multidimensional sequence [3]. Specifically, the RNN-RBM combines recurrent neural networks (RNNs), which can capture temporal interactions, and restricted Boltzmann machines (RBMs), which model conditional distributions.

RBMs have the disadvantage of having a gradient that is untractable to compute: the gradients of the loss with respect to the model parameters must be estimated by using a method such as contrastive divergence or Gibbs sampling. To obtain a tractable gradient, the RNN-NADE architecture replaces the RBM with a neural autoregressive distribution estimator (NADE) [13], which calculates the joint probability of a vector of binary variables $\mathbf{v} = [v_1, v_2, \cdots, v_n]$ (here used to represent the set of notes that are being played simultaneously) using a series of conditional distributions:

$$p(\mathbf{v}) = \prod_{i=1}^{|\mathbf{v}|} p(v_i | \mathbf{v}_{<i})$$

with each conditional distribution given by

$$\mathbf{h_i} = \sigma(\mathbf{b}_h + W_{:,<i}\mathbf{v}_{<i})$$
$$p(v_i = 1 | \mathbf{v}_{<i}) = \sigma(\mathbf{b}_{v_i} + V_{i,:}\mathbf{h_i})$$
$$p(v_i = 0 | \mathbf{v}_{<i}) = 1 - p(v_i = 1 | \mathbf{v}_{<i})$$

where $\mathbf{b}_v$ and $\mathbf{b}_h$ are bias vectors and W and V are weight matrices. Note that $\mathbf{v}_{<i}$ denotes the vector composed of the first $i-1$ elements of $\mathbf{v}$, $W_{:,<i}$ denotes the matrix composed of all rows and the first $i-1$ columns of W, and $V_{i,:}$ denotes the ith row of V. Under this distribution, the loss has a tractable gradient, so no gradient estimation is necessary.

In the RNN-NADE, the bias parameters $\mathbf{b}_v$ and $\mathbf{b}_h$ at each timestep are calculated using the hidden activations $\hat{\mathbf{h}}$ of an RNN, which takes as input the output $\mathbf{v}$ of the network at each timestep:

$$\hat{\mathbf{h}}^{(t)} = \sigma(W_{v\hat{h}}\mathbf{v}^{(t)} + W_{\hat{h}\hat{h}}\hat{\mathbf{h}}^{(t-1)} + \mathbf{b}_{\hat{h}})$$
$$\mathbf{b}_v^{(t)} = \mathbf{b}_v + W_{\hat{h}b_v}\hat{\mathbf{h}}^{(t)}$$
$$\mathbf{b}_h^{(t)} = \mathbf{b}_h + W_{\hat{h}b_h}\hat{\mathbf{h}}^{(t)}$$

The RNN parameters $W_{v\hat{h}}$, $W_{\hat{h}\hat{h}}$, $\mathbf{b}_v$, $W_{\hat{h}b_v}$, $\mathbf{b}_h$, $W_{\hat{h}b_h}$, as well as the NADE parameters W and V are all trained using stochastic gradient descent.

2   Translation Invariance and Tied Parallel Networks

One disadvantage of distribution estimators such as RBM and NADE is that they cannot easily capture relative relationships between inputs. Although they can readily learn relationships between any set of particular notes, they are not structured to allow generalization to a transposition of those notes into a different key. This is problematic for the task of music prediction and generation because the consonance or dissonance of a set of notes remains the same regardless of their absolute position.

As one example of this, if we represent notes as a one-dimensional binary vector, where a 1 represents a note being played, a 0 represents a note not being played, and each adjacent number represents a semitone (half-step) increase, a major chord can be represented as

$$\ldots 001000100100 \ldots$$

This pattern is still a major chord no matter where it appears in the input sequence, so

$$1000100100000,$$

$$0010001001000,$$

$$0000010001001,$$

all represent major chords, a property known as *translation invariance*. However, if the input is simply presented to a distribution estimator such as NADE, each transposed representation would have to be learned separately.

Convolutional neural networks address the invariance problem for images by convolving or cross-correlating the input with a set of learned kernels. Each kernel learns to recognize a particular local type of feature. The cross-correlation of two one-dimensional sequences $\mathbf{u}$ and $\mathbf{v}$ is given by

$$(\mathbf{u} \star \mathbf{v})_n = \sum_{m=-\infty}^{\infty} \mathbf{u}_m \mathbf{v}_{m+n}.$$

Crucially, if one of the inputs (say $\mathbf{u}$) is shifted by some offset δ, the output $(\mathbf{u} \star \mathbf{v})_n$ is also shifted by δ, but otherwise does not change. This makes the operation ideal for detecting local features for which the relevant relationships are relative, not absolute.

For our music generation task, we can obtain transposition invariance by designing our model to behave like a cross-correlation: if we have a vector of notes $\mathbf{v}^{(t)}$ at timestep t and a candidate vector of notes $\hat{\mathbf{v}}^{(t+1)}$ for the next timestep, and we construct shifted vectors $\mathbf{w}^{(t)}$ and $\hat{\mathbf{w}}^{(t+1)}$ such that $\mathbf{w}_i^{(t)} = \mathbf{v}_{i+\delta}^{(t)}$ and $\hat{\mathbf{w}}_i^{(t+1)} = \hat{\mathbf{v}}_{i+\delta}^{(t+1)}$, then we want the output of our model to satisfy

$$p(\hat{\mathbf{w}}^{(t+1)}|\mathbf{w}^{(t)}) = p(\hat{\mathbf{v}}^{(t+1)}|\mathbf{v}^{(t)}).$$

2.1   Tied Parallel LSTM-NADE

In order to achieve the above form of transposition invariance, but also handle complex temporal sequences and jointly-distributed output, we propose dividing

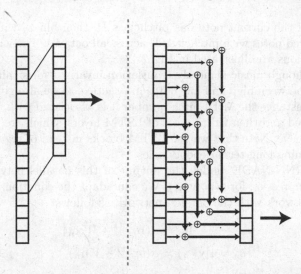

Fig. 2. Illustration of the windowing and binning operations. The thick-outlined box represents the current note. On the left, a local window around the note is extracted, and on the right, notes from each octave are binned together according to their pitch-class. For clarity, an octave is represented here by four notes; in the actual implementation octaves are of size 12.

the music prediction task into a set of *tied parallel networks*. Each network instance will be responsible for a single note, and will have tied weights with every other network instance. In this way, we ensure translation invariance: since each instance uses the same procedure to calculate its output, if we shift the inputs up by some amount δ, the output will also be shifted by δ. Our task is thus to divide the RNN-NADE architecture into multiple networks in this fashion, while still maintaining the ability to model notes conditionally on past output.

In the original RNN-NADE architecture, the RNN received the entire note vector as input. However, since each note now has a network instance that operates relative to that note, it is no longer feasible to give the entire note vector $\mathbf{v}^{(t)}$ as input to each network instance. Instead, we will feed the instance two input vectors, a local window $\mathbf{w}^{(n,t)}$ and a set of bins $\mathbf{z}^{(n,t)}$. The local window contains a slice of the note vector $\mathbf{v}^{(t)}$ such that $\mathbf{w}_i^{(n,t)} = \mathbf{v}_{n-13+i}^{(t)}$ where $1 \le i \le 25$ (giving the window a span of one octave above and below the note). If the window extends past the bounds of $\mathbf{v}$, those values are instead set to 0, as no notes are played above or below the bounds of $\mathbf{v}$. The content of each bin $\mathbf{z}_i$ is the number of notes that are played at the offset i from the current note across all octaves:

$$\mathbf{z}_i^{(n,t)} = \sum_{m=-\infty}^{\infty} \mathbf{v}_{i+n+12m}^{(t)}$$

where, again, $\mathbf{v}$ is assumed to be 0 above and below its bounds. This is equivalent to collecting all notes in each pitchclass, measured relative to the current note.

For instance, if the current note has pitchclass D, then bin $\mathbf{z}_2$ will contain the number of played notes with pitchclass E across all octaves. The windowing and binning operations are illustrated in Fig. 2.

Finally, although music is mostly translation-invariant for small shifts, there is a difference between high and low notes in practice, so we also give as input to each network instance the MIDI pitch number it is associated with. These inputs are concatenated and then fed to a set of LSTM layers, which are implemented as described above. Note that we use LSTM blocks instead of regular RNNs to encourage learning long-term dependencies.

As in the RNN-NADE model, the output of this parallel network instance should be an expression for $p(v_n|\mathbf{v}_{<n})$. We can adapt the equations of NADE to enable them to work with the parallel network as follows:

$$\mathbf{h_n} = \sigma(\mathbf{b}_h^{(n,t)} + W\mathbf{x_n})$$
$$p^{(t)}(v_n = 1|\mathbf{v}_{<n}) = \sigma(b_v^{(n,t)} + V\mathbf{h_n})$$
$$p^{(t)}(v_n = 0|\mathbf{v}_{<n}) = 1 - p^{(t)}(v_n = 1|\mathbf{v}_{<n})$$

where $\mathbf{x_n}$ is formed by taking a 2-octave window of the most recently chosen notes, concatenated with a set of bins. This is performed in the same manner as for the input to the LSTM layers, but instead of spanning the entirety of the notes from the previous timestep, it only uses the notes in $\mathbf{v}_{<n}$. The parameters $\mathbf{b}_h^{(n,t)}$ and $b_v^{(n,t)}$ are computed from the final hidden activations $\mathbf{y}^{(n,t)}$ of the LSTM layers as

$$b_v^{(n,t)} = b_v + W_{yb_v}\mathbf{y}^{(n,t)}$$
$$\mathbf{b}_h^{(n,t)} = \mathbf{b}_h + W_{yb_h}\mathbf{y}^{(n,t)}$$

One key difference is that b_v is now a scalar, not a vector, since each instance of the network only produces a single output probability, not a vector of them. The full output vector is formed by concatenating the scalar outputs for each network instance. The left side of Fig. 3 is a diagram of a single instance of this parallel network architecture. This version of the architecture will henceforth be referred to as Tied-Parallel LSTM-NADE (TP-LSTM-NADE).

2.2 Bi-Axial LSTMs

A downside to the architecture described in Sect. 2.1 is that, in order to apply the modified NADE, we must use windowed and binned summaries of note output. This captures some of the most important relationships between notes, but also prevents the network from learning any precise dependencies that extend past the size of the window. As an alternative, we can replace the NADE portion of the network with LSTMs that have recurrent connections along the note axis. This combination of LSTMs along two different axes (first along the time axis, and then along the note axis) will be referred to as a "bi-axial" configuration, to differentiate it from bidirectional configurations, which run recurrent networks both forward and backward along the same axis.

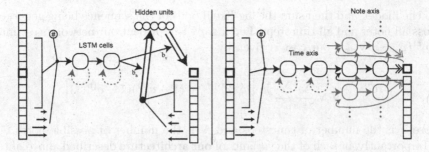

Fig. 3. On the left, schematic of a network instance for the tied parallel LSTM-NADE network. This instance is responsible for producing the output probability for the note indicated with the thick-outlined box. On the right, a schematic of an instance in the bi-axial LSTM network, showing a single instance of the time-axis network and three note-steps of the note-axis network. For each network, we concatenate a window of the note's vicinity, bins, and MIDI note number of the current note. Concatenations are indicated by lines connected by a solid black circle. Dashed arrows represent time-delayed connections, blue arrows represent recurrent connections, thick double-line-arrows represent the modified NADE estimation, and double-headed arrows indicate sampling a binary value from its corresponding probability.

In each "note-step", these note-axis LSTM layers receive as input a concatenation of two sources: the activations of the final time-axis LSTM layer for this note, and also the final output of the network for the previous note. The final activations of the note-axis LSTM will be transformed into a probability $p^{(n,t)}(v_n = 1|\mathbf{v}_{<n})$ using softmax activation. Note that, just as each note has a corresponding tied-weight time-axis LSTM network responsible for modeling temporal relationships for that single note, each timestep has a corresponding tied-weight note-axis LSTM network responsible for modeling the joint distribution of notes in that single timestep. Sequentially running the network for each note in a timestep allows us to determine the full conditional distribution for that timestep. This modification to the architecture is shown on the right side of Fig. 3, and will be referred to as Bi-Axial LSTM (BALSTM).

2.3   Training and Generation

We demonstrate our architecture by applying it to a polyphonic music prediction task, as described in Boulanger-Lewandowski et al. [3]. We train our network to model the conditional probability distribution of the notes played in a given timestep, conditioned on the notes played in previous timesteps. Specifically, we interpret the output of the nth tied-weight network instance at timestep t as the probability for playing note n at t, conditioned on previous note choices. Training our model thus amounts to maximizing the log-likelihood of each training sequence under this conditional distribution.

To calculate the log-likelihood of a given sequence, since we already know the notes that are chosen at all timesteps, we can use those notes as the inputs into the model, and then sum the log-likelihoods of the sequence being generated across all notes and all timesteps. Letting $q^{(n,t)}$ represent our network's estimate of $p^{(t)}(v_n = 1|\mathbf{v}_{<n})$, our cost is given by

$$C = -\frac{1}{TN} \sum_{t=1}^{T} \sum_{n=1}^{N} \ln\left[v_n^{(t)} q^{(n,t)} + (1 - v_n^{(t)})(1 - q^{(n,t)})\right],$$

where T is the number of timesteps and N is the number of possible notes.

Importantly, in each of the variants of our architecture described above, interaction between layers flows in a single direction; i.e. the LSTM time-axis layers depend only on the chosen notes, not on the specific output of the note-axis layers. During training, we already know all of the notes at all timesteps, so we can accelerate our training process by processing each layer independently: first preprocessing the input, then running it through the LSTM time-axis layers in parallel across all notes, and finally either using the modified NADE or the LSTM note-axis layers to compute probabilities in parallel across all timesteps. This massively parallel training process is ideal for training on a GPU or on a multi-core system.

Once trained, we can sample from this trained distribution to "compose" novel sequences. In this case, we do not know the entire sequence in advance. Instead, we must run the full network one timestep at a time. At each timestep, we process the input for that timestep, advance the LSTM time-axis layers by one timestep, and then generate the next timestep's notes. To do this, we sample from the conditional distribution as it is being generated: for each note, we choose $v_n^{(t)}$ from a Bernoulli distribution with probability $q^{(n,t)}$. Then, this choice is used to construct the input for the computation of $q^{(n+1,t)}$. Once all of the notes have been processed, we can advance to the next timestep.

The bottleneck of sampling from the distribution before processing the next timestep makes generation slower on GPUs or multi-core systems, since we can no longer parallelize the activation of note-axis computation. However, this can be mitigated somewhat by generating multiple samples simultaneously.

3 Experiments

3.1 Quantitative Analysis

We evaluated two variants of the tied-weight parallel model, along with a non-parallel model for comparison:

- *LSTM-NADE:* Non-parallel model consisting of an LSTM block connected to NADE as in the RNN-NADE architecture. We used two LSTM layers with 300 nodes each, and 150 hidden units in the NADE layer.
- *TP-LSTM-NADE:* Tied-parallel LSTM-NADE model described in Sect. 2.1. We used two LSTM layers with 200 nodes each, and 100 hidden units in the modified NADE layer.

Table 1. Log-likelihood performance for the non-transposed prediction task. For LSTM-NADE, TP-LSTM-NADE, and BALSTM, the two values represent the best and median performance across 5 trials. Data for other models is reproduced from Boulanger-Lewandowski et al. [3] and Vohra et al. [25].

Model	JSB Chorales	MuseData	Nottingham	Piano-Midi.de
Random	−61.00	−61.00	−61.00	−61.00
RBM	−7.43	−9.56	−5.25	−10.17
NADE	−7.19	−10.06	−5.48	−10.28
RNN-RBM	−7.27	−9.31	−4.72	−9.89
RNN (HF)	−8.58	−7.19	−3.89	−7.66
RNN-RBM (HF)	−6.27	−6.01	−2.39	−7.09
RNN-DBN	−5.68	−6.28	−2.54	−7.15
RNN-NADE (HF)	−5.56	−5.60	−2.31	−7.05
DBN-LSTM	−3.47	−3.91	−1.32	−4.63
LSTM-NADE	−6.00, −6.10	−5.02, −5.03	−2.02, −2.06	−7.36, −7.39
TP-LSTM-NADE	−5.88, −5.92	−4.32, −4.34	−1.61, −1.64	−5.44, −5.49
BALSTM	−5.05, −5.86	−3.90, −4.41	−1.55, −1.62	−4.90, −5.00

Table 2. Log-likelihood performance for the transposed prediction task. The two values represent the best and median performance across 5 trials.

Model	JSB Chorales	MuseData	Nottingham	Piano-Midi.de
LSTM-NADE	−9.04, −9.16	−5.72, −5.76	−3.65, −3.70	−8.11, −8.13
TP-LSTM-NADE	−5.89, −5.92	−4.32, −4.33	−1.61, −1.64	−5.44, −5.49
BALSTM	−5.08, −5.87	−3.91, −4.45	−1.56, −1.71	−4.92, −5.01

- *BALSTM:* Bi-axial LSTM with windowed+binned input, described in Sect. 2.2. We used two LSTM layers in the time-axis direction with 200 nodes each, and two LSTM layers in the note-axis direction with 100 nodes each.

We tested the ability of each model to predict/generate note sequences based on four datasets: *JSB Chorales*, a corpus of 382 four-part chorales by J.S. Bach; *MuseData*[1], an electronic classical music library, from CCARH at Stanford; *Nottingham*[2], a collection of 1200 folk tunes in ABC notation, consisting of a simple melody on top of chords; and *Piano-Midi.de*, a classical piano MIDI database. Each dataset was transposed into C major or C minor and segmented into training, validation, and test sets as in Boulanger-Lewandowski et al. [3]. Input was provided to our network in a piano-roll format, with a vector of length 88 representing the note range from A0 to C8.

---

[1] www.musedata.org.
[2] ifdo.ca/~seymour/nottingham/nottingham.html.

Dropout of 0.5 was applied to each LSTM layer, as in Moon et al. [15], and trained using RMSprop [21] with a learning rate of 0.001 and Nesterov momentum [19] of 0.9. We then evaluated our models using three criteria. Quantitatively, we evaluated the log-likelihood of the test set, which characterizes the accuracy of the model's predictions. Qualitatively, we generated sample sequences as described in Sect. 2.3. Finally, to study the translation-invariance of the models, we evaluated the log-likelihood of a version of the test set transposed into D major or D minor. Since such a transposition should not affect the musicality of the pieces in the dataset, we would expect a good model of polyphonic music to predict the original and transposed pieces with similar levels of accuracy. However, a model that was dependent on its input being in a particular key would not be able to generalize well to transpositions of the input.

Table 1 shows the performance on the non-transposed task. LSTM-NADE, TP-LSTM-NADE, and BALSTM correspond to the architectures described here, where the two values represent the best and median performance of each architecture, respectively, across 5 trials. Data for baseline models is reproduced from Vohra et al. [25] (for RNN-DBN and DBN-LSTM) and Boulanger-Lewandowski et al. [3] (for all other models). In particular, "Random" shows the performance of choosing to play each note with 50% probability, and the other architectures are variations of the original RNN-RBM architecture, which we do not describe thoroughly here.

Our tied-parallel architectures (BALSTM and TP-LSTM-NADE) perform noticeably better on the test set prediction task than did the original RNN-NADE model and many architectures closely related to it. Of the variations we tested, the BALSTM network appeared to perform the best. The TP-LSTM-NADE network, however, appears to be more stable, and converges reliably to a relatively consistent cost. Both tied-parallel network architectures perform comparably to or better than the non-parallel LSTM-NADE architecture.

The DBN-LSTM model, introduced by Vohra et al. [25], has superior performance when compared to our tied-parallel architectures. This is likely due to the deep belief network used in the DBN-LSTM, which allows the DBN-LSTM to capture a richer joint distribution at each timestep. A direct comparison between the DBN-LSTM model and the BALSTM or TP-LSTM-NADE models may be somewhat uninformative, since the models differ both in the presence or absence of parallel tied-weight structure as well as in the complexity of the joint distribution model at each timestep. However, the success of both models relative to the original RNN-RBM and RNN-NADE models suggests that a model that combined parallel structure with a rich joint distribution might attain even better results.

Note that when comparing the results from the LSTM-NADE architecture and the TP-LSTM-NADE/BALSTM architectures, the greatest improvements are on the MuseData and Piano-Midi.de datasets. This is likely due to the fact that those datasets contain many more complex musical structures in different keys, which are an ideal case for a translation-invariant architecture. On the other hand, the performance on the datasets with less variation in key is somewhat less impressive.

In addition, as shown in Table 2, the TP-LSTM-NADE/BALSTM architectures demonstrate the desired translation invariance: both parallel models perform comparably on the original and transposed datasets, whereas the non-parallel LSTM-NADE architecture performs worse at modeling the transposed dataset. This indicates that the parallel models are able to learn musical patterns that generalize to music in multiple keys, and are not sensitive to transpositions of the input, whereas the non-parallel model can only learn patterns with respect to a fixed key. Although we were unable to evaluate other existing architectures on the transposed dataset, it is reasonable to suspect that they would also show reduced performance on the transposed dataset for reasons described in Sect. 2.

3.2   Qualitative Analysis

In addition to the above experiments, we trained the BALSTM model on a larger collection of MIDI pieces with the goal of producing novel musical compositions. To this end, we made a few modifications to the BALSTM model.

Firstly, we used a larger subset of the Piano-Midi.de dataset for training, including additional pieces not used with prior models and pieces originally used as part of the validation set. To allow the network to learn rhythmic patterns, we restricted the dataset to pieces with the 4/4 time signature. We did not transpose the pieces into a common key, as the model is naturally translation invariant and does not benefit from this modification. Using the larger dataset maximizes the variety of pieces used during training and was intended to allow the network to learn as much about the musical structures as possible.

Secondly, the input to the network was augmented with a "temporal position" vector, giving the position of the timestep relative to a 4/4 measure in binary format. This gives the network the ability to learn specific temporal patterns relative to a measure.

Thirdly, we added a dimension to the note vector $\mathbf{v}$ to distinguish rearticulating a note from sustaining it. Instead of having a single 1 represent a note being played and a 0 represent that note not being played, we appended an additional 1 or 0 depending on whether that note is being articulated at that timestep. Thus the first timestep for playing a note is represented as 11, whereas sustaining a previous note is represented as 10, and resting is represented as 00. This adjustment enables the network to play the same note multiple times in succession. On the input side, the second bit is preprocessed in parallel with the first bit, and is passed into the time-axis LSTM layers as an additional input. On the output side, the note-axis LSTM layers output two probabilities instead of just one: both the probability of playing a note, and the probability of rearticulating the note if the network chooses to play it. When computing the log-likelihood of the sequence, we penalize the network for articulating a played note incorrectly, but ignore the articulation output for notes that should not be played. Similarly, when generating a piece, we only allow the network to articulate notes that have been chosen to be played.

Fig. 4. A section of a generated sample from the BALSTM model after being trained on the Piano-Midi.de dataset, converted into musical notation.

Qualitatively, the samples generated by this version of the model appear to possess complexity and intricacy. Samples from the extended BALSTM model demonstrate rhythmic consistency, chords, melody, and counterpoint not found in the samples from the RNN-RBM and RNN-NADE models (as provided by Boulanger-Lewandowski et al. [3]). The samples also seem to possess consistency across multiple measures, and although they frequently change styles, they exhibit smooth transitions from one style to another.

A portion of a generated music sample is shown in Fig. 4. Samples of the generated music for each architecture can be found on the author's website.[3]

4   Future Work

One application of the music prediction task is improving automatic transcription by giving an estimate for the likelihood of various configurations of notes [18]. As the music to be transcribed may be in any key, applying a tied parallel architecture to this task might improve the results.

A noticeable drawback of our model is that long-term phrase structure is absent from the output. This is likely because the model is trained to predict and compose music one timestep at a time. As a consequence, the model is encouraged to pay more attention to recent notes, which are often the best predictors of the next timestep, instead of on planning long-term structures. Modeling this structure will likely require incorporating a long-term planning component into the architecture. One approach to combining long-term planning with recurrent networks is the Strategic Attentive Writer (STRAW) model, described by Vezhnevets et al. [24], which extends a recurrent network with an action plan, which it can modify incrementally over time. Combining such an approach with a RNN-based music model might allow the model to generate pieces with long-term structure.

We also noticed that the network occasionally appears to become "confused" after playing a discordant note. This is likely because the dataset represents only a small portion of the overall note-sequence state space, so it is difficult to recover from mistakes due to the lack of relevant training data. Bengio et al. [2] proposed a scheduled sampling method to alleviate this problem, and a similar modification could be made here.

Another potential avenue for further research is modeling a latent musical style space using variational inference [11], which would allow the network to model distinct styles without alternating between them, and might allow the network to generate music that follows a predetermined musical form.

5   Conclusions

In this paper, we discussed the property of translation invariance of music, and proposed a set of modifications to the RNN-NADE architecture to allow it to

[3] https://www.cs.hmc.edu/~ddjohnson/tied-parallel/.

capture relative dependencies of notes. The modified architectures, which we call Tied-Parallel LSTM-NADE and Bi-Axial LSTM, divide the music generation and prediction task such that each network instance is responsible for a single note and receives input relative to that note, a structure inspired by convolutional networks.

Experimental results demonstrate that this modification yields a higher accuracy on a prediction task when compared to similar non-parallel models, and approaches state of the art performance. As desired, our models also possess translation invariance, as demonstrated by performance on a transposed prediction task. Qualitatively, the output of our model has measure-level structure, and in some cases successfully reproduces complex rhythms, melodies, and counterpoint.

Although the network successfully models measure-level structure, it unfortunately does not appear to produce consistent phrases or maintain style over a long period of time. Future work will explore modifications to the architecture that could enable a neural network model to incorporate specific styles and long-term planning into its output.

Acknowledgments. We would like to thank Dr. Robert Keller for helpful discussions and advice. We would also like to thank the developers of the Theano framework [20], which we used to run our experiments, as well as Harvey Mudd College for providing computing resources. This work used the Extreme Science and Engineering Discovery Environment (XSEDE) [23], which is supported by National Science Foundation grant number ACI-1053575.

References

1. Bellgard, M.I., Tsang, C.P.: Harmonizing music the boltzmann way. Connect. Sci. **6**(2–3), 281–297 (1994)
2. Bengio, S., Vinyals, O., Jaitly, N., Shazeer, N.: Scheduled sampling for sequence prediction with recurrent neural networks. In: Advances in Neural Information Processing Systems, pp. 1171–1179 (2015)
3. Boulanger-Lewandowski, N., Bengio, Y., Vincent, P.: Modeling temporal dependencies in high-dimensional sequences: application to polyphonic music generation and transcription. In: Proceedings of the 29th International Conference on Machine Learning (ICML-2012), pp. 1159–1166 (2012)
4. Eck, D., Schmidhuber, J.: A first look at music composition using LSTM recurrent neural networks. Istituto Dalle Molle Di Studi Sull Intelligenza Artificiale (2002)
5. Fernández, J.D., Vico, F.: AI methods in algorithmic composition: a comprehensive survey. J. Artif. Intell. Res. **48**, 513–582 (2013)
6. Greff, K., Srivastava, R.K., Koutnk, J., Steunebrink, B.R., Schmidhuber, J.: LSTM: A search space odyssey. arXiv preprint arXiv:1503.04069 (2015)
7. Hild, H., Feulner, J., Menzel, W.: HARMONET: a neural net for harmonizing chorales in the style of JS Bach. In: NIPS, pp. 267–274 (1991)
8. Hochreiter, S., Schmidhuber, J.: Long short-term memory. Neural comput. **9**(8), 1735–1780 (1997)
9. Kaiser, Ł., Sutskever, I.: Neural GPUs learn algorithms. arXiv preprint arXiv:1511.08228 (2015)

10. Kalchbrenner, N., Danihelka, I., Graves, A.: Grid long short-term memory. arXiv preprint arXiv:1507.01526 (2015)
11. Kingma, D.P., Welling, M.: Auto-encoding variational Bayes. arXiv preprint arXiv:1312.6114 (2013)
12. Krizhevsky, A., Sutskever, I., Hinton, G.E.: Imagenet classification with deep convolutional neural networks. In: Advances in Neural Information Processing Systems, pp. 1097–1105 (2012)
13. Larochelle, H., Murray, I.: The neural autoregressive distribution estimator. In: International Conference on Artificial Intelligence and Statistics, pp. 29–37 (2011)
14. Lewis, J.P.: Creation by refinement and the problem of algorithmic music composition. In: Music and Connectionism, p. 212 (1991)
15. Moon, T., Choi, H., Lee, H., Song, I.: RnnDrop: a novel dropout for RNNs in ASR. In: Automatic Speech Recognition and Understanding (ASRU) (2015)
16. Mozer, M.C.: Induction of multiscale temporal structure. In: Advances in Neural Information Processing Systems, pp. 275–275 (1993)
17. Nierhaus, G.: Algorithmic Composition: Paradigms of Automated Music Generation. Springer Science & Business Media, Verlag (2009)
18. Sigtia, S., Benetos, E., Cherla, S., Weyde, T., Garcez, A.S.d., Dixon, S.: An RNN-based music language model for improving automatic music transcription. In: International Society for Music Information Retrieval Conference (ISMIR) (2014)
19. Sutskever, I.: Training recurrent neural networks. Ph.D. thesis, University of Toronto (2013)
20. Theano Development Team: Theano: A Python framework for fast computation of mathematical expressions. arXiv e-prints abs/1605.02688, May 2016. http://arxiv.org/abs/1605.02688
21. Tieleman, T., Hinton, G.: Lecture 6.5-rmsprop: Divide the gradient by a running average of its recent magnitude. In: COURSERA: Neural Networks for Machine Learning 4 (2012)
22. Todd, P.M.: A connectionist approach to algorithmic composition. Comput. Music J. **13**(4), 27–43 (1989)
23. Towns, J., Cockerill, T., Dahan, M., Foster, I., Gaither, K., Grimshaw, A., Hazlewood, V., Lathrop, S., Lifka, D., Peterson, G.D., et al.: XSEDE: accelerating scientific discovery. Comput. Sci. Eng. **16**(5), 62–74 (2014)
24. Vezhnevets, A., Mnih, V., Osindero, S., Graves, A., Vinyals, O., Agapiou, J., et al.: Strategic attentive writer for learning macro-actions. In: Advances in Neural Information Processing Systems, pp. 3486–3494 (2016)
25. Vohra, R., Goel, K., Sahoo, J.: Modeling temporal dependencies in data using a DBN-LSTM. In: IEEE International Conference on Data Science and Advanced Analytics (DSAA), 2015. 36678 2015, pp. 1–4 (2015)

Mixed-Initiative Creative Drawing
with *webIconoscope*

Antonios Liapis[✉]

Institute of Digital Games, University of Malta, Msida, Malta
antonios.liapis@um.edu.mt

Abstract. This paper presents the *webIcononscope* tool for creative
drawing, which allows users to draw simple icons composed of basic
shapes and colors in order to represent abstract semantic concepts. The
goal of this creative exercise is to create icons that are ambiguous enough
to confuse other people attempting to guess which concept they represent. *webIcononscope* is available online and all creations can be browsed,
rated and voted on by anyone; this democratizes the creative process
and increases the motivation for creating both appealing and ambiguous
icons. To complement the creativity of the human users attempting to
create novel icons, several computational assistants provide suggestions
which alter what the user is currently drawing based on certain criteria
such as typicality and novelty. This paper reports trends in the creations
of *webIcononscope* users, based also on feedback from an online audience.

1 Introduction

The creativity in human thought processes, design practices or engineering has
been a topic of fascination since ancient times [17]. In recent years both philosophy and the cognitive sciences have allowed us to better understand and study
the process of being creative. Creativity is no longer perceived solely as an activity of reclusive geniuses who conceptualize completely new theories or inventions,
but also under the prism of an every-day, social form of creativity [5]. The last
30 years have seen a rise in popularity of the latter form of creativity (little-c
creativity) both in the commercial innovation sector and in educational settings.
Lateral thinking, i.e. the process of solving seemingly unsolvable problems or
tackling non-trivial tasks through an indirect, non-linear, creative approach [4],
is a skill that can be taught. The development of an educational curriculum
around the collaborative, improvisational creativity of students in groups has
gathered a strong support [2,21]. According to a survey of European teachers
[2], "schools promote a number of factors which favour creativity, such as learners' empowerment and open-mindedness, to rather a surprising extent" but "tend
to promote other important creativity enhancing factors, such as risk-taking and
mixing academic work and play, to a lesser degree".

To better integrate play into academic work, as noted above, the teaching
process increasingly includes games. While commercial games in the right context
can increase learner motivation and engagement [16,26], it is also valuable to

© Springer International Publishing AG 2017
J. Correia et al. (Eds.): EvoMUSART 2017, LNCS 10198, pp. 144–159, 2017.
DOI: 10.1007/978-3-319-55750-2_10

design games with the constraints of classroom use in mind, such as a limited play time or the need for short pauses for discussion. Such games are often aligned with an educational outcome or even an explicit topic. For instance, *Crystal Island* [20] tackles the topic of biology, as players interact with sick inhabitants of an island and attempt to find the solution to their ailments. In the realm of creativity support, many analog games promote creativity, e.g. *LEGO* bricks or the card game *Once Upon a Time* (Atlas Games 2012). However, there are few attempts at digital games designed to foster creativity in the classroom.

This paper describes *webIconoscope*, an online publicly available version of the Iconoscope game which was explicitly designed for fostering creativity within an educational setting [7]. In Iconoscope learners play in a group, attempting to draw icons which the other players will not be able to identify easily. Icons created via Iconoscope are evaluated by the other players, who vote which of the 3 possible concepts (described in words) is represented in the icon. The winner of a round of Iconoscope is the one with the most ambiguous icon, i.e. the icon with an equal number of correct and incorrect guesses. Ambiguous icons allow multiple interpretations from the viewers, and is an important disruptor which prompts lateral thinking [22]. Ambiguity results in cognitive dissonance between image and associated concept(s), or on the visual level itself (e.g. in optical illusions); this requires a creative, playful reading of the image. The Iconoscope game is played on Android tablets in groups of 4 or more players; each player uses a tablet to draw icons and passes it around the table during the voting phase. Iconoscope was deployed in educational institutions along with other creativity-oriented applications and games such as *4Scribes* [9]. Due to security and privacy concerns in educational settings, icons created with Iconoscope were only accessible to other users in the same educational institution. Instead, *webIconoscope* allows the anonymous use of both icon drawing and voting, making the icons publicly available and allowing anyone to engage with user-created content. This increases the application's publicity but more importantly allows for a broader evaluation of the user-created icons by a broader group of people. Compared to Iconoscope, *webIconoscope* lacks face-to-face interaction and feedback (relying instead on impersonal quantitative feedback such as the number of correct guesses) and the chance for immediate wins or losses. On the other hand, by redesigning the evaluation of ambiguity for potentially numerous votes and by enhancing the interface for audience feedback and presentation of results (e.g. as a leaderboard), *webIconoscope* transforms the short game sessions of Iconoscope into a broader, more social and public showcase of human (and human-computer) creativity.

2   Mixed-Initiative Co-Creation

With the pervasiveness of the digital world in every aspect of people's lives, diverse computer-aided design tools have emerged. Mixed-initiative tools are a special case of computer-aided design, where the computer takes on a more proactive role [8,27]. Mixed-initiative tools rely on both a human initiative and a computational initiative to perform the creative tasks and take the creative decisions. Likening the creative process to a conversation, Novick and Sutton [14] identify three types of initiative: the *task initiative* (who introduces the problem), the *outcome initiative* (who decides whether the problem has been solved), and the *speaker initiative* (who decides whose turn it is to speak).

Mixed-initiative interaction has been extensively explored for game design tasks such as level creation. There is a breadth of level design tools with different degrees of computational initiative, for instance showing optional suggestions in *Sentient Sketchbook* [11] or guiding the creative process with some indirect human guidance in interactive evolution tasks e.g. in [10]. In other cases user creations act as a goal for the computer to approximate, such as recreating a user's rough sketch in higher resolution [12]. In *Tanagra* [24], the computer attempts to "fill in the gaps" left by the user, while obeying user-specified constraints.

Besides designing the functional properties of games, mixed-initiative tools have also been used for freeform creative tasks. Examples where such creative tasks are part of a game setting can be found in *Petalz* [18] and *Artefacts* [15], where the core game mechanic is interactive evolution [25] (IEC) of flowers and blocks respectively. In Petalz, evolved flowers are posted on one's public gallery (their "balcony"); players can view their Facebook friends' balconies, like and comment on specific flowers, and sell their flowers at a marketplace for in-game currency. Artefacts is a sandbox creation game where players evolve 3D blocks into interesting shapes, combining them into complex 3D "sculptures". Artefacts lacks the social mechanics of Petalz (e.g. ownership of a balcony, sharing and liking, marketplace) and is closer to IEC in evolutionary art and music where users select which pieces will evolve without an external purpose or motivation. Among such interactive evolutionary art projects, of special note is *PicBreeder* [23] and *DrawCompileEvolve* [28]. Both systems allow users to submit an evolved image to a common public gallery, thus inviting others to rate how much they like the image (using a 5-star rating scale) or evolve the icon further via IEC. Users evolving each others' images allows for shared ownership of the output as well as negotiations of an image's meaning; both factors are important for little-c creativity to emerge [3]. DrawCompileEvolve allows for more human initiative than traditional IEC, as users seed evolution from their own drawn images.

Both *webIconoscope* and its predecessor Iconoscope follow a mixed-initiative approach to user interaction, with the computational initiative presenting optional suggestions (similarly to *Sentient Sketchbook*) but at the user's request (similarly to *Tanagra*). The computational initiative appears as assistants with profile pictures, names and implied personalities, thus strengthening the analogy of a conversation with the computer. In terms of collaborative creativity, *webIconoscope* borrows from the principles of Petalz and PicBreeder, with a public gallery that allows users to engage with each others' work (although they can not edit them further). Finally, *webIconoscope* goes beyond freeform creative exploration projects such as Artefacts and PicBreeder as it motivates users to guess the concept represented in the image and has a leaderboard for the most ambiguous images. This gives more purpose to the interaction with existing artifacts and a clearer framing of the goals of the creative process of new artifacts.

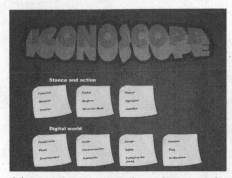
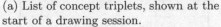
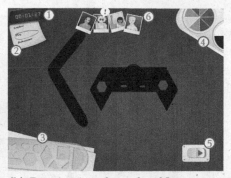

(a) List of concept triplets, shown at the start of a drawing session.

(b) Drawing interface of *webIconoscope*, with the interface elements annotated.

Fig. 1. The interface of *webIconoscope*, embedded on a webpage as a Flash application.

3   The *WebIconoscope* Interface

At the highest-level, *webIconoscope* is a drawing tool, not unlike Microsoft Paint or OpenOffice Draw. However, *webIconoscope* has more clear-cut goals for the player, with a target concept that must be drawn and the aim of creating ambiguous icons that can represent more than one concept.

Typical Use Case of *webIconoscope***:** In a sample use case of *webIconoscope*, a user starts by selecting a language of their choice, reading through the instructions page and inserting their username and an e-mail in case they want to participate in competitions[1]. Once this initial setup is completed, the user selects a triplet of concepts: each triplet is displayed on a post-it note (see Fig. 1a). When the user selects a triplet, they must also choose which concept they wish to draw among the three. The goal of the user is to create ambiguous icons which could be mis-interpreted as the other two concepts. The concepts are chosen by pedagogy experts to include thematically coherent but opposing ideas (e.g. "Protest", "Conform", "Sit on the Fence") or concepts that are semantically similar (e.g. "Freedom", "Play",' "Enthusiasm"). This way, the user is challenged to find the relationships on the semantic, thematic or visual level which can be exploited to create ambiguous icons. Once the player chooses a concept to draw, they are taken to the drawing interface where they can create their icon; once they are happy with their creation, or at the end of 5 minutes, the drawing is finished and uploaded to the database. The player has an option of choosing another concept to draw; if they do not, they are taken to the gallery page where they can survey their own and others' creations, vote for which concept is represented in each icon and rate the icons in terms of appeal.

---

[1] The launch of *webIconoscope* was followed by a competition running for 3 months; similar competitions are planned for the future to increase the use of *webIconoscope*.

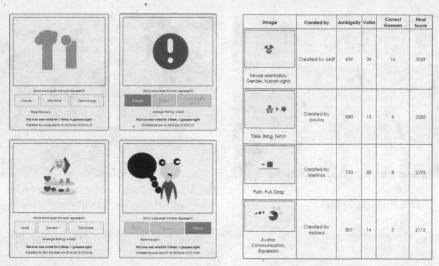

(a) A sample of the gallery interface.   (b) First 4 entries of the leaderboard.

Fig. 2. The website elements of *webIconoscope*. (Color figure online)

Drawing Interface: The drawing interface of *webIconoscope* looks like a drawing table (see Fig. 1b), showing a clock and the concepts to be drawn on a post-it note (1, 2 respectively in Fig. 1b). Drawing in *webIconoscope* is limited to the placement of pre-made abstract shapes, shown as a stencil (3). New shapes appear at the center of the screen in a neutral gray color: the user can then move, rotate, scale or recolor the shape as they desire. The shapes are mostly basic geometric shapes (squares, rhombi, circles, hexagons, triangles) and some more memorable shapes (star, heart). Shapes can be recolored via the palette (4): the palette has a small number of colors (mostly primary and secondary), as well as black and white. A button (5) allows users to end the drawing session before the five minutes are over. The interface includes portraits for all the computational assistants (6) which will be described in Sect. 4.

Website Interface: The main difference of *webIconoscope* from the multiplayer tablet-based digital game Iconoscope [7] is that the former is embedded in a website[2] which allows for many anonymous users to draw new icons and to survey previously created icons. All icons created through the *webIconoscope* interface are stored in a database alongside information on the concept represented and other interaction data. All icons in the database are shown in a gallery (see Fig. 2a), where any user can vote for which concept is represented by each icon (using the same concept triplet as the one used while drawing the icon). For each icon, the creator's name, the number of guesses by other users and the number of correct guesses are also displayed, as an invitation for the

---

user to guess correctly. The user can also rate the icon from 1 to 5 stars, based on how much they 'like' the icon. Since the goal of the created icons is ambiguity rather than appeal, the rating interface (and stars) is smaller and underplayed. Once a user has provided feedback on an icon, their selection is highlighted and "locked". Similarly, once they provide a star rating, that section is replaced by the average rating for this icon; users can not change their votes or ratings.

To promote competition, a page on the website displays a leaderboard of the top 10 icons (see Fig. 2b) along with their creator's name, the concept triplets they could be representing, and metrics on user feedback. These metrics include the number of votes, the number of correct guesses, the ambiguity score and a final score used to rank the top 10 icons. The ambiguity score A (calculated via Eq. 1) rewards icons with an equal number of correct and incorrect guesses, and also rewards a balance between the two wrong options. The final score F (Eq. 2) rewards icons with high ambiguity but also favors icons with more votes; the rationale being that more votes not only denote popularity but are more difficult to "get right" in terms of balanced correct and incorrect guesses.

$$A = 1000 - 500 \cdot \left(|1 - 2 \cdot \tfrac{c}{t}| + |1 - 2 \cdot \tfrac{i_{max}}{i}| \right) \tag{1}$$

$$F = log(t) \cdot A \tag{2}$$

where t is the total number of votes; c the number of correct votes; i the number of total incorrect votes; i_{max} the number of votes for the wrong option with the most votes. If $t = 0$ (i.e. there are no votes) then both A and F scores are 0.

4   Computational Assistants

In order to provide a creative stimulus, a set of *computational assistants* were added to the solitary drawing task of *webIconoscope*. The computational assistants show the user mutations of their current icon; the user can choose one of the four suggested alternatives to replace their icon and continue drawing (see Fig. 4) or discard all suggestions. The user can ask a computational assistant for suggestions at any point during the icon drawing process via its portrait on the drawing screen. All assistants are shown and can be interacted with in the drawing interface (see 6 in Fig. 1b); to motivate the use of assistants, every few seconds a random assistant's portrait swings while a dialog balloon pops up.

When selected, all computational assistants in *webIconoscope* perform a short evolutionary sprint, starting from an initial population consisting of mutated copies of the user's icon. Mutation can clone an existing shape in the icon (moving it to a random position) or remove a random shape. Moreover, during mutation every shape in the icon has a chance to be moved, rotated, scaled, recolored, or changed into another shape (e.g. a circle changing into a square).

Each assistant has a unique name and portrait (see Fig. 3), and they search the space differently: *Chaotic Kate* merely performs 10 random mutations to 4 copies of the user's icon, *Mad Scientist* performs novelty search [6] to diversify the population, *Typical Tom* and *Progressive Petra* attempt to respectively approach and deviate from a typical icon for this concept.

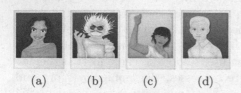

Fig. 3. Assistant profiles: Chaotic Kate (a), Mad Scientist (b), Typical Tom (c), Progressive Petra (d).

Fig. 4. Suggestions by Chaotic Kate.

Chaotic Kate performs the equivalent of a random walk and does not evaluate the quality of the content it produces in any way. The remaining three assistants choose the most promising individuals via fitness-proportionate roulette wheel selection. Icons evolve based on mutation alone and their fitness is computed based on the distance between phenotypes (icons). Since there is no obvious way of evaluating image diversity, a random distance metric based on difference in types, colors and positions of shapes is chosen: the five distance metrics are shown in Eqs. (3)–(7) and illustrated in Fig. 5. Equation (3) evaluates the number of colors that are not shared by both icons and Eq. (4) the number of shape types (e.g. circle, square) not shared by both icons. Equation (5) evaluates the number of different shape types and colors between icons, penalized by the number of shapes that share both shape type and color in both icons. This assumes that different shapes and colors are both perceptually and semantically different, but the same shapes with the same color (e.g. a red star) are important in carrying the meaning from one icon to the next (regardless of size or number of shapes). Equation (6) evaluates the average distance between all icons of one shape with all icons of the other shape: it largely rewards shapes placed in similar positions (also near each other) in both icons. Finally, Eq. (7) evaluates the difference in how "grouped" the shapes in each icon are.

$$d_c(i,j) = \frac{D_c(i,j)}{S_c(i,j)+D_c(i,j)} \tag{3}$$

$$d_s(i,j) = \frac{D_s(i,j)}{S_s(i,j)+D_s(i,j)} \tag{4}$$

$$d_{c,s}(i,j) = \frac{D_c(i,j)}{S_c(i,j)+D_c(i,j)} + \frac{D_s(i,j)}{S_s(i,j)+D_s(i,j)} + 10 \cdot S_{c,s}(i,j) \tag{5}$$

$$d_d(i,j) = \frac{1}{N(i) \cdot N(j)} \sum_{k=1}^{N(i)} \sum_{l=1}^{N(j)} d(\boldsymbol{p}_k, \boldsymbol{p}_l) \tag{6}$$

$$d_g(i,j) = |G(i) - G(j)| \tag{7}$$

(a) Icon 1   (b) Icon 2   (c) d_d calculation   (d) d_g calculation

Fig. 5. Distance metrics calculation for icons 1 and 2 (a and b). The icons do not share colors so $d_c = 1$ ($D_c = 3$ and $S_c = 0$); they share the half-circle shape type so $d_s = \frac{3}{4}$ since $D_s = 3$ (circle, rectangle, diamond) and $S_s = 1$ (half-circle). (c) shows the calculation of d_d, which is the average distance of all shapes in icon 1 to all shapes in icon 2 (black lines). (d) shows grouping G of icon 1 (average distance of green lines) and icon 2 (average distance of blue lines); their absolute difference is d_g. (Color figure online)

where $D_c(i,j)$ and $S_c(i,j)$ the number of colors not common and common (respectively) in icons i and j; $D_s(i,j)$ and $S_s(i,j)$ the number of shape types not common and common (respectively); $S_{c,s}(i,j)$ the number of combinations of color and shape common in the two icons; $N(i)$ the number of shapes in icon i; $d(\boldsymbol{p}_k, \boldsymbol{p}_l)$ the Euclidean distance between the centers of shape k and shape l; $G(i)$ the average Euclidean distance of all shapes in icon i.

These distance metrics are very lightweight computationally, especially when used one at a time: in comparison, using pixel-based distance of two images with 1200 by 900 pixels (as the ones shown in Fig. 5) would be impossible to compute in real-time, not to mention use for evolution. Choosing one metric, however, comes at the cost of expressivity and accuracy of evaluations: for instance, using d_d means the shapes and colors will remain the same in all suggestions (excluding random mutations). The obvious benefit of one distance metric is the reduced effort which allows for almost real-time generation of suggestions. A welcome side-effect of a random distance metric, however, is that it is almost impossible for the user to anticipate the resulting artifacts since at times they feature different shapes, at times different colors, and at times different positioning.

In the case of novelty search [6] (performed by the Mad Scientist), fitness is calculated based on the average distance of the individual with other members of the population and a novelty archive. The novelty archive initially contains the user's icon and in every generation the fittest (most novel) individual in the population is added to it. By diverging from the novelty archive, evolution maintains a memory of where the search has been and attempts to deviate from both historical (via the novelty archive) and current (via the current population) areas of the search space. While novelty search [6] traditionally considers a subset of the population and archive when computing novelty, the small sizes of both the population and the archive allows us to consider all individuals.

In the case of typicality search (performed by Typical Tom and Progressive Petra), fitness is calculated based on the distance from a *typical* icon for this concept. Typical icons are inserted into the database by experts and include simple but characteristic icons (e.g. a red heart for "kindness" or green triangles for "nature"). If no typical icon is found for a concept, a random one is created

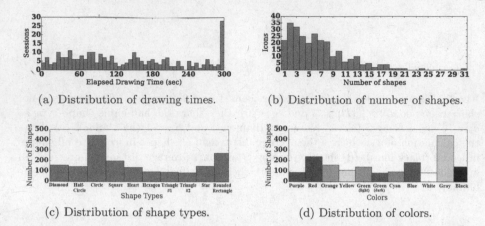

(a) Distribution of drawing times. (b) Distribution of number of shapes.

(c) Distribution of shape types. (d) Distribution of colors.

Fig. 6. Patterns of user interaction and final icons in *webIconoscope*. (Color figure online)

and added to the database instead. Typical Tom attempts to minimize the distance between the evolving icon and the typical icon: depending on the distance function, this may mean that for "kindness" at least one shape should become a heart (for d_s) or red (for d_c). Progressive Petra attempts to maximize the distance between the evolving icon and the typical icon, and so for "kindness" it might eliminate all instances of red hearts from the icon (for $d_{c,s}$).

5 Results

The results reported will discuss the degree of use of *webIconoscope* in terms of icons created via the drawing interface (in Sect. 5.1) and the response of the audience via the gallery (in Sect. 5.3). Moreover, the mixed-initiative aspect of *webIconoscope* will be evaluated in terms of the usefulness of its computational assistants (in Sect. 5.2).

5.1 Icons Created

Since its first launch in September of 2015 until the time of writing (October 2016), 275 valid icons have been created. Valid icons, in this case, have at least one shape and do not feature inappropriate content; the database was manually checked to ensure this. Of these 275 icons, most were created during 2015 due to a competition in schools and an extensive publicity push on social media.

Looking at patterns of *webIconoscope* use, most users chose English concepts (48%) with Greek concepts coming in a close second (45%) and German concepts at 7%. The primary reason for extensive use of Greek concepts was a publicity push in Greek schools. In general, the duration of drawing sessions in *webIconoscope* varied widely (see Fig. 6a) from 2 s to the full 5 minutes. The average drawing time was 142 s (standard deviation of 92 s).

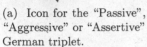

(a) Icon for the "Passive", "Aggressive" or "Assertive" German triplet.
(b) Icon for the "Nature', "Mankind" or "Technology" triplet.
(c) Icon for the "Sexual Orientation", "Gender" or "Social Rights" triplet.

Fig. 7. Sample icons with the fewest shapes, the most shapes and the most colors. (Color figure online)

Regarding the patterns of the final icons themselves, there was a large deviation in the number of shapes in each icon as shown in the distribution of Fig. 6b. While icons have an average of 6.4 shapes (standard deviation of 4.8), a nontrivial number of icons had only 1 shape (8%). This points to users who spend little time and effort drawing the simplest icons. On the other hand, certain icons were quite elaborate, with as many as 24 and 31 shapes. Regarding the types of shapes favored, Fig. 6c shows their distribution in all icons. Circles were the most popular shape; the thin rounded rectangle was interestingly the second most popular, as it seems to be used as a line[3]. It is surprising that shapes with more semantic associations (i.e. hearts and stars) were not used frequently. Finally, as shown in Fig. 6d the most prevalent color in the icons is gray, which is the default color (25% of all shapes), followed by red (13%). Again, gray shapes point to users that did not spend much time drawing[4]. In terms of color variety, most icons had one color (22%), three colors (20%) or two colors (18%), although some icons had as many as eight (2%) or nine (1%) colors.

To illustrate the variety in icons created via *webIconoscope*, Fig. 7 shows a sample of the icons, i.e. an icon with the fewest shapes (one), an icon with the most shapes (31) and an icon with the most colors (9). The large star of Fig. 7a plays with popular and historical symbols involving stars and uses a strong color (red) and a large size to show the "Aggressiveness" of the symbol; red color is often associated with passion and anger ("seeing red"). Figure 7b uses simple shapes (primarily circles) to make a composite shape: that of a human with blue triangles acting as blades (possibly). Green circles hint at "Nature", the human figure hints at "Mankind" and finally the blue triangles could hint at "Technology" (as they seem to replace human hands). Figure 7c is a multi-colored but highly abstract icon, juxtaposing a large red circle to a small hexagon. The size difference is obvious but the shape difference less so (making it intriguing). A group of multi-colored circles surround the large circle, perhaps in a threatening or idolizing fashion. It is not immediately obvious how

---

[3] *webIconoscope* does not allow users to draw lines.
[4] Gray is not a color that users can pick, so any gray shapes were never re-colored.

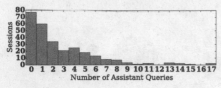
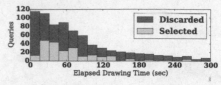

(a) Assistant queries per session.   (b) Timing of assistant queries.

Fig. 8. Assistant use.

this composition relates to any of the three concepts, although it could be a commentary on exclusion (for the "Social Rights" concept) or inequality (e.g. for "Gender").

5.2 Assistant Contributions

Most drawing sessions of *webIconoscope* (201 out of 275) included one or more queries to computational assistants, with users viewing their suggestions. This finding is surprising, as in the tablet-based Iconoscope game users forgot to consult the assistants. There are two reasons for this shift in user behavior: (a) the assistants in *webIconoscope* are more animated, swaying and showing a speech bubble from time to time, and (b) drawing in *webIconoscope* is a single-user experience (often at home rather than a classroom) with fewer distractions.

While assistants were queried at least once in 201 sessions, there were often more than one queries per session. With a total of 747 queries in those 201 sessions, this averages to 3.7 queries per session. As shown in Fig. 8a, in most sessions assistants were queried once, possibly as a test; however in some sessions they were queried extensively (up to 17 times). Due to the many sessions with only one assistant query, one can assume that users either liked the assistants' suggestions or didn't; those that queried an assistant once likely did not appreciate the suggestions and did not query it again.

There did not seem to be a big difference between queries to different assistants: 21% of queries were to Chaotic Kate, 30% to the Mad Scientist, 29% to Typical Tom and 20% to Progressive Petra. It is suspected that users chose whichever assistant was animating at the time (which was randomly chosen).

A relevant analysis for assistant queries is the timing when such a query was made: Fig. 8b shows the distribution of elapsed drawing time when assistants were queried. It is obvious that most users used assistants early in the drawing process, in the first 30–100 s; as in earlier findings in Sentient Sketchbook [11], the suggestions are often used as inspiration in early stages of the icon design when a blank canvas causes creative block. Assistants rarely get queried late in the process (in part due to the fact that few drawing sessions lasted the full 5 minutes). This is contrary to the use of Sentient Sketchbook suggestions, where many designers used the suggestions to fine-tune a design to e.g. reach perfect scores in game balance. This is likely because the evolutionary algorithms

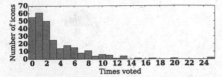

(a) Distribution of times each icon was voted for.

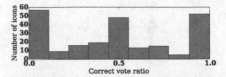

(b) Distribution of the ratio of correct votes over all votes of the same icon.

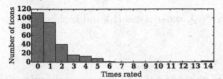

(c) Distribution of times each icon was rated.

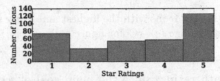

(d) Number of stars rated in each interaction with *webIconoscope*.

Fig. 9. Audience feedback to *webIconoscope* icons.

(which mostly reward divergence) and the high-level and context-agnostic distance metrics result in visually "noisy" suggestions which often break patterns of the users' icons. Thus, querying assistants at the final stages of the design process is almost guaranteed to "break" the user's nigh-final icon, which is undesirable. It is important to note that while assistants were queried often, their suggestions were selected to replace the user's icon in 102 of those sessions. However, in those 102 sessions, the assistants' suggestions were selected 230 times, i.e. more than twice per session on average. There was only a slight bias towards selecting suggestions from the Mad Scientist (29% of all selected suggestions); the least popular was Progressive Petra (21%). As noted above there was a divide between users who liked the suggestions and those who didn't. Users who queried assistants once actually selected a suggestion only in 32% of the sessions; by comparison, users who queried assistants 2 to 4 times selected a suggestion at least once in 53% of the sessions and those who queried assistants over 4 times selected a suggestion at least once in 67% of the sessions. Obviously, those who found the suggestions appealing queried the assistants more often and were more likely to take advantage of their suggestions. Interestingly, the ratio of selected suggestions versus queried assistants does not change regardless whether users make one, two or many queries: there is roughly a 1 in 3 chance that suggestions were found appropriate any time a user queried a computational assistant.

5.3 Audience Feedback

Since September 2015, there were 984 interactions with the feedback section of *webIconoscope*. Out of those, 935 were votes on which concept is represented while 347 rated an icon's appeal. The downplayed role of rating (with smaller stars and text located under the voting buttons) explains the lower engagement

(a) Icon for the "Avatar", (b) Icon for the "Sexual (c) Icon for the "Danger", "Communication", "Ex- Orientation", "Gender" or "Safety" or "Protecting the pression" triplet. "Social Rights" triplet. young" Greek triplet.

Fig. 10. Icons with the highest ambiguity score (a), most votes (b) and highest ratings on average (c). (Color figure online)

with rating; since the main goal of *webIconoscope* was to create ambiguity in icons, the large proportion of guessing interactions was the desired outcome. As with the drawing sessions of *webIconoscope*, the majority of interactions (89%) were made during 2015 due to dissemination in schools and a competition.

As shown in Fig. 9a, many of the icons were voted for once (22% of all icons) or not at all (20%). Many of the icons, therefore, did not entice users to attempt to guess their concept, although the lower use of *webIconoscope* in 2016 could also explain the few votes: icons created in 2016 were not seen by many users. While most icons received few (if any) votes, 28% of the icons received 5 or more votes. The icon with the most votes (25) is shown in Fig. 10b; it also has the highest final score, largely due to this large number of votes, i.e. t in Eq. (2).

Regarding correct versus incorrect guesses in user's votes, Fig. 9b shows the distribution of that ratio. Since many icons received a few votes (i.e. one or two), in many cases all votes were incorrect (24% of all icons) or all votes were correct (22%); similarly, in many cases (21%) there was an equal number of correct and incorrect votes. This prevalence of all-correct or all-incorrect guesses, coupled with many icons being voted once or not at all, led to ambiguity scores of 500 (40% of icons) and 0 (19%) due to no votes. Only 3% of the icons reached ambiguity scores between 950 and 1000 (the maximum value). Among icons with maximum ambiguity, the icon in Fig. 10a has the most votes (8): this means that the icon has 4 correct votes and 2 votes for each of the wrong concepts.

In terms of ratings, Fig. 9d shows the distribution of all ratings to icons in *webIconoscope*; as expected most ratings have extreme values (5 stars or 1 star) since people often give feedback when they really like or dislike something. As shown in Fig. 9c many icons were rated once or not at all, which is not surprising since many of the newer icons created during 2016 were not seen by many users. One icon was rated 14 times (shown in Fig. 10b). Among the icons with an average rating of 5 stars, the one with the most ratings is shown in Fig. 10c. The eye (formed by the half-circle) hints at surveillance (perhaps "Safety" or "Protecting the young"); the red color of the iris and the eyebrow (formed by the rounded rectangle) makes the expression angry which hints at "Danger".

6   Discussion and Future Work

The online deployment of *webIconoscope* allowed a large number of users to draw icons and leave them for other users to appraise. The 275 icons in the *webIconoscope* database feature a broad range of visual styles and topics (i.e. concept triplets). Icons were often fairly simple, with a few shapes and colors; however, even a few shapes (such as the six shapes in Fig. 10c) can be quite effective at creating a strong visual message. Some icons proved more popular than others, not only in terms of their average ratings but also in terms of the number of times other users interacted with them (either to guess the concept or to rate it). Unfortunately, audience interactions during 2016 was not proportional to the number of icons created during this period, leading to many of the latest icons having few if any votes or ratings. Another competition or publicity push could help increase attention and engagement with the recently created icons.

In terms of the computational assistants, their representation as human-like portraits and their periodic movement made them enticing to interact with. However, based on Sect. 5.2 the suggestions provided by the computational assistants were often not deemed appropriate and thus were not selected to replace the user's icon. The 'visual difference' metrics used to drive 3 of the 4 assistants were admittedly quite simplistic, favoring fast computation over accurate evaluation of perceptual differences. The role of assistants was primarily that of a random stimulus [1] which could break the user's frame of reference [22] and drawing practices. Under this prism, the fact that they were selected to replace a user's icon at least once in 102 of 275 sessions should be considered an achievement. On the other hand, there are many possible improvements in the generative processes of computational suggestions in order to increase the appeal, usefulness, and co-creative potential of *webIconoscope*. An obvious improvement could be the combination of all 5 distance metrics into a Euclidean distance: however this would increase the computational cost as well as introduce imbalances due to different value ranges in the distance metrics. Another improvement could be in the mutation operators, where rather than manipulating each shape individually, mutation can create groups (e.g. by color, by shape or by proximity) and apply scaling, recoloring, or cloning operators on all shapes in the group. This would increase the consistency and the semantic attributes of the user's icon, as a group of shapes is likely more than the sum of its parts.

While the analysis of the users' contributions (both in terms of icons drawn and votes on others' icons) is extensive in this paper, there is potential for more in-depth evaluation of the data collected via *webIconoscope*. The icons drawn can be analyzed via unsupervised machine learning techniques such as Non-Negative Matrix Factorization to find clusters with similar shapes and colors [13]. Moreover, the semantic associations of icons can be learned by applying supervised machine learning (e.g. Deep Learning) so that the system is able to predict the semantic association of an icon by its appearance alone. Finally, similar supervised learning methods can be used to learn a mapping between the icon and the average audience rating. Such a computational model of visual aesthetics for icons can then be used as an objective function for a computational assistant targeting value [19] (rather than typicality and novelty which are targeted currently), or used as a constraint on minimum predicted audience rating which tests all computational suggestions before they are presented to the user.

7   Conclusion

This paper presented the interface of *webIconoscope* and its outcomes on the first year of its deployment. Its online availability led to a frequent use of the creative aspect (with 275 user-created icons) as well as the evaluation aspect (with 984 instances of audience feedback) of *webIconoscope*. The included computational assistants were designed to act as a disruptor to the user's frame of reference while drawing; although they were frequently queried, the rather simplistic way in which they target visual diversity led users to often discard their suggestions. Overall, *webIconoscope* managed to collect a rich dataset of visual depictions of abstract concepts (see Figs. 7 and 10) which can be exploited further to analyze visual aesthetics, or learn computational models of icon ambiguity or appeal.

Acknowledgment. The author would like to thank Serious Games Interactive and the FP7 ICT project C2Learn (project No: 318480) for the implementation of Iconoscope. The ongoing research has received funding from the European Union's Horizon 2020 research and innovation programme under grant agreement No: 693150.

References

1. Beaney, M.: Imagination and Creativity. Open University, Milton Keynes (2005)
2. Cachia, R., Ferrari, A., Kearney, C., Punie, Y., Van, W., Berghe, D., Wastiau, P.: Creativity in schools in Europe: a survey of teachers (2009). http://ipts.jrc.ec.europa.eu/publications/pub.cfm?id=2940. Accessed Nov 2016
3. Chappell, K., Craft, A.R., Rolfe, L., Jobbins, V.: Humanizing creativity: valuing our journeys of becoming. Int. J. Educ. Arts **13**(8) (2012)
4. De Bono, E.: Lateral Thinking: Creativity Step by Step. Harper Collins, New York (2010)
5. Jeffrey, B., Craft, A.: The universalization of creativity. In: Craft, A., Jeffrey, B., Leibling, M. (eds.) Creativity in Education. Continuum, London (2001)
6. Lehman, J., Stanley, K.O.: Abandoning objectives: evolution through the search for novelty alone. Evol. Comput. **19**(2), 189–223 (2011)
7. Liapis, A., Hoover, A.K., Yannakakis, G.N., Alexopoulos, C., Dimaraki, E.V.: Motivating visual interpretations in iconoscope: designing a game for fostering creativity. In: Proceedings of the Foundations of Digital Games Conference (2015)
8. Liapis, A., Yannakakis, G.N.: Boosting computational creativity with human interaction in mixed-initiative co-creation tasks. In: Proceedings of the ICCC Workshop on Computational Creativity and Games (2016)
9. Liapis, A., Yannakakis, G.N., Alexopoulos, C., Lopes, P.: Can computers foster human users' creativity? Theory and praxis of mixed-initiative co-creativity. Digit. Cult. Educ. (DCE) **8**(2), 136–152 (2016)
10. Liapis, A., Yannakakis, G.N., Togelius, J.: Limitations of choice-based interactive evolution for game level design. In: Proceedings of AIIDE Workshop on Human Computation in Digital Entertainment (2012)

11. Liapis, A., Yannakakis, G.N., Togelius, J.: Sentient sketchbook: computer-aided game level authoring. In: Proceedings of the 8th Conference on the Foundations of Digital Games, pp. 213–220 (2013)
12. Liapis, A., Yannakakis, G.N., Togelius, J.: Sentient world: human-based procedural cartography. In: Machado, P., McDermott, J., Carballal, A. (eds.) EvoMUSART 2013. LNCS, vol. 7834, pp. 180–191. Springer, Heidelberg (2013). doi:10.1007/978-3-642-36955-1_16
13. Lim, C.U., Liapis, A., Harrell, D.F.: Discovering social and aesthetic categories of avatars: a bottom-up artificial intelligence approach using image clustering. In: Proceedings of the International Joint Conference of DiGRA and FDG (2016)
14. Novick, D., Sutton, S.: What is mixed-initiative interaction?. In: Proceedings of the AAAI Spring Symposium on Computational Models for Mixed Initiative Interaction (1997)
15. Patrascu, C., Risi, S.: Artefacts: minecraft meets collaborative interactive evolution. In: Proceedings of the IEEE Conference on Computational Intelligence and Games (CIG) (2016)
16. Pirius, L.K., Creel, G.: Reflections on play, pedagogy, and world of warcraft. EDUCAUSE Q. **33** (2010)
17. Cairns, H.: Plato: The Collected Dialogues. Princeton University Press, New Jersey (1961)
18. Risi, S., Lehman, J., D'Ambrosio, D.B., Hall, R., Stanley, K.O.: Combining search-based procedural content generation and social gaming in the petalz video game. In: Proceedings of the Artificial Intelligence and Interactive Digital Entertainment Conference (2012)
19. Ritchie, G.: Some empirical criteria for attributing creativity to a computer program. Mind. Mach. **17**(1), 67–99 (2007)
20. Rowe, J., Shores, L., Mott, B., Lester, J.: Integrating learning, problem solving, and engagement in narrative-centered learning environments. Int. J. Artif. Intell. Educ. **21**(1–2), 115–133 (2011)
21. Sawyer, K.: Educating for innovation. Thinking Skills Creativity **1**, 41–48 (2006)
22. Scaltsas, T., Alexopoulos, C.: Creating creativity through emotive thinking. In: Proceedings of the World Congress of Philosophy (2013)
23. Secretan, J., Beato, N., D'Ambrosio, D.B., Rodriguez, A., Campbell, A., Folsom-Kovarik, J.T., Stanley, K.O.: Picbreeder: a case study in collaborative evolutionary exploration of design space. Evol. Comput. **19**(3), 373–403 (2011)
24. Smith, G., Whitehead, J., Mateas, M.: Tanagra: reactive planning and constraint solving for mixed-initiative level design. IEEE Trans. Comput. Intell. AI Games **3**(3), 201–215 (2011)
25. Takagi, H.: Interactive evolutionary computation: fusion of the capabilities of EC optimization and human evaluation. Proc. IEEE **89**(9), 1275–1296 (2001). Invited paper
26. Watters, A.: Legos for the digital age: students build imaginary worlds (2011). http://blogs.kqed.org/mindshift/2011/03/legos-for-the-digital-age-students-build-imaginary-worlds/. Accessed Nov 2016
27. Yannakakis, G.N., Liapis, A., Alexopoulos, C.: Mixed-initiative co-creativity. In: Proceedings of the 9th Conference on the Foundations of Digital Games (2014)
28. Zhang, J., Taarnby, R., Liapis, A., Risi, S.: DrawCompileEvolve: sparking interactive evolutionary art with human creations. In: Johnson, C., Carballal, A., Correia, J. (eds.) EvoMUSART 2015. LNCS, vol. 9027, pp. 261–273. Springer, Cham (2015). doi:10.1007/978-3-319-16498-4_23

Clustering Agents for the Evolution
of Autonomous Musical Fitness

Róisín Loughran$^{(\boxtimes)}$ and Michael O'Neill

Natural Computing Research and Applications Group, University College Dublin,
Dublin, Ireland
roisin.loughran@ucd.ie

Abstract. This paper presents a cyclical system that generates
autonomous fitness functions or *Agents* for evolving short melodies.
A grammar is employed to create a corpus of melodies, each of which is
composed of a number of segments. A population of Agents are evolved to
give numerical judgements on the melodies based on the spacing of these
segments. The fitness of an individual Agent is calculated in relation to
its clustering of the melodies and how much this clustering correlates
with the clustering of the entire Agent population. A preparatory run is
used to evolve Agents using 30 melodies of known 'clustering'. The full
run uses these Agents as the initial population in evolving a new best
Agent on a separate corpus of melodies of random distance measures.
This evolved Agent is then used in combination with the original melody
grammar to create a new melody which replaces one of those from the
initial random corpus. This results in a complex adaptive system creating new melodies without any human input after initialisation. This
paper describes the behaviour of each phase in the system and presents
a number of melodies created by the system.

Keywords: Algorithmic composition · Grammatical Evolution · Clustering · Self-adaptive system · Autonomous fitness function

1 Introduction

Boden has suggested three ways in which computers can display creativity [2]:

- Combining novel ideas
- Exploring the limits of conceptual space
- Transforming established ideas that enable the emergence of unknown ideas

Grammar based evolutionary methods such as Grammatical Evolution (GE) [4]
offer an interesting parallel to such processes. The 'combination of ideas' concept can be likened to the crossover operator used in evolutionary systems, while
'exploration' can be likened to the mutation operator. The use of grammars in
GE can facilitate the third idea of 'transformation' listed above. Thus we propose
that grammar-based evolutionary systems are well-suited to creative tasks such

© Springer International Publishing AG 2017
J. Correia et al. (Eds.): EvoMUSART 2017, LNCS 10198, pp. 160–175, 2017.
DOI: 10.1007/978-3-319-55750-2_11

as melody writing. The creation of melodies offers a particularly difficult evolutionary computational challenge as there is no absolute correct answer; judging whether one melody is better than another is inherently a subjective matter. Often, this problem is addressed by using a human as a fitness function, using a set of known musical rules or comparing the music to a given style or genre. Each of these methods are based on the assumption that human-made music is best — and consequently is what is being searched for. But there already is an abundance of music being created (by humans) that follow such rules, with more being created every day. In looking at algorithmic composition as a computational problem, we are given an opportunity to consider it from a different angle. Assuming that the music created by machines must automatically be judged in human terms is an assumption that has the potential to limit the capabilities of any computationally creative system [16]. As Boden stated [3]:

> 'The ultimate vindication of AI-creativity would be a program that generated novel ideas which initially perplexed or even repelled us, but which was able to persuade us that they were indeed valuable'.

As long as autonomous creative systems are focussed on human judgement, this will be impossible to realise. For these reasons this paper proposes a system that evolves melodies, not according to any musically-derived fitness function, but by developing an autonomous fitness function that is created in response to the system itself.

The proposed system creates a population of *Agents* that result in a numerical output for a melody, evolved using a given melody corpus. An Agents' fitness is not based on any musical quality but on how well each individual Agent clusters the melody corpus in relation to the average clustering of the entire population of Agents. As such, it mimics the social phenomenon of agreement: those Agents that conform to the population are given better fitness than those that do not. In this way an individual Agent does not have any merit on its own — its performance can only be measured in relation to the overall behaviour of the population. Once an Agent is evolved, it is used in a further evolutionary run as a fitness function to create a new melody. This new melody replaces one of those in the original corpus and the process is repeated. This results in a cyclical process of Agents used to create melodies that are in turn used to create new Agents. Thus we present a complex adaptive system that continuously updates in response to its own behaviour without any human influence.

The following section describes relevant literature in evolutionary systems applied to melodic composition. Section 3 describes each stage of the system. Experimental results and the behaviour of each stage of the system is described in Sect. 4. Conclusions and proposed future work are given in Sect. 5.

2   Previous Work

This study proposes a novel method for using GE to evolve melodies, focussing on developing an autonomous fitness function that has no a priori musical knowledge or preference. In recent years, a number of EC methods have been applied to the problem of algorithmic composition. Genetic Algorithms (GA) have been applied in the systems GenJam to evolve real-time jazz solos [1], GenNotator to manipulate musical compositions using a hierarchical grammar [23] and to create four-part harmony from music theory [9]. More recently, adapted GAs have been used with local search methods to investigate human virtuosity in composing with unfigured bass [18], with a grammar to augment live coding in creating music with Tidal [10], and with non-dominated sorting in a multi-component generative music system that could generate chords, melodies and an accompaniment with two feasible-infeasible populations [20]. Genetic Programming (GP) has been used to recursively describe binary trees as genetic representation for the evolution of musical scores. The recursive mechanism of this representation allowed the generation of expressive performances and gestures along with musical notation [7]. Interactive Grammatical Evolution (GE) has been used for musical composition with promising results [21]. GE has also been used recently with autonomous fitness functions based on statistical measures of tonality and the Zipf's distribution of musical attributes [13,14]. These studies found the musical representation created by the grammar and the combination of individuals from the final population could be as important as the fitness function. Some studies have addressed the problematic issue of determining musical subjective fitness by removing it from the evolutionary process entirely. GenDash was an early developed autonomous composition system that used random selection to drive the evolution [25]. Others used only highly fit individuals within the population from initialisation and then used the whole population to create melodies [1,8].

The evolution of a population of individual Agents (or 'Critics' or similar terminology) that adjudicate a melody in some way has been proposed in a number of notable studies. The concept of populations co-evolving in a composer-critic paradigm was presented in [24]. This modelled the production of birdsong in nature by co-evolving males who composed songs along with female critics who decided, based on these songs, who to choose as a mate for the next generation. An evaluation framework consisting of a number of critics was proposed in [19]. This study induced a set of critics from a set of musical examples after first specifying specific musical criteria. The system then created music and was evaluated by a set of human listeners. A distributed population of autonomous composing agents is described in [17], which co-evolved agents with repertoires of melodies according to a measured 'sociability'. This sociability was measured in terms of similarity of the agent's repertoires; individual melodies survived or were altered depending on reinforcement feedback between co-evolving agents. This study differs from the proposed method as it is the correlation of a individual Agent's clustering of melodies to that of the (single) population that is measured in this system rather than a direct similarity measure between melodies.

A notable study demonstrated that in Computationally Creative Evolutionary systems, it is only important that the decision of fitness need be defensible; what makes one creative item better than another may not be what a human would choose but it must be a sensible, defensible and reproducible choice by the computer program. In other words there must be a logical and explainable method in assigning fitness measures. This was investigated using the idea of a preference function by measuring qualities such as specificity, transivity and reflexivity to determine the choice of a system in a number of subjective tasks [6]. Such a measure may not agree with what a human may choose as the best but, most importantly, it agrees with itself. This preference function chooses one item over another due to a logical system of comparing between items and determining a decisive preference. We try to build on this idea in the system proposed. Creating a fitness function (or Agent) based on a dynamic measurement of the system rather than a typical human measure is a key idea in the proposed system.

It has been proposed that using a pre-specified objective is not necessarily the best approach to searching. This theory suggests that searching for novelty is a better method when considering a problem, that good solutions can be found when looking for a different solution or when searching for no particular solution at all [12,22]. Such a theory fits very well in searching any creative space. A musician may not know exactly what piece of music they are trying to create when they start, they work through ideas, changing their process and hence their output as they observe what they are creating. Furthermore it was discussed in [16] that using human-based fitness measures may not be ideal in generative music; the prevalent and consistent adjudication of autonomously generated music against human opinion or measures determined from human-created music theory may in fact be limiting the potential of systems that could create music outside of such constraints. It is for such reasons that the current system develops an autonomous fitness function that is based purely on self-referential organisation of melodies by the system as it develops; the proposed method constitutes a complex, adaptive system that recursively amends an initialised corpus of melodies in response to the continually updating fitness measure. The steps in this process are detailed in the following section.

3 Method

The main focus of the system is in the creation of a fitness function that can be used to search through a population of melodies and 'adjudicate' them by means other than using known musical qualities. To do this a population of fitness functions is evolved, which from henceforth will be referred to as *Agents* in this work. Each Agent is used to cluster a population of constructed melodies. An overall clustering measure from the Agent population is calculated and each individual Agents' fitness is measured in relation to how much it agrees with this general clustering. The form of the Agents is specific to the melodies created for these experiments; Agents are constructed as a linear combination of the distances between each segment of the given melodies. The full proposed system is cyclical: a corpus of melodies is used in the evolution of an Agent which is used to evolve a new melody to be included in the original corpus, and the cycle repeats. This section describes the representation of the melodies, the representation of the Agents and how they are used together as the system evolves.

3.1 Melody Representation

Each Agent is evolved according to its correlation with the population in the clustering of a selection of melodies. Throughout the system, melodies are created using a previously developed system for composing short melodies with GE. A full description of this system and the results obtained can be found in [15]. The grammar used is based on:

```
<piece>::= <seg><seg><seg><seg><seg><seg>
<seg>::= <event><event><event><event><event>
<event>::= <style>,<oct>,<pitch>,<dur>
<style>::= <n>|<n>|<n>|<n>|<n>|<n>|<chd>|<chd>|<chd>|<chd>|<turn>|<arp>
<chd>::= <in>,0,0|<in>,<in>,0|<in>,<in>,<in>
<turn>::= <dir>,<len>,<dir>,<len>,<stp>
<in> ::= 3|4|5|7|5|5|7|7
<len>::= <stp>|<stp>,<stp>|<stp>,<stp>,<stp>|<stp>,<stp>,<stp>,<stp>
<dir>::= down|up
<stp>::= 1|1|1|1|1|1|2|2|2|2|2|2|2|2|3
<oct>::= 3|4|4|4|4|5|5|5|5|6|6
<pitch>::= 0|1|2|3|4|5|6|7|8|9|10|11
<dur>::= 1|1|1|2|2|2|4|4|4|8|8|16|16|32
```

This grammar creates a melody `<piece>` containing six segments, each comprising of a number of musical events. Each `<event>` can either be a single note (`<n>`), a chord, a turn or an arpeggio. A single note is described by a given pitch, duration and octave value. A chord is given these values but also either one, two or three notes played above the given note at specified intervals. A turn results in a series of notes proceeding in the direction up or down or a combination of both. Each step in a turn is limited to either one, two or three semitones. An arpeggio is similar to a turn except it allows larger intervals and longer durations. The application of this grammar results in a series of notes each with a given pitch and duration. The inclusion of turns and arpeggios allows a variation in the number of notes played, depending on the production rules chosen by the grammar.

The use of this grammar results in MIDI melodies split into six segments. A pitch vector for each segment is found by expanding the segment to give the pitch value at each demisemiquaver. This vector is normalised by setting the first value to zero (and transposing the remaining pitches accordingly) resulting in a pitch contour for each melodic segment. These contours can then be compared directly. In this manner, a distance can be measured between each of the segments of a given melody. The distances between all six segments in a melody can be represented numerically in a 6 by 6 matrix. This matrix is symmetrical about the diagonal, allowing it to be collapsed into a single vector of length 15 (i.e. $5+4+3+2+1$) that represents the distance between each pair of segments. It is from these distances that an Agent measures a given melody.

3.2   Agent Construction

As described above, each melody may be represented by a vector of 15 distance values. Each Agent is formed to syntactically result in a linear combination of these 15 measured distances. This is realised using GE with the following grammar:

```
<expr> ::= <O><D1><O><D2><O><D3><O><D4><O><D5><O><D6><O><D7><O><D8>
           <O><D9><O><D10><O><D11><O><D12><O><D13><O><D14><O><D15>
<O> ::= <op><scalar>
<op> ::= + | - | *
<scalar> ::= 1 | 2 | 3 | 4 | 5
```

This very simple grammar takes a linear combination of each of the 15 distance measures used to represent a given melody. Thus any Agent will output a single numerical value for each melody. The fitness function used to evolve a given Agent is based on a measure of how the individual Agent clusters the melodies in relation to how the current population of Agents cluster the melodies. Thus the Agent has no merit from its own output, but only in relation to the way in which it performs in respect to the rest of the population. This idea forms the crux of the proposed system. It can be summarised in the following steps:

- Each Agent clusters the corpus of melodies
- The most prevalent clustering of all melodies across all Agents is noted
- Each Agent is assigned fitness according to how well it correlates or 'agrees' with the overall clustering

As each Agent produces a numerical result for each melody, clustering in one dimension is only to be considered. This was implemented using Jenks natural breaks optimisation technique [11]. This method determines the best division of data between classes by seeking to minimise the average standard deviation of each element from the class mean while maximising each class' mean from those of the other classes. We consider between two and six clusters and choose the option with the best return of variance. To acquire a measure of classification we consider whether or not each pair of the melodies are found to be in the same class, assigning a value of 1 if they are and 0 otherwise. This results in a matrix of 1s and 0s again symmetrical about the diagonal (if melody 2 is in the same class as melody 3 then the reciprocal is also true). As an illustrative example, consider the arrangement [a, a, b, b, c, c] — a corpus of six individuals that should be logically clustered into three groups of two elements. If these values are indexed 1–6, all possible pairings between these six individuals can be represented by the 15-length vector:

$$[(1,2),(1,3),(1,4),(1,5),(1,6),(2,3),(2,4),(2,5),(2,6),(3,4),(3,5),(3,6),(4,5),(4,6),(5,6)]$$

If we consider '1' indicates both indexed elements are in the same cluster and '0' implies not, the clustering of the 6-element list of letters should be:

$$[1, 0, 0, 0, 0, 0, 0, 0, 0, 1, 0, 0, 0, 0, 1]$$

A similar vector can be calculated for any known clustering of one-dimensional data — such as that produced by each Agent on a corpus of melodies.

A population of 100 Agents will thus result in 100 binary-valued vectors. These are summed across all 100 Agents and normalised within the limits [0, 1] to give the *Population Clustering*. The mean squared error of the 'cluster result' of the given Agent to this Population Clustering is calculated and assigned as the fitness value of that Agent. By minimising this fitness the Agent that conforms with the majority clustering of the population is assigned best fitness in the evolutionary run. The experiments in Sect. 4 describes two implementations of this phase of the system — the preparatory run and the full system implementation. In the preparatory run this Population Clustering is replaced by an ideal clustering known by creating melodies of three specific shapes; in the full evolutionary system it is updated with each new Agent population at each generation.

3.3   Evolving a Melody

Once a best Agent has been evolved, this can then be used to evolve a new melody. GE is employed with the grammar described above in Sect. 3.1 and this best Agent as a fitness function to evolve a new melody. Each Agent is a linear combination of the distances measured from the six segments within the given melody. Hence the fitness of each melody is calculated as:

$$\text{fitness}_{melody} = abs(\text{Agent}(\text{distances}_{melody}))$$

Minimising this fitness will result in the melody with the lowest absolute output, as measured by the Agent, being deemed the best melody. This melody replaces one of the randomly generated melodic representations. After 30 full cycles of the system, the corpus has been re-populated with melodies created by the system.

3.4   Full System

As detailed in the following section, an preparatory phase is used with melodies of known shape to create an initial population of Agents. Once this is completed, a full run of the system can be undertaken using a randomly initialised corpus of melodies. The saved Agents from the preparatory phase are used as the initial population in evolving a best Agent, which is then used to evolve a best melody, which is subsequently used to update the corpus. A graphical overview of the system is shown in Fig. 1.

The phases in these experiments are all based on evolutionary runs. The parameters shown in Table 1 (chosen as those typical in the literature) are common to all experiments; generation and population size can vary and are stipulated in each relevant section.

4   Experimental Results

This section describes the behaviour of each individual section of the system. A selection of melodies produced and described below can be found at http:// ncra.ucd.ie/Site/loughranr/evo_2017.html.

Table 1. EC parameters common to each evolutionary phase.

Parameter	Value
Selection	Tournament (size 2)
Crossover rate	0.7
Mutation rate	0.01
Initial genome length	100
Elite size	1

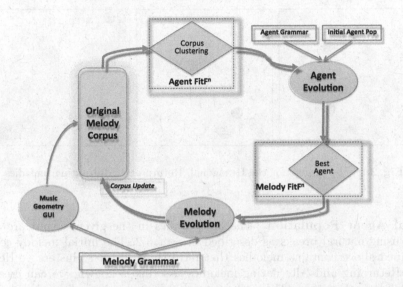

Fig. 1. Graphical overview of the full proposed system.

4.1 Preparatory Cycle

This proposed system is cyclical and self-referential, each stage of the cycle responding to the current state of the system. To start the system, however, we initialised a population of melodies that conformed to three pre-defined shapes, thus having a predictable clustering. This initial melody corpus was used to create an initial population of Agents, which can be shown to possess the required (or typical) clustering ability. These initialisation steps are described below.

Initial Melody Corpus. To create the initial population of Agents, a preparatory training corpus of 30 melodies conforming to the three shapes in Fig. 2 was created. The melodies were created to contain six segments, using the grammar described above in Sect. 3.1. Each of the three shapes can be represented by measuring the distance between each of the six given points of the shape, again resulting in a 15-point vector. Melodies can then be evolved towards one of these shapes using the mean-squared error between the melodic contour representation and that specified by the graphical shape. The evolutionary strategy used to evolve the melodies was based on a population-based hill-climbing strategy entitled the *Music Geometry GUI*, described in full in [15]. This strategy applied variable neighbourhood search as a series of operators of increasing complexity until an improvement was found. Using this method, an initial corpus of 30 melodies was evolved containing 10 Hexagonal, 10 Returning and 10 Alternating shapes.

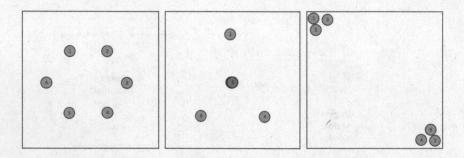

Fig. 2. Shape targets for the Hexagonal, Return and Alternating melodies

Initial Agent Population. The 30 melodies in the given corpus are clustered using natural breaks as described in Sect. 3.2. The initial melody corpus described above contains melodies that form three distinct clusters — Hexagonal, Returning and Alternating melodies. As this is known, we can create a target 'ideal' clustering 435-length vector (30×30 matrix again reduced along the diagonal) for the evolving Agents. The fitness measure of each Agent is calculated as the mean-squared error from its clustering vector to this ideal target clustering vector. This was run independently 100 times with a population of 100 over 50 generations. A plot of the population average and best fitnesses achieved over these 100 runs is shown in Fig. 3. This shows a typical evolutionary response with a steady reduction in both average and best fitness over successive generations. Each Agent consists of a linear combination of each of the 15 distance measures in each melody e.g.

```
-1D1+3D2*4D3-0D4+2D5-2D6+5D7*4D8*4D9*1D10+4D11+5D12-5D13-3D14-1D15.
```

These 100 best evolved Agents were examined and it was found that, although they achieve similar fitness results, no two were syntactically identical. Hence we can be confident there is enough diversity among these Agents to use them as an initial population for an evolutionary run. Evolving 100 unique solutions to a problem indicates that it is a simple problem with many local optimal solutions. Such a challenge is ideal for initialising a problem such as this: the aim is to create a diverse group of individuals that have some ability relevant to the domain, yet are not specialised.

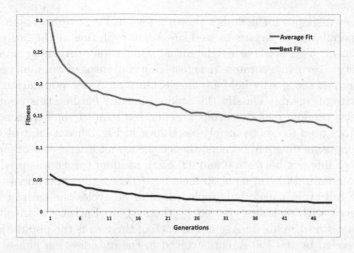

Fig. 3. Average vs. best fitness of the evolution of the initial Agents averaged over 100 runs.

· Of the 100 best Agents it was found that 30 achieved a perfect fitness score of 0. Further to such fitness measures, it is possible to examine the actual clustering of these Agents on the data. Using the original training (ordered as 10 Hexagonal, 10 Alternating, 10 Returning) the ideal clustering order (of the indexes of the melodies) is naturally:

[(0,1,2,3,4,5,6,7,8,9) (10,11,12,13,14,15,16,17,18,19) (20,21,22,23,24,25,26,27,28,29)]

In comparison to this ideal cluster pattern, 83 melodies were clustered otherwise by the best Agents. Considering 30 melodies over 100 experiments, this results in a Clustering Accuracy of 97.23%. As a test to the generality of these Agents, the data corpus was rearranged to be ordered [Hexagonal, Alternating, Returning, Hexagonal, Alternating, Returning...] resulting in ideal clustering of

[(0,3,6,9,12,15,18,21,24,27) (1,4,7,10,13,16,19,22,25,28) (2,5,8,11,14,17,20,23,26,29)]

When compared it was found that 83 (different) melodies were mis-clustered by the Agents, again leading to a Clustering Accuracy of 97.23%. This indicates that this initial population of Agents can generalise and are robust to re-arranging the melody corpus. These Agents were then used to cluster unseen data in a full run of the system.

4.2 Full Cycle

Each full cycle of the system evolves one best Agent, which is used to create one best melody, that replaces one of the melodies from the original corpus. Hence in each cycle the corpus is different only by one melody, and the evolving population of Agents is initialised *each time* from the population created in the preparatory run. The Agents are evolved according to an agreement among the population as to what way to cluster the current corpus of melodies — hence the system learns through self-organisation.

The known shapes for melodies shown above in Fig. 2 were chosen to create an initial population of Agents to seed the Agent evolution at the beginning of a full experiment. This is to ensure the initial population of Agents contains some useful ability — evolving towards the clustering consensus of randomly generated Agents does not necessarily hold any merit. Once an Agent population has been created that can reliably classify the ordered melody corpus, this population is saved. A new random melody corpus is generated to start the experiments. The graphical tool used to create the shapes shown in Fig. 2 has a limit of length 42 (between opposing diagonal corners), hence the target integer vectors were filled with random integers between 0 and 42. Each random target was used in a GE run to create an initial corpus of 30 melodies to start the experiments.

In each full cycle a melody is replaced; after 30 cycles the entire corpus has been replaced with melodies created by the system. Each cycle involves a full evolutionary run to create an Agent. For each of these runs the population was of size 100 (seeded by the 100 Agents evolved in the initialisation phase) run over 100 generations. A plot of the best and average fitnesses achieved across 100 generations is shown in Fig. 4. This plot shows a rapid decrease in best fitness which tapers off around generation 16 and becomes more stable after generation 50. The average fitness remains higher which indicates the population is still diverse at the end of a run. Only a single best Agent is chosen as a fitness function for the following Melody run.

The average of the best fitnesses across 30 Melody runs is also shown in Fig. 4. While on average the best fitness does taper off over generations it is clear that there are large differences in the values obtained (note the $\log_{10}$ scale) resulting in large peaks particularly towards the beginning of a run. These differences are a result of the syntax of the Agent (fitness function). For example Agent:

-1D1+3D2*4D3-0D4+2D5-2D6+5D7*4D8*4D9*1D10+4D11+5D12-5D13-3D14-1D15.

contains several multiplication terms which will lead to very large results for a number of melodies. As the system always evolves with minimising fitness, these melodies will be eliminated over time but these wide ranges are an inherent part of the system. Such large differences made it infeasible to include the average fitnesses on the same plot.

4.3   Melodies

This paper presents a study based around algorithmic composition, although composing 'good' melodies or compositions better than other, more focussed systems was not the main aim of the study. This study is based on the concept of creating a self-adaptive cyclical system creating subjective fitness functions that make decisions in a justified, reproducible and explainable manner. Nevertheless, it is an implemented compositional system and as such we offer the reader a number of short melodies available at http://ncra.ucd.ie/Site/loughranr/evo_2017.html. Hex1, Hex2, Alt1, Alt2, Return1 and Return2 are examples of melodies created from the shapes shown in Fig. 2 used to create the initial Agent population. Each melody is played twice to help the ear recognise the given shape. The Hex melodies

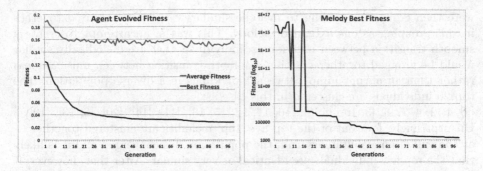

Fig. 4. Average vs. best fitness of the evolution of the evolving Agents and the best fitness of the evolving Melodies averaged over 30 runs.

display a series of slightly altering segments that return to the original segment such as in an hexagonal (or ideally circular) shape, the Alt melodies alternate between two distinct styles of segments and the Return melodies start with a segment, change it somewhat and return to the original a number of times. For completeness we have included two sample Random melodies.

It is worth noting that the average fitness obtained when creating the preparatory corpus of melodies was 5.69 whereas the average fitness obtained when creating the Random melodies was 22. This is as to be expected as, although the randomly created distances were defined to be within the ranges of those of the shapes, the relative distances between them were not controlled to be graphically meaningful and so may not have been possible to achieve. It was observed by listening to the melodies from the initial Random corpus that some of these melodies compensated for this by simply repeating individual segments. Thus there is a wide variety within the Random melodies with some containing elements of repetition such as Random1, whereas others such as Random2 contain no such discernible patterns. This is acceptable in the proposed experiment, as all that is required to start the experiment is a varied population of melodies represented in segments. Whether or not they now conform to specific shapes is of no consequence once the full evolutionary run is started.

Four short Evolved melodies are included as a demonstration of the output of the final system. Each is again played twice in succession. From listening to the final outputs, it is clear that some of the final melodies have kept remnants of the original shapes whereas others have not. Evol45 for example clearly displays an Alternating pattern. Evol0 on the other hand (the first melody created by the system) does not display any such pattern. Thus the best Melodies, and subsequent best Agents, have learned from a corpus that was originally full of well patterned melodies, yet such knowledge is sometimes apparent in the output and sometimes not.

Melody20 is a longer composition created from the final (melody) population of the last cycle. For this composition the top 20 melodies were selected according to the fitness given by the current Agent. A distance metric was calculated between each of these melodies by calculating the Levenshtein distance between the pitch contours (at each time-step) of each pair of melodies. These 20 melodies were then grouped according to melodic similarity and concatenated together in this grouping as one extended melody. This was in an attempt to create a smooth transition between individuals; individuals that are melodically similar would be grouped together creating a smooth transition between musical ideas rather than an abrupt jump between individuals in a diverse population. The top 20 individuals are clustered in this case as:

[[0, 1, 2, 4, 7, 9], [3, 10], [5, 8], [16, 17, 12, 15, 18], [11], [14], [20], [6, 13], [19]]

Unsurprisingly, four out of the top five 'top' melodies are clustered together — at the end of an evolutionary run, the population has converged and it is likely that the top few individuals are identical or very similar. After that, however, it is evident that the melodies are grouped more in relation to similarity than in fitness. There is, however, some similar content audible in a number of individuals. Much of the content in the top individual can be heard to re-surface many times throughout this composition. This is to be expected in an evolutionary run and we have previously considered it to be of benefit in using evolutionary methods for composition as this 'similar yet different' aspect of parts of a melody can lead to variation on a theme, which is known to be a pleasant quality in music [14]. In future studies we plan to investigate more interesting ways of traversing through the population for the creation of well-formed compositions.

4.4   Discussion

An important concept throughout an experiment such as this is to consider how data or knowledge is being transferred through the system. The initial Agent population is created using a clearly patterned database (melodies evolved to emanate a certain shape) but such patterns are not directly input into the system again at any point. The population was created in such as way as to ensure the individuals had some meaning — that each Agent was not completely naive but was created to start with some innate ability. Thus the system is not provided with an explicit target for the fitness function; the initial population is created from a preparatory run using a known target and the ability learned from this preparatory run is maintained within the initial population and propagated through the system in an indirect manner. The system adapts to its own response in creating new Agents and subsequent new melodies once the full cycle is started. Hence the system is self-sustainable and runs without any external human input, once it has been initialised.

In each cycle, the population was initialised using the same Agents created in the preparatory run, with all other parameters remaining the same; the only difference between cycles is one different melody in the corpus. The stochastic elements of EC methods combined with this one change resulted in a different Agent population and hence new melody in each cycle. This demonstrates that even in controlled experiments with very few degrees of freedom, evolutionary methods still have the power to develop numerous varied results that satisfy the proposed criteria. This search ability and flexibility is one of the reasons EC methods are so suitable for the development of creative systems.

Furthermore, the transformation of knowledge or ability throughout an experiment such as this, is reminiscent of Boden's third suggestion of how computers may be creative as discussed in the Introduction: the 'transformation' of ideas. The proposed system transforms ideas and information many times throughout its operation: in the transformation from genome-phenome through the use of the grammar, in the passing of clustering ability through from the preparatory step to the evolution of the clustering Agents and in the cyclical employment of the evolved Agent in evolving new melodies. Knowledge transformation is a key concept throughout this proposed system for it is through the transformation and abstraction of data, knowledge and ideas that true creativity can emerge.

The Lovelace Test for creativity states that for a system A with output o and human architect H, the system can only be deemed to be creative if H cannot explain how A created o [5]. While on the surface this may seem an easy test to pass, on closer inspection it is remarkably difficult — if at all possible. By default the architect (or programmer) — assuming they understand their own code — will be able to explain how the system created the resultant output. The system proposed in this study is certainly still explainable, but it was created in a way that the knowledge gained and behaviour displayed by the system was abstracted an extra level away from the human architect. We hope that further studies into conceptual, knowledge transferring and self-organising systems may assist in the development of computational creative systems or creative AI.

Although this system does produce melodies (discussed in Sect. 4.3), at this stage of development we have not conducted any human-evaluations on these melodies. The system at the moment represents a theoretical computational study; the focus of this study is on the description and proposal of the ideas within the system rather than an adjudication of the output. Furthermore, we acknowledge that at the moment, the melodies produced are short and not particularly impressive. We believe this can be improved in the next phase of the system that will run in a similar cyclical and adaptive manner, but could be employed with an improved grammar that can create more sophisticated melodies. As a compositional system we are interested in gauging human response to the results. Future work on more sophisticated version of the system will involve a comprehensive survey-based set of human evaluations.

5   Conclusions

This paper presents a cyclical algorithmic compositional system based on mutually dependent runs of GE. The system creates a best Agent, evolved using a corpus of melodies, which is subsequently used as a fitness function to create a new melody that replaces a melody in the corpus and the process is repeated. This best individual Agent is evolved according to the way in which it clusters the melodies, and how much its clustering correlates with the average clustering of the population of Agents. Each initial population of Agents is populated with a set of Agents trained in a preparatory step on a corpus of melodies whose ideal clustering is pre-defined. In this way, the Agents take some learned ability and transfer it into the system. No other outside influence is introduced to the system once it has started. A number of melodies created during various stages of the system were presented.

Although an implemented system is presented accompanied by a number of produced melodies, we still consider this system to be of more theoretical than practical interest at this stage. The strength of the proposed system lies in the data transformation and knowledge transfer that is described throughout the paper. Immediate future work will consider further analysis of the running of the system over a number of cycles to determine if there is any predictable behaviour exhibited by the system and what this may imply. We plan to develop the system to include a more sophisticated grammar that would enable more interesting musical compositions to be created. The key challenge in this development will be to improve the compositional ability of such a system while explicitly maintaining the autonomy of the entire system. We feel that the strength and novelty of this study lie in the autonomy of a self-adaptive complex system applied to a subjective task and, as such, it is imperative that this autonomy is maintained and preserved in future implementations.

Acknowledgments. This work is part of the App'Ed (Applications of Evolutionary Design) project funded by Science Foundation Ireland under grant 13/IA/1850. We would like to thank the anonymous reviewers for their suggestions on how to improve the paper.

References

1. Biles, J.A.: Straight-ahead jazz with GenJam: a quick demonstration. In: MUME 2013 Workshop (2013)
2. Boden, M.A.: The Creative Mind: Myths and Mechanisms. Psychology Press, New York (2004)
3. Boden, M.A.: Computer models of creativity. AI Mag. **30**(3), 23 (2009)
4. Brabazon, A., O'Neill, M., McGarraghy, S.: Grammatical evolution. In: Brabazon, A., O'Neill, M., McGarraghy, S. (eds.) Natural Computing Algorithms, pp. 357–373. Springer, Heidelberg (2015)
5. Bringsjord, S., Bello, P., Ferrucci, D.: Creativity, the turing test, and the (better) lovelace test. In: Moor, J.H. (ed.) The Turing Test, pp. 215–239. Springer, Heidelberg (2003)
6. Cook, M., Colton, S.: Generating code for expressing simple preferences: moving on from hardcoding and randomness. In: Proceedings of the Sixth International Conference on Computational Creativity June, p. 8 (2015)
7. Dahlstedt, P.: Autonomous evolution of complete piano pieces and performances. In: Proceedings of Music AL Workshop. Citeseer (2007)
8. Eigenfeldt, A., Pasquier, P.: Populations of populations: composing with multiple evolutionary algorithms. In: Machado, P., Romero, J., Carballal, A. (eds.) Evo-MUSART 2012. LNCS, vol. 7247, pp. 72–83. Springer, Heidelberg (2012). doi:10.1007/978-3-642-29142-5_7
9. Göksu, H., Pigg, P., Dixit, V.: Music composition using genetic algorithms (GA) and multilayer perceptrons (MLP). In: Wang, L., Chen, K., Ong, Y.S. (eds.) ICNC 2005. LNCS, vol. 3612, pp. 1242–1250. Springer, Heidelberg (2005). doi:10.1007/11539902_158

10. Hickinbotham, S., Stepney, S.: Augmenting live coding with evolved patterns. In: Johnson, C., Ciesielski, V., Correia, J., Machado, P. (eds.) EvoMUSART 2016. LNCS, vol. 9596, pp. 31–46. Springer, Cham (2016). doi:10.1007/978-3-319-31008-4_3

11. Jenks, G.F.: The data model concept in statistical mapping. In: International Yearbook of Cartography, vol. 7(1), pp. 186–190 (1967)

12. Lehman, J., Stanley, K.O.: Efficiently evolving programs through the search for novelty. In: Proceedings of the 12th Annual Conference on Genetic and Evolutionary Computation, pp. 837–844. ACM (2010)

13. Loughran, R., McDermott, J., O'Neill, M.: Grammatical evolution with zipf's law based fitness for melodic composition. In: Sound and Music Computing Conference, Maynooth (2015)

14. Loughran, R., McDermott, J., O'Neill, M.: Tonality driven piano compositions with grammatical evolution. In: 2015 IEEE Congress on Evolutionary Computation (CEC), pp. 2168–2175. IEEE (2015)

15. Loughran, R., McDermott, J., O'Neill, M.: Grammatical music composition with dissimilarity driven hill climbing. In: Johnson, C., Ciesielski, V., Correia, J., Machado, P. (eds.) EvoMUSART 2016. LNCS, vol. 9596, pp. 110–125. Springer, Cham (2016). doi:10.1007/978-3-319-31008-4_8

16. Loughran, R., O'Neill, M.: Generative music evaluation: why do we limit to 'human'?. In: Computer Simulation of Musical Creativity (CSMC), Huddersfield, UK (2016)

17. Miranda, E.R.: On the evolution of music in a society of self-taught digital creatures. Digit. Creativity 14(1), 29–42 (2003)

18. Munoz, E., Cadenas, J., Ong, Y.S., Acampora, G.: Memetic music composition. IEEE Trans. Evol. Comput. 20(1), 1–15 (2016)

19. Pearce, M., Wiggins, G.: Towards a framework for the evaluation of machine compositions. In: Proceedings of the AISB 2001 Symposium on Artificial Intelligence and Creativity in the Arts and Sciences, pp. 22–32. Citeseer (2001)

20. Scirea, M., Togelius, J., Eklund, P., Risi, S.: MetaCompose: a compositional evolutionary music composer. In: Johnson, C., Ciesielski, V., Correia, J., Machado, P. (eds.) EvoMUSART 2016. LNCS, vol. 9596, pp. 202–217. Springer, Cham (2016). doi:10.1007/978-3-319-31008-4_14

21. Shao, J., McDermott, J., O'Neill, M., Brabazon, A.: Jive: a generative, interactive, virtual, evolutionary music system. In: Chio, C., et al. (eds.) EvoApplications 2010. LNCS, vol. 6025, pp. 341–350. Springer, Heidelberg (2010). doi:10.1007/978-3-642-12242-2_35

22. Stanley, K.O., Lehman, J.: Why Greatness Cannot Be Planned: The Myth of the Objective. Springer, Heidelberg (2015)

23. Thywissen, K.: GeNotator: an environment for exploring the application of evolutionary techniques in computer-assisted composition. Organised Sound 4(02), 127–133 (1999)

24. Todd, P.M., Werner, G.M.: Frankensteinian methods for evolutionary music. In: Griffith, N., Todd, P.M. (eds.) Musical Networks: Parallel Distributed Perception and Performace, p. 313. MIT Press, Cambridge (1999)

25. Waschka II, R.: Composing with genetic algorithms: GenDash. In: Miranda, E.R., Biles, J.A. (eds.) Evolutionary Computer Music, pp. 117–136. Springer, Heidelberg (2007)

EvoFashion: Customising Fashion Through Evolution

Nuno Lourenço[✉], Filipe Assunção, Catarina Maçãs, and Penousal Machado

CISUC, Department of Informatics Engineering, University of Coimbra,
Pólo II - Pinhal de Marrocos, 3030 Coimbra, Portugal
{naml,fga,cmacas,machado}@dei.uc.pt

Abstract. In today's society, where everyone desires unique and fashionable products, the ability to customise products is almost mandatory in every online store. Despite of many stores allowing the users to personalize their products, they do not always do it in the most efficient and user-friendly manner. In order to have products that reflect the user's design preferences, they have to go through a laborious process of picking the components that they want to customise. In this paper we propose a framework that aims to relieve the design burden from the user side, by automating the design process through the use of Interactive Evolutionary Computation (IEC). The framework is based on a web-interface that facilitates the interaction between the user and the evolutionary process. The user can select between two types of evolution: (i) automatic; and (ii) partially-automatic. The results show the ability of the framework to promote evolution towards solutions that reflect the user aesthetic preferences.

Keywords: Evolutionary algorithm · Fashion design · Interactive evolutionary computation · Product customisation

1 Introduction

In today's society many online stores allow customers to personalise their products during their purchase so that each individual can carry unique, stylish and distinctive items [1]. This customisation usually includes letter engraving and selection of colours and/or materials to specific parts of the product. Nevertheless, this is often a lingering process, where the buyer is asked endless questions about his/her preferences. Additionally, the customisation of the product can have a direct impact in other aspects of the item, such as its price or availability, making the process even more difficult and not user friendly. In order to alleviate the design burden from the customers side, and ease the customisation of products, we propose a framework based on Evolutionary Algorithm (EA). We propose an automatic tool to measure the user's preferences, capable of speeding the process of personalising items and adjusting the products parameters. To allow an effortless integration with online stores, we show the viability of the approach by proposing a web framework that allows the customisation of fashion products.

© Springer International Publishing AG 2017
J. Correia et al. (Eds.): EvoMUSART 2017, LNCS 10198, pp. 176–189, 2017.
DOI: 10.1007/978-3-319-55750-2_12

Despite the wide range of practical applications of this type of frameworks, we choose the evolution of shoe design as a case-study. Our experiments consider the use of an automatic and user-guided fitness function to guide the evolutionary process. Results show the viability of the proposed framework, evincing that the combination of automatic and user-guided fitness functions is advantageous, since it alleviates the customisation burden from the user, and, at the same time, allows the emergence of unique pieces specially tailored for the user's taste.

The remainder of the paper is organised as follows. Section 2 details on methodologies applied to the customisation and optimisation of products based on EA. Our proposal is detailed in Sect. 3, followed by the experimental analysis (Sect. 4). Finally, Sect. 5 gathers the main conclusions and points towards future work.

2 Related Work

EAs [2] are computational models that have their origins in the early 1960's. At the time, and inspired by theories of natural-selection (Charles Darwin) and of Mendelian inheritance (Gregor Mendel), researchers started to develop different techniques to evolve artificial systems. These theories are based on the idea that the best adapted individuals have higher chances of survival and, as such, higher chances of reproducing and passing their characteristics to the next generation of individuals. As generations pass, individuals will become more adapted to the environment. EAs have been successfully applied to different domains from science to art. In the next paragraphs, focus will be given to works within the scope of this paper, i.e., the application of EAs in works concerning industry and design optimisation and customisation.

There are several examples of the application of EAs to optimise the shape and efficiency of an object. In architecture, EAs can be applied to improve the energy efficiency and design of a house. In [3], Caldas and Norford use Evolutionary Computation (EC) to evaluate several aspects concerned with the energetic performance of buildings, such as window placing and size. Still focusing on architectural efficiency aspects, in [4] Besserud and Cotten, evolved the optimal structure of a 300-meter tower, targeting the maximization of sun rays incidence for the installation of photovoltaic panels. Concerning the evolution of designs, Michalek et al. [5] automatically generate the layout of floorplans, and Juan et al. [6] propose a system that enables house customisation by combining case-based reasoning with a Genetic Algorithm (GA).

Several other works related with the optimisation and customisation of products are present in the literature. For example, in [7] the design of antennas is addressed, with the objective of increasing the efficiency of the design process. In [8–10] several methodologies for design of layouts are introduced. Morcillo et al. [8] used the principles and techniques of EC together with Fuzzy Logic to automatically obtain the layout of a single-page design. Morcillo's approach was not able to learn different styles, and in 2014 O'Donovan et al. [9] developed an approach to automatically create graphic design layouts through a new energy-based model derived from design principles. With this method, it was possible to arrange a set of elements in a particular style, previously learned from the analysis of designs presented to the system. In [10] the layout of a set of album images is evolved.

In the majority of the aforementioned works, it is possible to define a fitness function that measures how far we are from our objective. However, when dealing with the design of wearables or with customisable products that are to be purchased by masses, the definition of quality is subjective to the buyer and to its notion of aesthetics, making it much harder to define. With the increasing importance of online shopping, the application of EC to allow and facilitate the customisation and personification of products, have emerged. Whilst the majority of the previous works use automatic fitness assigning schemes, when dealing with fashion it is common to involve the customer in the evolutionary process. Such methodologies are normally referred to as Interactive Evolutionary Computation (IEC) [11] approaches, where the user is involved in the evolutionary process to define the fitness of a certain candidate solutions.

An example of the application of IEC to fashion is in the combination of clothing parts to generate new pieces. In [12] Kim et al. used IEC to generate unique dress designs from separated parts, previously classified in three categories. Through user interaction, it is possible to set the fitness and create outputs that reflect personal preferences. Khajeh et al. [13] also applied this method to enhance the creativity of the designer in order to produce novel sets of clothes design. In their work, clothes components and fabric patterns were designed and encoded separately. Then, through user interaction and the application of fashion design rules, the system combines the components and the fabric patterns. The system proved to be efficient in the generation of fashion designs, minimising the cost and time according to the user preferences. Another important aspect in fashion is concerned with reduction of fabric waste; in [14] and [15] the nesting problem is addressed. Still focusing on fashion, but more specifically on shoe design, in [16], Shimoyama et al. use EC to optimise the sole structure of a shoe. In [17] Dasan, aided by IEC evolves shoe models according to the user preference.

A problem inherent to the majority of IEC approaches concerns user weariness and consistency, i.e., it is difficult for the user to efficiently rank and compare all the individuals in large populations, maintaining the same classification criteria throughout the entire process. Therefore, in the current work we aim at developing a framework that reduces the need for interaction, but at the same time allows customers to taylor their products. Similarly to other approaches, the user interacts with the system to feed his (the preferences. However, in addition to common IEC approaches, we follow the same guidelines behind the work by Machado et al. [18]. The user will not need to classify every single candidate solution and will be able to set filter values, reducing the necessary number of clicks and interaction. An archive where the user can store the best solutions according to his/her preferences is also provided.

Front		Side		Sole		Lining		Zipper Tape		Heel		Straps		Straps Tips		Hardware	
3	11	4	22	0	2	0	1	0	1	4	2	0	2	0	0	0	1

Fig. 1. On top, the genotype of a candidate solution; On the bottom, the resulting phenotype.

3   Framework

EvoFashion is a customisation framework that uses principles of IEC to promote the personalisation of fashion products. Despite the wide range of applicability in fashion domains, we will focus our attention on the evolution of the design of shoe models due to the availability of an engine capable of rendering them[1]. In the following sub-sections we detail the main components of the developed EA and the web interface.

3.1   Representation

Solutions are encoded as a set of integers with the same length as the number of parameters allowed in the customisation of a certain product. Each gene (i.e., integer) has a value in the $[0, num\_possibilities]$ interval, where $num\_possibilities$ is the maximum number of different possibilities for that parameter.

Our experiments will focus on the customisation of a specific shoe model, with 9 customisable parts. Each part allows personalisation of two properties: material and colour. Consequently, the candidate solutions will be composed by $9 \times 2 = 18$ integers. An example of a possible solution along with its phenotypic representation is depicted in Fig. 1.

3.2   Genetic Operators

To promote the evolution and the proper exploration of the problem domain we rely on recombination and mutation. We use uniform crossover to recombine two parents. Firstly we create a random mask of the same size of the genotype, and then swap the genetic material according to the previously generated mask. Regarding the mutation operator, we apply a per gene mutation to the candidate solutions, which allows the algorithm to change, from generation to generation, a percentage of the genes to other valid ones.

---

[1] In the current work we have used the my-swear platform to render the generated shoes, which can be found in https://www.my-swear.com/.

3.3   Fitness Assessment

The main goal of the framework is to promote the evolution of candidate solutions towards regions of the search space that the customer sees as aesthetic. To accomplish that, practitioners often use IEC systems, asking the user to rank the evolved solutions. However, this might lead to some disparities, as the user needs to look at every single evolved solution and compare it with the remaining ones. For that reason, we have used principles of partially interactive computation, which combines automatic fitness components, such as filters, with the interaction between the framework and the user. Three automatic fitness components are used, and they work as filters for the colour, material and price of the evolved models. Further details are presented in the next paragraphs.

Colour – the user defines a colour that he/she likes and the goal of the evolution is to promote the convergence of all parts of the generated shoes to that colour. For that, the Root Mean Square Error (RMSE) is used to compute the distance between a snapshot of the generated shoe models and the target colour, defined by the user;

Material – similar to the colour, but for materials, i.e., the objective is to converge to an individual where all shoe parts are made of the material selected by the user. As with colour, the evolution is guided using an error that is to be minimised and represents the percentage of shoe parts that are not made using the selected material;

Price – by selecting a price range the framework promotes the emergence of candidate solutions that are within that interval, using the RMSE to evaluate the distance between the evolved shoe models' price and the closer bound of the target price range.

From the previous, we propose the following model to compute the quality of each individual:

$$fitness = \frac{1}{1 + automatic\_fitness} + user\_input,$$

where,

$$automatic\_fitness = colour + material + price.$$

That is, the fitness is made of two distinct parts: (i) the automatic evolution towards user selected criteria; and (ii) a user input that works as a weight for ranking the individuals. Both parts are optional, but at least one of the fitness components must be provided, otherwise all individuals would be assigned the same fitness value. In the first part, if one of the automatic components is not defined its RMSE value is 0.

3.4   Archive

The archive works as a repository for the individuals that the user finds aesthetically appealing. If, during the evolutionary process, the user finds a good

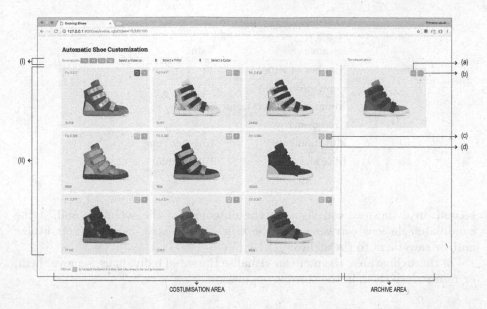

Fig. 2. The Customisation Area, on the left side, is divided in two sub-areas: the evolutionary process management (I) and the visualisation of the population (II). In the last, the user can increase the quality of individual shoes (d) and/or add them to the archive (c). In the Archive Area, the user can: (a) add the saved shoes back to the population; and (b) remove archive members.

solution, he can store it in the archive and re-introduce it later in the population. The archive also helps avoiding convergence, by increasing the diversity at the population level; for that, if solutions start being too similar, without having achieved the desired result, the user just needs to repopulate the generation with archived members.

3.5 Interface

To enable the visualisation and exploration of the different individuals and to allow the interaction with the user, we developed a web-based application. This application allows the user to define different filters, such as colour or price, to improve the fitness of the individuals by clicking on them, and to store them in the archive. Figure 2 shows the different areas of the interface and buttons functionality.

To simplify navigation, the interface is a single page divided into two main areas: the Customisation Area and the Archive Area. This way, the user is always aware of the status of the evolutionary process and can easily manage the archive and change the filters.

The Customisation Area is divided into two sub-areas. In the first one, the user can manage the evolution, by increasing the number of generations or modifying the filters parameters (such as colour, material or target price). In the

Table 1. Experimental parameters.

Parameter	Value
Number of runs	30
Population size	12
Number of generations	50
Crossover rate	70%
Mutation rate	10% per gene
Elite size	1 individual

second area, the user can visualise the outcomes of the evolution and, if the candidate solutions correspond to the user's preferences, increase their fitness and/or save them to the archive.

In the archive area, the user can visualise the saved individuals, remove them, or reintroduce them in the current population.

4   Experimental Results

We will use EvoFashion to conduct two different experiments. First, we will address the ability of the automatic fitness components to promote evolution towards a target colour, material, price or a combination of multiple parameters. Next, the automatic fitness components will be combined with user interaction, through IEC.

4.1   Experimental Setup

Table 1 details the framework parameters used in the experiments conducted in the following sections. All the values, except the number of generations, are kept fixed throughout the experiments. When performing tests using interactive fitness the number of maximum generations is not fixed, as the stopping criteria depends on the user preferences. In order to prevent user exhaustion, and to allow the majority of the population to be displayed in the screen at the same time, we keep the population size small, i.e., 12 individuals. For that reason, we use high mutation and crossover rates so that the individuals in each generation have noticeable differences. This enables a faster convergence towards feasible solutions, i.e., the emergence of individuals that are considered of high quality by the user. Parent selection is performed using roulette wheel.

The shoe models that are going to be optimised allow the customisation of the following parts: front, side, sole, lining, zipper tape, heel, straps and strap tips, and hardware. For each shoe part it is possible to define two parameters: colour and material.

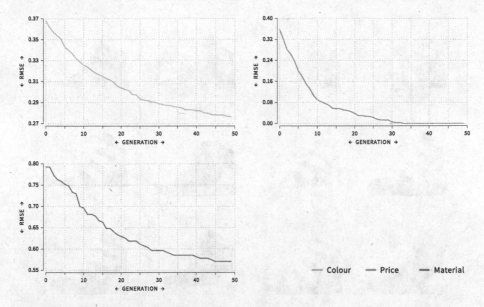

Fig. 3. Evolution of the RMSE of the best individuals across 50 generations for each of the automatic fitness components: colour, price and material. The results are averages of 30 independent runs. (Color figure online)

4.2   Automatic Evolution

The first experiments focus only on the use of the automatic fitness components. Therefore, we have analysed the ability of each fitness function component to promote the emergence of candidate solutions that depict the desired characteristics for a specific target value. The targets for colour, price and material were set to $\#dae415$ (■), > 7000 and crocodile, respectively.

Figure 3 shows the evolution of the RMSE of the best individuals across 50 generations for each of the automatic fitness components. The results are averages of 30 independent runs. A perusal analysis of the results confirms that in all scenarios the system is capable of promoting the convergence towards regions of the search space that have the desired characteristics, i.e., yellow shoe models, with prices above 7000 or with the majority of the shoe parts made of crocodile, respectively for the evolution guided by colour, price and material.

A more detailed analysis of the charts reveal that when evolving towards a target colour (in this case, $\#dae415$ ■) the convergence speed is lower than in the remaining experiments. Moreover, while in the experiments concerning evolution based on price and material evolution attains results close to the best possible solution, with colour this seems not to be the case. In the evolution towards crocodile materials, the best possible solution has a RMSE of $5/9$ because only 4 shoe parts out of the 9 can be made of crocodile. However, it is expected that RMSE values attained with colour are higher than those reached by the remaining metrics. The reason for that is related to the values that each of

Fig. 4. Example of the initial population of one evolutionary run focusing the evolution towards a target colour: #*dae*415 (■). Individuals are ordered from left to right and from top to bottom according to their fitness values. (Color figure online)

Fig. 5. Example of the last population for the same run of Fig. 4. Individuals are ordered from left to right and from top to bottom according to their fitness values. (Color figure online)

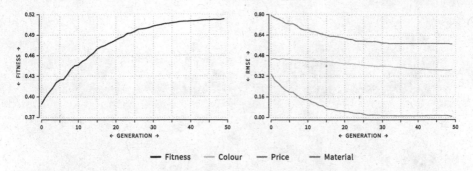

Fig. 6. On the left, the evolution of the fitness of the best individuals across 50 generations with a combination of the automatic fitness components: colour (*#dae*415), price (> 7000) and material equal crocodile. On the right, the evolution of each of the automatic fitness components. The results are averages of 30 independent runs.

the variables can assume and how the error is computed: in the colour guided evolution, the error is computed as a distance to a target colour, which is specified by the user. But, the specified colour may not be allowed to all shoe parts, or may have slightly different RGB colour values depending on the used material. As such, the goal of this fitness component is not to attain a value of 0, but rather an approximation of the input colour.

Figures 4 and 5 show, respectively, the initial and last populations of a randomly chosen run aiming at converging the colours of the different shoe parts towards *#dae*415. Candidate solutions are ordered from the best to the worst individuals in terms of RMSE, i.e., the first individuals are those closer to having all shoe parts with the same colour, and equal to the target one. Looking at the images of the shoe models it is possible to observe that in the initial generation the shoes depict a great variance, having different colours and materials. Additionally, only 2 out of 12 shoes depict parts with yellow colours. On the contrary, in the last generation, 9 out of the 12 shoes have at least one part of the shoe with a colour similar to the target one, being that the first 4 presented models are almost entirely yellow.

From the above conducted experiments we conclude that when we only aim at tackling one of the automatic fitness components the framework is able to find adequate solutions. In the upcoming paragraphs, we will analyse the ability of the framework to obtain shoe models subjected to several restrictions. More precisely, we will promote the evolution towards models that are yellow, have a price above 7000 and are made of crocodile.

Figure 6 depicts the results obtained when using all automatic fitness components. From the left figure it is possible to acknowledge that across generations evolution is promoted. To explain how better solutions are being attained, we need to look at the image on the right, which shows the decomposition of the automatic fitness function into the different components. The fitness value is to be maximised and the errors of the different fitness components are to be minimised. Thus, in the left chart the higher the values the better, and on the right hand side, the lower the better.

Fig. 7. On the left, the best candidate solutions of the 6 evolutionary runs that obtained the highest fitness results. On the right, the opposite, i.e., the best candidate solutions of the 6 evolutionary runs that obtained the lowest fitness results. Results are from the 30 evolutionary runs guided using the combined fitness function. (Color figure online)

An analysis of the right panel shows that the automatic fitness components that have a greater contribution to the fitness evolution are price and material, which attain results closer to the best possible: 0 in the price error, and 5/9 in the material error (only 4 of the 9 shoe model parts can be made of crocodile). The colour evolution is much slower, ending with an error of approximately 0.4, which is not far from the initial 0.45. However, looking at the candidate solutions of the evolutionary runs that reach the best and worst results (see Fig. 7) it is possible to see that even the solutions from the runs that obtain the worst results have at least one of the shoe parts coloured in yellow. As such, we conclude that the colour objective is also being accomplished. Because of the used material, the RGB components in the crocodile material are different than the ones of the selected colour by the user. Therefore, it is impossible to get shoe parts painted with the selected colour, due to conflicting evolutionary objectives. As a curiosity, the shoe with the highest fitness in Fig. 7 (top left corner) was capable of achieving all evolutionary objectives: it has a price of 7450, all shoe parts use crocodile, and the shoe model is entirely yellow.

4.3   Partially Interactive Evolution

Another aspect that needs to be analysed is the ability of the user to interact with the framework. To accomplish this task, we developed an easy to use web-interface (Sect. 3.5). The user sees the results of the evolutionary in a grid layout, and may provide feedback regarding the quality of each of the generated solutions. For that, the user presses the improve fitness button as many times as he/she likes, i.e., each time the button is pressed, the fitness value is increase by an a priori defined value. Additionally, if the user finds a shoe design aesthetically pleasing, he/she can store it on the archive. Notwithstanding, the user can combine his/her own preferences with the aforementioned automatic fitness components.

To demonstrate that the framework is able to comply with the user preferences, following the steps described above, please refer to the video posted in https://vimeo.com/cdvlab/evofashion.

5   Conclusions and Future Work

Motivated by the increase in online shopping and by the desire of customers to personalise the products they buy we propose a partially interactive evolutionary computation framework for the customisation of fashion products. The framework is based on a web-interface which allows the user to visualise the evolutionary process and select the products he/she finds aesthetically pleasing. In addition to user attribute fitness, the framework also lets the user to specify filter values, which are used to automatically guide evolution. There is an archive where best solutions can be saved, and later re-introduced in the evolutionary process. The archive has two main goals; first, it allows the customer to keep track of the most aesthetic solutions; secondly, it avoids the loss of population diversity.

Results show the ability of the framework to successfully customise fashion products. In concrete, we have targeted the personalisation of shoe models. In the first experiments, focus was given to the evolution only considering the automatic fitness components, i.e., the evolution of candidate solutions towards a specific colour, price range or material. Then, tests regarding the evolution of shoe models that satisfy more than one of the previous conditions were conducted. In both scenarios it was possible to obtain the expected results, and the algorithm successfully converged to regions of the search space where the shoe model had the desired characteristics.

After assessing the ability of the automatic fitness function components, we performed experiments where the user was asked to evaluate the generated shoe models. Results show that, by using an interactive fitness function, the user is able to promote solutions with the defined parameters and that, when his/her needs are satisfied by a given solution, he/she can increase its fitness, promoting the shoe model characteristics through the next generations.

Next steps to expand on this work will focus on experimenting with the customisation of different fashion products, using different rendering engines.

Additionally, we will look into methodologies for automatically updating the archive. An example of such approaches is the work of Vinhas et al. [19], where the an archive of candidate solutions is updated according to the new solutions' novelty degree, i.e., for individuals to be added to the archive they have to be different from those that are already in the archive, and their quality must be superior to a given minimum threshold. Approaches for automatically initialising the first population from the archive, as well as other automatic fitness function metrics will also be investigated.

Acknowledgments. All shoe images are copyright of MYSWEAR (https://www. my-swear.com/), and are used as fair use for academic purposes only. We gratefully acknowledge the support of NVIDIA Corporation for the donation of a Titan X GPU. We would also like to thank Tiago Martins for all the patience making the charts herein presented.

References

1. Pine, B.J.: Markets of One: Creating Customer-Unique Value Through Mass Customization. Harvard Business Press, Brighton (2000)
2. Eiben, A.E., Smith, J.E.: Introduction to Evolutionary Computing, vol. 3. Springer, Heidelberg (2003)
3. Caldas, L.G., Norford, L.K.: A genetic algorithm tool for design optimization. In: Proceedings of the 1999 Conference of the Association for Computer-Aided Design in Architecture (ACADIA 1999), Salt Lake City, UT (1999)
4. Besserud, K., Cotten, J.: Architectural genomics (2008)
5. Michalek, J., Choudhary, R., Papalambros, P.: Architectural layout design optimization. Eng. Optim. **34**(5), 461–484 (2002)
6. Juan, Y.K., Shih, S.G., Perng, Y.H.: Decision support for housing customization: a hybrid approach using case-based reasoning and genetic algorithm. Expert Syst. Appl. **31**(1), 83–93 (2006)
7. Hornby, G.S., Globus, A., Linden, D.S., Lohn, J.D.: Automated antenna design with evolutionary algorithms. In: AIAA Space, pp. 19–21 (2006)
8. Gonzalez-Morcillo, C., Mártin, V.J., Vallejo, D., Castro-Schez, J.J., Albusac, J.: Gaudii: an automated graphic design expert system. In: IAAI (2010)
9. O'Donovan, P., Agarwala, A., Hertzmann, A.: Learning layouts for single-pagegraphic designs. IEEE Trans. Visual. Comput. Graphics **20**(8), 1200–1213 (2014)
10. Geigel, J., Loui, A.: Using genetic algorithms for album page layouts. IEEE Multimedia **10**(4), 16–27 (2003)
11. Takagi, H.: Interactive evolutionary computation: fusion of the capabilities of EC optimization and human evaluation. IEEE Proc. **89**(9), 1275–1296 (2001)
12. Kim, H.S., Cho, S.B.: Application of interactive genetic algorithm to fashion design. Eng. Appl. Artif. Intell. **13**(6), 635–644 (2000)
13. Khajeh, M., Payvandy, P., Derakhshan, S.J.: Fashion set design with an emphasis on fabric composition using the interactive genetic algorithm. Fashion Text. **3**(1), 1–16 (2016)
14. Crispin, A., Clay, P., Taylor, G., Bayes, T., Reedman, D.: Genetic algorithm coding methods for leather nesting. Appl. Intell. **23**(1), 9–20 (2005)

15. Shiyou, Y., Guangzheng, N., Yan, L., Renyuan, T.: Shape optimization of pole shoes in harmonic exciting synchronous generators using a stochastic algorithm. IEEE Trans. Magn. **33**(2), 1920–1923 (1997)
16. Shimoyama, K., Seo, K., Nishiwaki, T., Jeong, S., Obayashi, S.: Design optimization of a sport shoe sole structure by evolutionary computation and finite element method analysis. Proc. Inst. Mech. Eng. Part P J. Sports Eng. Technol. **225**(4), 179–188 (2011)
17. Dasan, A.: The evolutionary design of generative shoes. http://arandasan.co.uk/Evolve (2012). Accessed 06 Nov 2016
18. Machado, P., Romero, J., Cardoso, A., Santos, A.: Partially interactive evolutionary artists. New Gener. Comput. **23**(2), 143–155 (2005)
19. Vinhas, A., Assunção, F., Correia, J., Ekárt, A., Machado, P.: Fitness and novelty in evolutionary art. In: Johnson, C., Ciesielski, V., Correia, J., Machado, P. (eds.) EvoMUSART 2016. LNCS, vol. 9596, pp. 225–240. Springer, Cham (2016). doi:10.1007/978-3-319-31008-4_16

A Swarm Environment for Experimental Performance and Improvisation

Frank Mauceri and Stephen M. Majercik$^{(\boxtimes)}$

Bowdoin College, Brunswick, ME 04011, USA
{fmauceri,smajerci}@bowdoin.edu

Abstract. This paper describes Swarm Performance and Improvisation (Swarm-PI), a real-time computer environment for music improvisation that uses swarm algorithms to control sound synthesis and to mediate interactions with a human performer. Swarm models are artificial, multi-agent systems where the organized movements of large groups are the result of simple, local rules between individuals. Swarms typically exhibit self-organization and emergent behavior. In Swarm-PI, multiple acoustic descriptors from a live audio feed generate parameters for an independent swarm among multiple swarms in the same space, and each swarm is used to synthesize a stream of sound using granular sampling. This environment demonstrates the effectiveness of using swarms to model human interactions typical to group improvisation and to generate organized patterns of synthesized sound.

Keywords: Interactive human/computer music improvisation · Swarm intelligence

1 Introduction

Real-time, interactive music generation systems often draw inspiration from the interactions of improvising performers. Group improvisation provides examples of spontaneous music generation involving player autonomy as well as reaction and responsiveness between players. Generative systems for music provide possibilities for complex, real-time, pattern generation analogous to improvisation. For the purpose of interactivity, these systems can be designed to accept human input in real-time. However, this is not a sufficient condition for creating interactivity analogous to group improvisation.

Interaction in group improvisation happens in several musical dimensions (e.g. changes of timbre, thematic material, dynamics, phrasing) and addresses different time frames (e.g. instant changes of texture or mood; short term motivic call and response; recapitulation of remembered materials). In artificial systems, the mapping between the human player and the computer needs to address multiple parameters and should have consequences over multiple time scales.

Audience recognition of the interactions among players is an essential element of the drama of performance. By witnessing interactions among players,

© Springer International Publishing AG 2017
J. Correia et al. (Eds.): EvoMUSART 2017, LNCS 10198, pp. 190–200, 2017.
DOI: 10.1007/978-3-319-55750-2_13

the audience understands that decisions are made in the moment. Similarly, in artificial systems, the interface between player and computer should allow the audience to witness the exchange of information and discover its significance.

Swarm-PI is a real-time computer environment for music improvisation that uses swarm algorithms to control sound synthesis and to mediate interactions with a human performer. Other swarm-based interactive improvisation systems have been developed; we discuss some of these in Sect. 5. What distinguishes this environment from previous attempts to use swarms for interactive music systems is the introduction of a swarm avatar, i.e. a swarm that is analogous to acoustic features of the human performance. This swarm's parameters are continuously modulated by changes in the performer's sound. This avatar swarm, representing the performer, interacts with computer generated swarms.

In Sects. 2 and 3, we describe the goals that guided the design of Swarm-PI and discuss the nature of interactivity. We describe the details of Swarm-PI in Sect. 4. In Sect. 5, we discuss related work on swarm-based music systems. We discuss further work in Sect. 6, and we conclude in Sect. 7.

2   Design Goals

Our goal was to create an environment in which an improviser playing a conventional musical instrument could interact with a generative music system that responds to musical gestures, including changes in timbre. We constructed an interactive environment for improvisation to satisfy the following goals:

- The environment should facilitate the exploration of unfamiliar materials, relationships, and structures.
- The system should not presume any stylistic predisposition; the goal is not to simulate jazz (for example) or any other pre-existing music syntax.
- The observers (the performer and the audience) should be continually invited to anticipate the sequence of system states, although the system's complexity makes this sequence hard to predict.
- The performer's improvisational decisions and the environment must have meaningful consequences for each other.

3   Interactivity

Frequently, the creative work in a group of improvisors depends on interactions between performers who have independent, yet intersecting sets of artistic goals and strategies. In our environment, each swarm generates its own stream of sound, which develops according to the swarms unique rule set.

Improvisers' decisions often involve simple, immediate responses to one or more performers. In the course of playing together, organized behaviors emerge that are not the consequence of previously established global rules, but rather the consequence of local decisions. In this respect, the interactions of improvising performers resemble the behavior of swarm systems, where local interactions often result in surprising global patterns. For many performers, the emergence of unexpected patterns and relationships is a primary artistic goal. In improvised performance, the level of information is always in flux; artists constantly adjust their strategies in order to preserve the conditions of uncertainty and discovery.

In the introduction, it was noted that the audience's recognition of a responsive exchange between players was essential to the experience of interactivity. For that reason, a live musical performer was used for human input into Swarm-PI. The audience is thus given visual cues as well as auditory cues from the performer that are intended to facilitate the understanding of the audience. The performer exhibits listening and response in her comportment, gesture, and the expressive dimension of playing an instrument. Additionally, the authors intend that a display of the video representation of the swarms will be part of a performance, further facilitating the recognition of interaction between performer and computer. For example, an increase in the population size could be mirrored by an increase in the amplitude of the generated sound.

4   System Dynamics

The system dynamics of Swarm-PI arise from the interactions among a human avatar swarm and up to six computer generated swarms. A swarm's dynamics are not governed by a global controller or external force, but rather emerge from the interactions of the individual agents in the swarm (see Sect. 4.1). The self-organization and emergent behavior of the swarms make them appropriate for modeling creative production; each swarm is a virtual performer and its dynamics resemble the spontaneous autonomy of an improvised solo. The swarm can be described in terms of its location, density, velocity, and coherence; this allows for a nuanced mapping of swarm movements onto sound synthesis parameters. Swarms react to the location and movement of neighboring swarms, reflecting the exchange of information typical of group improvisation.

In Swarm-PI, the swarms interact with each other and are represented by real-time video animation in a 3-dimensional virtual space (Fig. 1). Audio from the human player is analyzed and acoustic descriptors are mapped onto a swarm avatar that interacts with the computer-generated swarms. The locations and movements of the computer-generated swarms are mapped to parameters for granular sound synthesis. The computer's reaction to the human player is mediated by the swarm representation of the performer and by the swarm rules governing the movement of swarm individuals. The performer is free to respond to the audio output of the system. Figure 2 provides an overview of the system in a configuration in which there is only one computer generated swarm. The Swarm-PI structure readily maps onto Blackwell's PQf architecture [5].

4.1   Swarm Movement

A swarm is composed of some number of virtual agents, typically between 50 and 100. The movement of individual agents in Swarm-PI is governed by rules

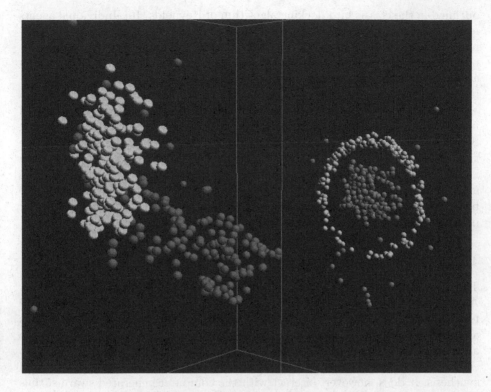

Fig. 1. Visual representation of swarms

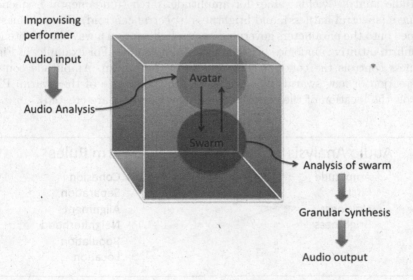

Fig. 2. Swarm-PI system overview

outlined in the classic bird flocking algorithm of Reynolds [10]. Individual agents move according to the location and velocity of other agents in a local neighborhood defined by a radius of perception. Three basic rules govern agent movements. An agent tries to match the average heading of its neighbors (*alignment*) and move toward the average location of its neighbors (*cohesion*), while avoiding collisions (*separation*). In our implementation, agents also have a *normal speed* and a *maximum speed*, and a degree of *random motion* can be introduced into their motions. With this basic model a wide variety of behaviors is observed. Swarms of agents can be coherent or chaotic; unexpected behaviors such as splitting and rejoining are possible. When more than one swarm is present, each with it own rules, a variety of interactions are possible including orbiting behaviors and predator/prey relationships. The swarm algorithm and video routines are based on Sayama's Swarm Chemistry [12] and written in Processing. The user interface and sound synthesis are implemented in Max/MSP.

4.2 Human Avatar

The human performer is introduced into the swarm space as a swarm avatar. An acoustical analysis of the performer (using Tristan Jehan's analyzer external for Max/MSP [8]) characterizes the sound of the player's improvisation. This analysis in turn defines parameters for the performer's swarm avatar. Like Rowe and Singer's stage work *A Flock of Words* [11], which uses acoustical analysis to control a swarm, the swarm avatar in Swarm-PI is not used directly for sound synthesis. It does, however, interact with the computer generated swarms, thus mediating between the human-improvised sound and the computer generated swarms, which generate their own sound (see Sect. 4.3).

Audio analysis yields values for amplitude, pitch (fundamental frequency), noisiness (spectral flatness), and brightness (spectral centroid). These values are mapped onto the parameters governing swarm behavior, in a way that is intended to confirm intuitive analogies between visuals and sound. For example, in Fig. 3, noisiness controls the cohesion and density of the swarm. Amplitude controls the creation of new swarm individuals and, thus, the size of the swarm. Pitch controls the location of the emitter (where new particles are created).

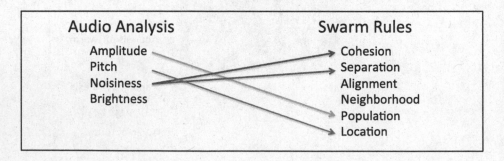

Fig. 3. Example mapping for swarm avatar

4.3 Sound Generation

The computer generated swarms, whose parameters are generated randomly, provide the basis for sound generation. It was noted in early experiments that these swarms settled into predictable patterns. Movements of the performer's avatar would perturb swarm movement and create interesting patterns, but additional changes in behavior were desired to prevent the repetition of these patterns over larger time scales. This was accomplished by cyclically modulating the swarm parameters using low frequency sine waves (e.g. 0.1 Hz).

Sonification of the swarms is accomplished via granulated audio sampling (using GMU Max externals developed at GMEM [1]). Sound granulation involves taking small audio samples (5–200 ms) and subjecting them to an amplitude envelope; constellations of these sound granules are assembled to synthesize complex timbres. Each sound grain is characterized by the location from which it is selected in a stored audio file, and by its envelope, duration, amplitude, playback speed, and pan position. The density of grain events is controlled by the inter-onset time between sound grains. Sequences of grains can be synchronous or asynchronous.

Parameter values for granular synthesis are obtained from an analysis of the swarm. Mean location, location variance, mean velocity, and variance of velocity are mapped onto the parameters of the granular synthesis engine. Again, mappings are chosen to make intuitive analogies between swarm behavior and changes in sound. For example, in Fig. 4, swarm density (location variance) correlates to granular density (inter-onset time and length), movement in the x-axis controls which sample is used and the pan position, movement in the y-axis affects transposition, and movement in the z-axis in conjunction with the population size affects amplitude.

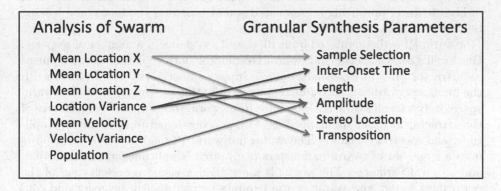

Fig. 4. Example mapping for granular synthesis

The parameter spaces of both the computer generated swarms and the synthesis technique are manifold, resulting in rich possibilities for mapping system dynamics onto sound. Initial performances used a live saxophone player and audio samplings from recordings of musical instruments for the granular synthesis. These performances demonstrated interesting interactions with the granulated sound generated by Swarm-PI. The performer found it engaging to control the swarm avatar and felt that the generated sound had a rich, complex, and continuously changing texture that was an interestingly stark contrast to his saxophone. He felt, however, that the system was slow to react to sudden changes or fast sequences of notes, and that the timbre of his saxophone needed to be better integrated into the ensemble of swarm improvisers. Finally, we felt that changes in the sound produced did not capture the vivid changes of organization displayed by swarms. Finding satisfying parameter mappings continues to be a challenge.

5   Related Work

Similarities between the dynamic of improvising musicians and that of swarms has inspired the creation of a number of swarm-based music generation systems. Three of these are Blackwell's: Swarm Music [4], Swarm Granulator [3], and Swarm Techtiles [6]. In Swarm Music, multiple swarms of particles move in a virtual 3-dimensional space, using to Reynolds' flocking algorithm [10], described in Sect. 4.1. Music from human performers influences swarm behavior by creating targets within the environment. Swarm particles are mapped into a virtual 3-dimensional music space, the axes of which represent volume, pitch, and pulse, thus providing a straightforward mapping from swarm to sound. Swarm Granulator is similar to Swarm Music, but operates at the sound grain level rather than the note level. Swarm Techtiles uses an entirely different approach. Audio samples from two channels are scaled to pixel values and the two streams are used to create a two-dimensional image with texture. Swarms search for areas of high texture, producing a granulated sound based on the image and leaving behind attractors to attract other swarms.

Swarm-PI is distinguished from Blackwell's systems in a number of respects. Blackwell's Swarm Music and Swarm Granulator both represent audio input as an attractor. Swarm-PI visualizes audio input as another swarm. In Swarm-PI, the human performer is therefore represented by this swarm avatar in the swarm space. In Blackwell's Swarm Granulator framework, the x, y, and z coordinates of the attractor location are mapped onto three acoustic parameters (pitch, amplitude, and spectral centroid). The avatar in Swarm-PI maps acoustic descriptors onto a larger set of swarm parameters (e.g., pitch, amplitude, spectral centroid, and spectral variance). The avatar is more than a visual representation of the performers sound; the avatar swarm exhibits its own swarm behavior and continually interacts with other, sound-generating swarms. These sound-generating swarms and the avatar together, visually represent the interactive ensemble; the avatar preserves the metaphor of an ensemble of equals, improvising together.

Another difference between Blackwell's approach and Swarm-PI is the method for translating swarms into granulated audio. Blackwell's Swarm Granulator and Swarm Techtiles use the position of individuals in the swarm for the calculation of sound grains. Swarm-PI uses a statistical analysis of the swarm to control parameters for sound granulation. This method sacrifices the specificity of representing swarm individuals, but is less computationally expensive and more robust in real-time applications.

Two other swarm-based systems, Jones' AtomSwarm [9] and an unnamed system by Beyls [2], are not currently configured for interaction with human performers. AtomSwarm employs a flocking mechanism similar to that of Reynolds, a system of virtual hormones that affects the behavior of an individual, and a genetic component that regulates the hormone mechanism and modifies the musical behavior of an individual. The cyclic modulation of swarm rules in Swarm-PI is similar to the hormonal cycles in AtomSwarm, both serving to promote more varied behavior.

In Beyls' swarm music system, swarms operate in a 2-dimensional space associated with a MIDI player. The behavior of swarm individuals is governed by various factors, including position, energy level, level of attraction/repulsion with respect to other individuals, an activation level, and a personality. Movement is controlled by a stress factor that is similar to the cohesion/separation factors in a flocking swarm. Individuals form clusters; sound is produced by the first agent in the cluster with the most energy, its personality determining the melody's pitch-intervals, durations, and velocities. The other agents in the cluster add to the melody only if their activation level is high enough, whereas all swarm individuals in Swarm-PI contribute to the sound produced.

6   Further Work

Swarm-PI realizes an interactive computer system inspired by group improvisation. One of the most attractive features of this system is the performer's ability to function as if improvising in a group situation, playing an instrument and interacting with the system using only musical gestures and cues. There are many directions for further work; we describe four that we are currently pursuing.

6.1   The "Edge of Creativity"

In improvised performance, the artist is constantly adjusting her strategies to try to maintain a balance between uncertainty and creation [7]. There must be enough uncertainty so that there is "room for" new musical ideas, but not so much uncertainty that nothing created can be sustained.

Complex dynamical systems, such as swarms, have become an influential concept in psychology and cognitive science, especially with reference to creativity [13,14]. Such systems exhibit three types of behavior that are analogous to different aspects of creativity: (1) self-organization, a phenomenon in which the elements of the system become organized or structured in a seemingly purposeful way despite the absence of a global controller, (2) emergent behavior, i.e. behavior that is not predictable from knowledge of the system's constituent parts, but instead arises from the operation of the system, and (3) phase changes, which are relatively abrupt transitions from one behavioral regime to another very different regime. These three characteristics suggest analogies to the organization of ideas into new patterns in the creative process (self-organization), the novel and unexpected qualities of creative ideas and artifacts (emergent behavior), and the nonlinear quality of the creative process (phase changes).

In particular, phase changes provide a way to model the efforts of the performer to maintain the balance between uncertainty and creation. In the computer generated swarms in Swarm-PI, some parameter values result in a fixed state, e.g. all the swarm members are trapped in a tiny sphere centered at a fixed point. Even if that point were moving, the behavior would, in a sense, still be fixed; only the center of the sphere is moving. Changing the parameter values, even smoothly, can transform a fixed state into an apparently random state, in which swarm members are evenly distributed in the virtual space.

In between these two regimes, however, it is possible to find regimes of behavior in which swarm members form structures, such as rings (see Fig. 1), which are stable for a period of time, but then dissolve and re-form as some other type of structure, such as a sphere. Two structures can collide and produce a new structure. There is both uncertainty and creation. There is no formula that one can use to produce this regime, but information flow appears to play a key role. There needs to be sufficient information flow—swarm members need to know what other swarm members are doing—to prevent fixed behavior and enable structure formation, but not so much information flow that the structures created are destroyed by new structures as soon as they form. Our hypothesis is that an analysis of the information flow in the swarm will allow us to determine the conditions necessary to produce and maintain a "sweet spot" between the fixed and random regimes, where uncertainty and creation are in a productive balance. This could be used to create a self-regulation mechanism for the computer generated swarms in Swarm-PI.

6.2   Modeling Memory

Human performers not only respond to the immediate environment, but also to things that happened in the immediate past and to things that they anticipate happening in the future. The novelty of emerging patterns is always measured against expectations held in memory. The neighborhood of interactions extends in time as well as in space.

The human player's interaction with Swarm-PI includes the player's memory and anticipation. Ideally, the swarm players would respond not only to their immediate environment but also to a span of "remembered" events. One method to implement system memory would be to have swarms lay persistent or recurring stigmergic traces or trails in the virtual swarm space. These traces, functioning as attractors or obstacles, would imprint the topology of the solution space with a map of past events. Blackwell uses stigmergic attractors rather than a swarm avatar to represent human input [4]. We would like to retain the advantages of using an avatar and add stigmergic mechanisms for introducing responsiveness over a range of time frames.

6.3 Self-reflective Feedback

Many models of creativity recognize evaluation as essential to the process; human players recognize the processes they are engaged in as productive or not and adjust their responses appropriately. Similar feedback for swarm players might be implemented using measures of information flow and pattern emergence, which could be used as prompts modulating mappings and swarming rules. Feedback mechanisms could also contribute to the organization of the system over longer time scales.

6.4 Evolution of Swarms

Measures of information flow and pattern emergence could potentially be used as fitness measures for the evolution of the computer generated swarms, producing additional interesting dynamics. The values for the rule parameters governing a particular swarm would be that swarm's genome. Exchanging some rule parameters between two swarms would result in two child swarms, similar to [12]. Mutation would introduce slight changes in the parameters. The two parent swarms would die, via age-based mortality, and be replaced by the child swarms. The lifespan of a swarm (with a suitable lower bound to prevent an early demise) would be one of the parameters described by the genome. This evolutionary scenario would produce system dynamics for each generation that are new, yet related to the dynamics of the previous generation.

7 Conclusion

Swarm-PI provides a framework for explorations of swarm-based interactive performance. The swarm avatar provides a flexible mechanism for mediating interactions between the human player and the improvisation system. Avatars exist in a rich and complex space; the parameter space is very large and avatars are capable of a wide range of behaviors. This supports the mapping of a detailed characterization of the human player's output, including non-musical gestures and cues, into the virtual space of the system. The complexity of the possible mappings provides the opportunity for subtlety in the system's internal dynamics and its reaction to input, and it affords users great flexibility in configuring the environment. We have sampled only a tiny fraction of the ways human output can be mapped onto avatars and the ways avatars can interact with the computer generated swarms. Further explorations of this complex space may uncover new modes of interaction between human and computer performers.

Acknowledgments. The authors would like to thank Bowdoin College students John Burlinson, Nicole Erkis, Grace Handler, Octavian Neamtu, and John Truskowski for their work.

References

1. Bascou, C., Pottier, L.: GMU, a flexible granular synthesis environment in Max/MSP. In: Proceedings of the Sound and Music Computing Conference (2005)
2. Beyls, P.: Interaction and self-organization in a society of musical agents. In: Proceedings of the ECAL 2007 Workshop on Music and Artificial Life (MusicAL 2007) (2007)
3. Blackwell, T.M.: Swarm granulation. In: Romero, J., Machado, P. (eds.) The Art of Artificial Evolution: A Handbook on Evolutionary Art and Music, pp. 103–122. Springer, Heidelberg (2008)
4. Blackwell, T.M.: Swarm music: improvised music with multi-swarms. In: Proceeding of the 2003 AISB Symposium on Artificial Intelligence and Creativity in Arts and Science, pp. 41–49 (2003)
5. Blackwell, T.M.: Evolutionary computer music. In: Miranda, E.R., Al Biles, J. (eds.) Swarming and Music, pp. 194–217. Springer, London (2007)
6. Blackwell, T.M., Jefferies, J.: Swarm tech-tiles tim. In: Rothlauf, F., et al. (eds.) EvoWorkshops 2005. LNCS, vol. 3449, pp. 468–477. Springer, Heidelberg (2005). doi:10.1007/978-3-540-32003-6_47
7. Borgo, D.: Sync or Swarm: Improvising Music in a Complex Age. Continuum Publishing, New York (2005)
8. Jehan, T.: Max/MSP (2012). http://web.media.mit.edu/~tristan
9. Jones, D.: AtomSwarm: a framework for swarm improvisation. In: Giacobini, M., et al. (eds.) EvoWorkshops 2008. LNCS, vol. 4974, pp. 423–432. Springer, Heidelberg (2008). doi:10.1007/978-3-540-78761-7_45
10. Reynolds, C.W.: Flocks, herds, and schools: a distributed behavioral model. SIGGRAPH Comput. Graph. **21**(4), 25–34 (1987)
11. Rowe, R., Singer, E.: Two highly-related real-time music and graphics performance systems. In: Proceedings of the International Computer Music Conference, pp. 133–140 (1997)
12. Sayama, H.: Swarm chemistry. Artif. Life **15**(1), 105–114 (2009)
13. Sulis, W., Combs, A.: Nonlinear dynamics in human behavior. In: Studies of Nonlinear Phenomena in Life Science, vol. 5. World Scientific (1996)
14. Zausner, T.: Process and meaning: nonlinear dynamics and psychology in visual art. Nonlinear Dyn. Psychol. Life Sci. **11**(1), 149–165 (2007)

Niche Constructing Drawing Robots

Jon McCormack[✉]

sensiLab, Faculty of Information Technology,
Monash University, Caulfield East, Australia
Jon.McCormack@monash.edu
http://jonmccormack.info

Abstract. This paper describes a series of experiments in creating autonomous drawing robots that generate aesthetically interesting and engaging drawings. Based on a previous method for multiple software agents that mimic the biological process of niche construction, the challenge in this project was to re-interpret the implementation of a set of evolving software agents into a physical robotic system. In this new robotic system, individual robots try to reinforce a particular niche defined by the density of the lines drawn underneath them. The paper also outlines the role of environmental interactions in determining the style of drawing produced.

1 Introduction

The idea of autonomous machine creativity has occupied human thought at least since the ancient Greeks, although the concept of creativity and its interpretation has changed significantly [25]. In more recent times, the field of computational creativity has sought a scientific understanding of various aspects of human creation and creativity, and a major goal of this discipline is to create autonomous creative artists [8]. However, the emphasis – and the standard benchmark – is human creativity, most often through the lens of classical and modernist art or established genres of design or music [19].

The research described here investigates the concept of autonomous machine creativity through the design and implementation of autonomous drawing robots. In these experiments, the motivation and inspiration is not the mimicry or replication of human drawing or human creativity.[1] The investigations described are motivated by a desire to understand *poesies* in a post-anthropocentric sense [24], one that considers creative expression more broadly than the examples exhibited by humans and human cultures. In this sense, human and machine creativity may be seen as synergistic or complementary, contributing to an on-going dialogue on the fundamentals of creativity, and expanding how we consider creativity in a post-anthropocentric worldview.

---

[1] Although of course humans may appreciate the aesthetics and creativity too!.

© Springer International Publishing AG 2017
J. Correia et al. (Eds.): EvoMUSART 2017, LNCS 10198, pp. 201–216, 2017.
DOI: 10.1007/978-3-319-55750-2_14

2   Background

The research described here relates to a class of drawing machines, known as *autonomous drawing robots* (often referred to as "Drawbots"), where one or more autonomous robots interact to create a drawing. Typically the resultant drawings are abstract, created through a pattern-making process rather than fig-urative representations drawn from captured images, although such distinctions normally represent a continuum of possibilities rather than a binary division. In some cases, many individual robots are involved in the generation of a single drawing. Much like the swarm [15] and multi-agent or individual-based mod-elling systems [14], these systems draw on the emergent interactions between multiple robotic agents and their environment.

Also important is the *performative* nature of autonomous, artistic drawing systems. People are fascinated by machines that appear to show intent, agency and autonomy – even if it is often alien or ambiguous. Drawbots play on this fascination to produce artefacts that provide a visual record of that agency and autonomy. Unlike their human counterparts robots are always eager and ready to perform: they do not need to wait until "the right moment" or search for inspiration in time or from place in order to create. Hence, many systems are readily exhibited "in the act of drawing", rather than exhibiting only the finished drawings they have created, as is the norm with conventional human artists.[2]

2.1   Related Work

The artist Jean Tinguely developed a series of autonomous drawing machines ("Méta-Matics"), first exhibited in Paris in 1959 [9]. Through these machines Tinguely deliberately expressed an ironic distance "from the abstract-gestural painting of his contemporaries". The visual form and kinetic behaviour of these machines fascinated audiences, explicitly prompting the question of the automa-tion of artists and artistic processes.

A extensive visual survey of drawing machines can be found in the *Draw-ing Machines Vimeo Channel*, moderated by Gary Warner (https://vimeo.com/channels/drawingmachines). Drawing machines are "autonomous machines and devices that make purposeful marks on surfaces, and leave drawings in the world as evidence of their agency" and while being a broader categorisation than autonomous drawing robots, capture much of the essence of human fascination with machine autonomy, intent and creativity.

Perhaps the most sustained study of modern autonomous artistic drawing robots can be found in the work of Leonel Moura, who has been developing draw-ing robots since 2001 [20,21]. Many of Moura's works draw upon concepts from biology, situated awareness, collective self-organisation and stigmergy found in

---

[2] This fascination does extend to conventional human artists using computers as well – recent David Hockney exhibitions have included numerous works painted using iPhone and iPad apps that are exhibited showing the paint strokes the artist made in real time, revealing the artist's drawing process as it occurred.

the collective behaviour of ants and other social insects. These robots use infrared (IR) and colour sensors to detect the marks locally and feed them autonomously.

The *Drawbots* system [2] was an investigation into autonomous drawing using evolutionary robotics as a way to remove "the personal signature of the artist" [3, Chap. 6]. In this work "implicit" fitness measures were defined that did not restrict or fix the type of marks the robot drawer should make, including an "ecological model" involving interaction between environment resource acquisition and expenditure through drawing. However, the results demonstrated only minimal creativity, and the authors concluded that fitness functions that embodied "artistic knowledge about 'aesthetically pleasing' line patterns" would be necessary if the robot were to make drawings worthy of exhibition.

The work described in this paper has its origins in software-based agent simulation, described in [17]. Greenfield has worked extensively with virtual drawing robots, principally modelled on Khepera robots [11,13]. In [12] he describes a series of experiments for autonomous "avoidance drawings" whereby virtual drawing robots generate complex patterns by generating random walks that avoiding the lines already drawn by themselves and other robots. This concept of self-avoiding walks has also been explored by Chappell, who described pattern-formation processes using self-interacting curves [5] and spatially rhythmic parametric curves described by Whewell equations that simulate the meandering of bends and rivers [6].

What many of these systems share in common is the idea of avoiding interactions with other lines, or stopping drawing if a line intersects with an existing line. Such rules give the drawings structure and assist in the perception of intent in robot behaviour.

2.2   Niche Construction

Niche construction in biology is the process whereby organisms modify or influence their local, heritable environment. Proponents argue for its importance in more fully understanding the feedback dynamics of evolutionary processes in nature [22]. Niche construction processes determine feedback systems that can modify the dynamics of the evolutionary process, because ecological and genetic inheritance co-influence the evolutionary process. Computational models of niche construction show that it can influence the inertia and momentum of evolution and introduce or eliminate polymorphisms in different environments [7].

In previous work, niche construction was employed to enhance the diversity and variation in a virtual autonomous drawing system (Fig. 1) [17]. In this work, a collection of software drawing agents were randomly placed on a blank canvas. Each agent consisted of a six-allele genome that determines the agent's behaviour. Individual alleles – normalised real numbers – determine drawing style (rate of curvature, propensity to meander), rate of fecundity, mortality and, importantly, a local preference for drawing density. As the agents move around the canvas, they leave a trail of ink. If an agent intersects with a line drawn by another agent or itself, it dies. At any time a line-drawing agent may spawn an offspring, with the resultant daughter agent's genome a mutated version of the

Fig. 1. Two sample outputs from the software niche constructing drawing system.

parent. Mutation is performed with a probability of $\frac{1}{L}$, where L is the genome length. So-called "creep mutation" is used, whereby a normally distributed random number with mean 0 is added to an allele selected for mutation.

The allele for local drawing density preference is used to determine what constitutes an acceptable "niche" for the agent. One can think of this as being an ideal value at which the agent is most "happy". The further away from this ideal value the less happy the agent is and the less likely it is to reproduce.[3] Even greater deviations from the preferred value may effect survival, as shown in Fig. 2.

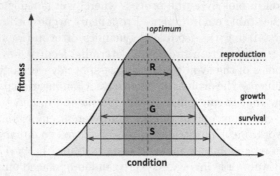

Fig. 2. An idealised agent viability curve for a single dimensional environmental condition (redrawn from [1]).

---

[3] The propensity to reproduce is also determined by a separate allele in the genome, which effectively controls the density of offspring lines created by the agent.

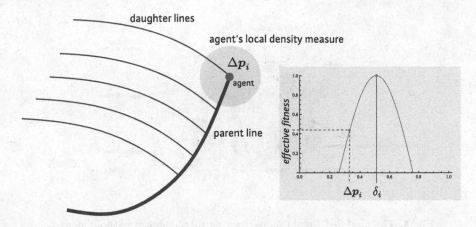

Fig. 3. The niche construction mechanism for drawing agents, who try to construct a niche of local density that satisfies their genetic preference.

For each agent, i, δ_i defines its preferred niche. Local density, defined as the ratio of inked to blank canvas per unit area, is measured over a small area surrounding the agent at each time step (Fig. 3). Proximity to the preferred niche determines the probability of reproduction, given by

$$Pr(rep) = f_i \cdot \cos^\omega(\texttt{clip}(2\pi(\Delta_{p_i} - \delta_i)), -\frac{\pi}{2}, \frac{\pi}{2}), \tag{1}$$

where Δ_{p_i} is the local density around the point p_i, the agent's position, ω a global parameter that varies the effective niche width, f_i is the agent's fecundity and $\texttt{clip}$ is a function that limits the first argument to the range specified by the next two. Being in a non-preferred niche similarly increases the probability of death.

3   Robot Design

Reinterpreting a software simulation as physical robot designs involves many constraints and challenges. The aim of this initial research was to use the niche construction method to control physical robot behaviour. More advanced aspects of the software simulation, such as self-reproduction were removed from the experiments reported here.

For the initial experiments we used a collection of modified Pololu *m3pi* robots.[4] The m3pi combines a *3pi* robot base with an expansion board that sits above the base (Fig. 4). The base board uses an Atmel ATmega328P microcontroller running at 20 MHz. This microcontroller handles low-level functionality such as driving the motors, poling the QTR-RC reflectance sensors (used to detect lines and line density, see Sect. 3.2 below) and writing to a small 8×2 character LCD display, used for diagnostic messages and debugging.

---

[4] https://www.pololu.com/product/2150.

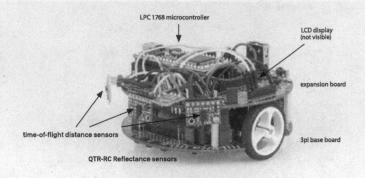

Fig. 4. The modified Pololu m3pi drawing robot (shown without pen).

The expansion board houses a significantly more powerful microcontroller (an 80 MHz ARM mbed NXP LPC1768). This microcontroller communicates with the base board via a serial protocol, commands include reading the reflectance sensors, writing to the LCD and driving the motors at different speeds (both forwards and backwards). Additionally, this microcontroller manages a series of three distance sensors that calculate the distance between the robot and any obstacles. The mbed system[5] supports a full C++ development environment, making it ideal for more complex robotic programming tasks.

Each robot is powered by four rechargeable NiCad batteries and will run continuously for up to 35 min on a full charge. While automated charging was investigated (where upon reaching a reserve power condition the robot seeks out the battery charger and automatically charges itself) for the experiments reported here we relied on manual charging for reliability purposes. A lasercut mounting stage allows mounting of a marker pen at the front of the robot. In this configuration, the pen is in permanent contact with the paper, so is always drawing (i.e. it is not possible for the robot to not draw when moving in this configuration).

3.1   Navigation and Collision Avoidance

Commands control the speed of the two geared motors that turn the wheels on the left and right side of the robot. The software allows continuous control of left and right motor speed independently, and in both directions (a differentially steered drive system). This independent control allows the robot to turn according to the differential speed between the left and right motors [16]. The robot has no turn counter on the wheels, nor IMU or Compass, so it has no accurate sense of its exact physical location within the drawing area. Three time-of-flight distance sensors provide real-time information on the proximity of the robot to

[5] See http://www.mbed.com for more details.

obstacles and allow it to react accordingly to avoid them. The sensors have an effective range of approximately 15 cm. A distance-proportionate Virtual Force Field (VFF) method [4] is used to avoid walls and other obstacles. Typically, the robots are set up in a rectangular "drawing arena" that houses a large paper floor surrounding by a low-height wall to stop the robots escaping. The obstacle avoidance algorithm successfully allows the robot to avoid bumping into the surrounding walls, other robots and obstacles.

When the obstacle avoidance task is not in effect (i.e. when the robot is not in close proximity to any obstacles), the robot is free to roam according to a meandering algorithm, similar to that used for the virtual niche constructing agents described in [17], but with modification (explained in Sect. 3.2). In the software version, a combination of alleles determined the agent's basic motion, based on constant rate of curvature and a degree of "irrationality".

For the robot implementation described here, a new movement algorithm was devised, based on a weighted sum of line following and meandering, determined by the suitability of the current niche to the individual's preference.

3.2   Niche Construction Implementation

As detailed in Sect. 2.2, the niche construction drawing algorithm revolves around an agent's local density preference δ_i measured against the actual line density in the immediate area of the robot Δ_{p_i}. To estimate the local density, each robot is fitted with a $5 \times$ QTR-RC reflectance sensor array, which measures the reflectance level directly below each sensor from an infrared light integrated into the sensor. The sensors are arranged around the front edge of the base board and cover a distance of approximately 6 cm. To calculate the local density the average reading across all sensors is read into a ring buffer and the buffer is updated proportionally to the forward speed of the robot (Fig. 5). Hence the average value of all values currently stored in the ring buffer gives the local density in the area directly passed under the robot. We also experimented with centre-weightings for the reflectance sensor array, which did not result in better perception of density.

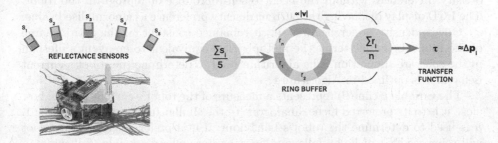

Fig. 5. Calculation of local density.

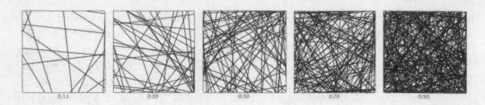

Fig. 6. Perceptual density samples used for calibration and calculation of the transfer function.

To map the reflectance sensor readings to a perceptual reading of line density a simple calibration sheet with five sample areas of pen-drawn lines of increasingly higher density was designed (Fig. 6). A computer-calculated density match of random lines drawn over a 6 cm × 6 cm area was used to create each sample. The test samples vary in density from normalised values of 0.1–0.9. The robot was moved directly over each sample and the raw sensor values were recorded, along with the cumulative averages while moving over the area of constant density. The calibration was repeated several times at different angles for each sample to remove any directional bias in the readings. These values were used to build a transfer function, τ, that maps raw sensor readings to a normalised perceptual measure of line density. The transfer function is applied to the readings obtained from the raw readings accumulated in the ring buffer to give a normalised reading of the local density of lines in the robot's immediate vicinity. This corresponds to Δ_{p_i} as detailed in Eq. 1.

Responding to a "niche". In the original niche constructing work, the preference of an agent to its niche was determined by an individual allele. The closer the agent to its preferred niche, the higher the probability of reproduction, which daughter lines spawned from the parent agent as it draws along the canvas. In the robotic version self-reproduction was not used, so the emphasis was placed on the individual agent *constructing* its niche. To do this, a density preference was set for each robot via a small potentiometer installed at the base of the robot. Changing the potentiometer allowed easy experimentation with different density preferences without the need to self-reproduce or reprogram the robot. The LCD display displaying the current density preference as a normalised value.

Once a density preference was set, it remained constant for the entire charge of the robot, in simple terms the metaphorical equivalent to an agent's lifespan in the software simulation. The algorithm used to determine the robot's current behaviour is outlined in Algorithm 1.

The variable μ (line 9) represents a measure of the robot's current fitness: how close it is to its preferred niche, analogous to the similar function shown in Eq. 1. μ is used to determine the robot's behaviour: if in a preferred niche, the robot will reinforce that niche by following existing lines where possible (following an existing line draws over the top without a lot of change to the local density). The proximity to the preferred niche determines how closely the robot follows the line, as expressed in the **lineFollow**(μ) function (line 11).

Algorithm 1. Robot Niche Constructing algorithm

1: $\epsilon \leftarrow 0.2$ ▷ Transition range for preferred niche
2: $\delta \leftarrow$ getDensityPreference()   ▷ Read potentiometer value from base of robot
3: **while** batteryHasCharge() **do**
4:      $[s_1 \ldots s_n] \leftarrow$ readDistanceSensors()   ▷ read ToF distances into $s_1 \ldots s_n$
5:      **if** objectDetected(s) **then**   ▷ check if an object detected by $s_1 \ldots s_n$
6:         steerToAvoid($\min(s)$)   ▷ Avoid collisions
7:      **else**
8:         $\Delta_{p_i} \leftarrow$ getLocalDensity()   ▷ Read local density via transfer function
9:         $\mu \leftarrow \cos^{0.5}(2\pi(\Delta_{p_i} - \delta))$ ▷ μ represents niche proximity
10:         **if** $\mu \geq (1 - \epsilon)$ **then**   ▷ in the required niche – so reinforce this niche
11:            lineFollow(μ)
12:         **else if** $\Delta_{p_i} < \delta$ **then**   ▷ not enough lines in this area
13:            steerInDirection(∇s)   ▷ turn in the direction of increasing density
14:            meander(p_i, μ)   ▷ try to find area of closer density match
15:         **else**   ▷ too much density in this area, steer to avoid
16:            steerInDirection($-\nabla s$)   ▷ turn in the direction of decreasing density
17:            meander(p_i, μ)   ▷ try to find area of closer density match
18:         **end if**
19:      **end if**
20: **end while**

The **meander()** function is based on a similar function to that developed for the software agents. In the original software agents [17], two alleles determined the agent's constant rate of curvature (σ) and "irrationality" (r). These values were used to calculate the rate of change of angular velocity:

$$\nabla\theta = \sigma + \texttt{fracSum}(p, k \cdot r)^{0.89r^2}, \tag{2}$$

where p is the agent's current position, k a constant known as the *octave factor*, and $\texttt{fractSum}$ a function that sums octaves of 2D noise [23]. For the robots, a similar method was used, based on an estimate of the robots relative position. The position vector p is calculated by estimation. To perform this estimate it is assumed that the robot's starting position and orientation – wherever that may be – defines the origin and direction of an orthogonal, 2D coordinate system, with the robot initially heading in the positive x direction (i.e. in the direction of the vector $[1, 0]$).

The current position, p, is calculated using Euler integration, based on the estimated velocity vector, v, derived from the differential speed of the left and right motors [16]. The **meander()** function takes two arguments: the robot's current position, p_i, and a degree of irrationality, r, a normalised value. In the algorithm shown, the irrationality (how much the agent meanders) is determined by the proximity to a preferred niche: the further away from the preferred niche, the higher the irrationality, hence the more the robot is likely to twist and turn, writhing to try and find a more preferred niche.

4   Results

The niche constructing drawing robots were exhibited in 2015 as part of the group exhibition ARTE@IJCAI, held at the Centre Cultural Borges in Buenos Aires, Argentina. The exhibition, curated by Luc Steels, highlighted the role of AI and machine creativity in art, and was a companion event to the International Joint Conference on Artificial Intelligence (IJCAI 2015).

In this exhibition, the robots were placed within a 4 m × 4 m drawing arena and left to draw with a fully charged battery (Fig. 7). The large size of the drawing area allowed the robots to roam freely without the need to constantly avoid the edges of the arena.

Drawings typically took several hours to complete, which required charging the robots several times during the course of the day. During recharges the density preference was changed in some of the drawings to achieve an overall variation in the drawing "style" of each completed drawing. Once a drawing was complete, it was hung on the gallery wall and a new drawing started. Over the course of the exhibition seven different large-scale drawings were completed.

Generally, people were fascinated by the drawing process as much as the resultant drawings. They described intent and autonomy in interviews carried out while the work was running, ascribing purposeful behaviours to the robots as they drew. Often people would place their hands or other objects into the drawing arena, causing any robot nearby to go scurrying away.

Changing the density preference had significant effects on the resultant drawings, with values in the middle range proving the most visually satisfactory. As the robot is unable to raise the pen from the canvas, it is always drawing, hence, always increasing density. However, after a time of drawing over an existing line, the density does not increase. This forces the robot into a conundrum: for a low density preference one should just keep drawing over existing lines (line following), but that places the robot typically over areas of high density that seeks to avoid. Low density preference robots tend to be "explorers", always trying to move away from areas where lines exist. Middle to high density preference robots tended towards being "reinforcers", endlessly drawing around existing lines in increasingly complex patterns, reminiscent of insect trails (Figs. 7 and 9).

As with the original software agents, "founder lines" (those drawn first) have significant impact on the resultant drawing, particularly when the robot agents seek to reinforce existing lines, since these become the main pathways that are reinforced by drawing lines around them or by line following. Of equal importance to the final results as each robot's programmed behaviour was its interaction with the environment. As already mentioned, people would interact with the robots, placing their hands in their pathway, or try to catch them (thankfully difficult given that the robot can move at up to 1 m/s). Moreover, inconstancies in the paper surface would often cause the robot to change course due to slippage or the pen getting caught up on the edges of the paper.

These kind of interactions point to the importance of the environment in determining overall behaviour, furthering the previously introduced concept of creative ecosystems [18]. In subsequent performances of the robots we

Fig. 7. Exhibition at ARTE@IJCAI in Argentina. The top image shows the drawing arena and a previously drawn image hanging on the wall. The bottom image shows detail of an image drawn by a robot with a niche density preference of 0.7. Notice how the robot has reinforced the niche by repeated drawing over earlier patterns.

Fig. 8. A "rock garden" version of the drawing arena (left) and the resultant drawing (right). Photo curtesy of Gary Warner.

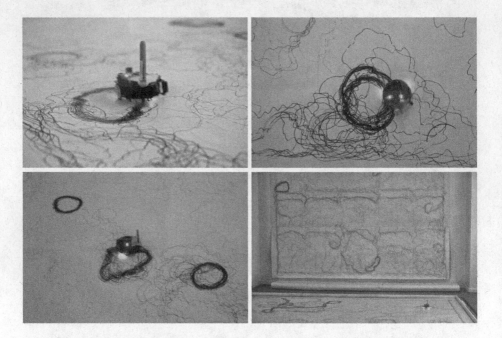

Fig. 9. Various stages of constructing a "niche" of high density.

experimented with placing rocks inside the drawing area, similar to that of the *kare-sansui* dry landscape zen rock gardens in Japan. As expected, the nature of the drawings reflected the interaction of the robots and their environment, finding niches in between rocks in spaces large enough that allowed reinforcement of a specific niche preference (Fig. 8).

5   Conclusions and Future Work

From the initial experiments reported here, it is clear that a number of interesting directions for further research present themselves. Some experimentation is in progress with a directional light sensor so a swarm of robots can use a light source as a navigational aid (in a similar way to some organisms use various tropisms). In the implementation described here, there is little robot-robot interaction beyond collision avoidance. We have begun to experiment with more sophisticated forms of robot-robot and robot-environment interactions. These robots use coloured light and audible sounds that the robots emit and react to (Fig. 10), creating the possibility of a sonic niche. Specific sounds or light frequencies can be used to identify groups – all members of a particular group use the same frequency signature – with different groups responding in different ways to the proximity of outsiders.

A behaviour currently under investigation is inspired by insect behaviour, specifically the defensive behaviour of the African bombardier beetle (*Stenaptinus insignis*) [10]. These beetles emit a hot, acrid spray when attacked. The bombardier is particularly intriguing as it is able to precisely direct its spray in the direction of the attacker. Inspired by these natural wonders, we have designed a number of insect-like robots that squirt different coloured inks,

Fig. 10. Further designs for the niche constructing drawing robots, developed by Nick Jones, based on the insect species *Stenaptinus insignis*. These insects fire a toxic acid spray when attacked. The microphone, which responds to sounds, can be seen in the middle of the front of the robot. These robots are currently under development.

Table 1. Comparison of software agent and physical robot implementations.

Feature	Software virtual agents	Physical robots
Emphasis	Collective behaviour	Individual drawing behaviour
Niche mechanism	Line density	Line density
Reproduction?	Yes	No
Behavioural mechanism	Genome	Decision tree
Behaviour change	Evolution	Environment
Drawing style	Complex structure	Niche reinforcement

as shown in the figure. In addition to microphones and small speakers, these robots have colour sensors underneath, allowing the possibility of more complex, colour-dependent niche construction. Experiments with these new, more complex niche constructing robots are currently ongoing.

In conclusion, in this paper we have described how the original, software-based niche constructing agent model can be interpreted in real physical robots. A summary of the major differences is outlined in Table 1. Obviously features such as self-reproduction, and hence evolution, are difficult to achieve in a physical robotic system, so as a first step, we have focused on the role of niche preference in the individual. Even with this limitation, the resultant drawing and robot behaviour show interesting results. While the style of the drawings is quite different between the software and physical robot versions, it is easy to see the important role that niche construction plays in determining the overall features of the drawings produced. In this respect there are similarities in the way niche construction provokes interesting and unexpected regions of organisation by trying to build niches of suitable density in both the software and physical robot versions.

Our new experiments that broaden the conditions under which the robots seek to niche construct (e.g. sound and colour), suggest that this method is a rich and useful technique for generating autonomous creative behaviour in artificial systems.

Acknowledgments. Nick Jones worked on the *Stenaptinus insignis* robots as an Industrial Design student in our lab. This research was supported by Australian Research Council Discovery Project grants DP1094064 and DP160100166.

References

1. Begon, M., Townsend, C., Harper, J.: Ecology: from individuals to ecosystems. Wiley-Blackwell, Maldon (2006)
2. Bird, J., Husbands, P., Perris, M., Bigge, B., Brown, P.: Implicit fitness functions for evolving a drawing robot. In: Giacobini, M. (ed.) EvoWorkshops 2008. LNCS, vol. 4974, pp. 473–478. Springer, Heidelberg (2008). doi:10.1007/978-3-540-78761-7_50

3. Boden, M.A.: Creativity and Art: Three Roads to Surprise. Oxford University Press, Chicago (2010)
4. Borenstein, J., Koren, Y.: Real-time obstacle avoidance for fast mobile robots. IEEE Trans. Syst. Man Cybern. **19**(5), 1179–1187 (1989)
5. Chappell, D.: Taking a point for a walk: Pattern formation with self-interacting curves. In: Greenfield, G., Hart, G., Sarhangi, R. (eds.) Bridges 2014 Conference Proceedings, pp. 337–340. Tessellations Publishing, Phoenix, Arizona (2014)
6. Chappell, D.: Sinuous meander patterns: a family of multi-frequency spatial RHYHMS. J. Math. Arts **9**(3–4), 63–76 (2015)
7. Day, R.L., Laland, K.N., Odling-Smee, J.: Rethinking adaptation: the niche-construction perspective. Perspect. Biol. Med. **46**(1), 80–95 (2003)
8. d'Inverno, M., McCormack, J.: Heroic vs collaborative AI for the arts. In: Proceedings of IJCAI 2015 (2015)
9. Dohm, K., Hoffmann, J.: Kunstmaschinen Maschinenkunst (Art Machines, Machine Art). Kehrer Verlag, bilingual (German/English) edn. (2008)
10. Eisner, T.: For Love of Insects. The Belknap Press of Harvard University Press, Cambridge (2003)
11. Greenfield, G.: A platform for evolving controllers for simulated drawing robots. In: Machado, P., Romero, J., Carballal, A. (eds.) EvoMUSART 2012. LNCS, vol. 7247, pp. 108–116. Springer, Heidelberg (2012). doi:10.1007/978-3-642-29142-5_10
12. Greenfield, G.: Avoidance drawings evolved using virtual drawing robots. In: Johnson, C., Carballal, A., Correia, J. (eds.) EvoMUSART 2015. LNCS, vol. 9027, pp. 78–88. Springer, Cham (2015). doi:10.1007/978-3-319-16498-4_8
13. Greenfield, G.: Robot paintings evolved using simulated robots. In: Rothlauf, F., et al. (eds.) EvoWorkshops 2006. LNCS, vol. 3907, pp. 611–621. Springer, Heidelberg (2006). doi:10.1007/11732242_58
14. Grimm, V., Railsback, S.F.: Individual-based Modeling and Ecology, Princeton Series in Theoretical and Computational Biology. Princeton University Press, Princeton (2005)
15. Kennedy, J., Eberhart, R.C., Shi, Y.: Swarm Intelligence. Morgan Kaufmann Publishers, San Francisco (2001)
16. Lucas, G.W.: An elementary model for the differential steering system of robot actuators. Technical report, The Rossum Project (2000). http://rossum.sourceforge.net/papers/DiffSteer/DiffSteer.html
17. McCormack, J.: Enhancing creativity with niche construction. In: Fellerman, H., Dörr, M., Hanczyc, M.M., Laursen, L.L., Maurer, S., Merkle, D., Monnard, P.A., Stoy, K., Rasmussen, S. (eds.) Artificial Life XII, pp. 525–532. MIT Press, Cambridge (2010)
18. McCormack, J.: Creative ecosystems. In: McCormack, J., d'Inverno, M. (eds.) Computers and Creativity, Chap. 2, pp. 39–60. Springer, Heidelberg (2012)
19. McCormack, J.: Aesthetics, art, evolution. In: Machado, P., McDermott, J., Carballal, A. (eds.) EvoMUSART 2013. LNCS, vol. 7834, pp. 1–12. Springer, Heidelberg (2013). doi:10.1007/978-3-642-36955-1_1
20. Moura, L.: Machines that make art. In: Herath, D., Kroos, C., Stelarc, (eds.) Robots and Art, Cognitive Science and Technology, pp. 255–269. Springer, Heidelberg (2016)
21. Moura, L., Pereira, H.G.: Man + Robots – Symbiotic Art. LxXL, Black and white facsimile edn. (2014)
22. Odling-Smee, J., Laland, K.N., Feldman, W.M.: Niche Construction: The Neglected Process in Evolution. Monographs in Population Biology. Princeton University Press, Princeton (2003)

23. Perlin, K.: Improving noise. ACM Trans. Graph. (TOG) **21**(3), 681–682 (2002)
24. Roudavski, S., McCormack, J.: Post-anthropcentric creativity. Digital Creativity **27**(1), 3–6 (2016)
25. Still, A., d'Inverno, M.: A history of creativity for future AI research. In: Pachet, F., Cardoso, A., Corruble, V., Ghedini, F. (eds.) Proceedings of the Seventh International Conference on Computational Creativity (ICCC 2016), pp. 147–154, June 2016

Automated Shape Design
by Grammatical Evolution

Manuel Muehlbauer[1]([⊠]), Jane Burry[1], and Andy Song[2]

[1] Spatial Information Architecture Laboratory (SIAL),
RMIT University Melbourne, Carlton 3000, Australia
{manuel.muehlbauer,jane.burry}@rmit.edu.au
[2] Evolutionary Computation and Machine Learning Group,
RMIT University Melbourne, Carlton 3000, Australia
andy.song@rmit.edu.au
http://www.sial.rmit.edu.au/
https://titan.csit.rmit.edu.au/e46507/ecml/

Abstract. This paper proposes a automated shape generation methodology based on grammatical genetic programming for specific design cases. Two cases of the shape generation are presented: architectural envelope design and facade design. Through the described experiments, the applicability of this evolutionary method for design applications is showcased. Through this study it can be seen that automated shape generation by grammatical evolution offers a huge potential for the development of performance-based creative systems.

Keywords: Shape generation · Design · Genetic programming

1 Introduction

This research presents a Genetic Programming (GP) approach to shape design using grammatical evolution to provide a methodology to extend the potential for the use of machine intelligence during early stages of architectural design processes. A method of this kind could greatly enhance efficiency and support decision making for architects. The opportunities and limitations of the proposed grammatical evolution for the design of architectural spaces is elaborated.

The main contribution of the presented research is the development of a methodology for grammatical evolution and the definition of the representation for the encoding of particular design spaces. It explores the potential of this process for the generation of simple envelopes - the exterior hull of a building, for architectural shapes, combined shapes for architectural envelopes and complex shapes for facade components.

During our research, different methods are used to develop a methodology for shape design by grammatical evolution. Strongly-typed tree-based GP [13] is used as the core. Shape-based nodes are used as main representations for the different design cases.

© Springer International Publishing AG 2017
J. Correia et al. (Eds.): EvoMUSART 2017, LNCS 10198, pp. 217–229, 2017.
DOI: 10.1007/978-3-319-55750-2_15

The paper reports the research outcomes which are architectural shapes produced with gradually increasing complexity during the development of case specific representations. During the process of developing representations for architectural shapes, the feasible of those representations are explored and discussed in the experiments section (Sect. 4) and discussions section (Sect. 5) of this paper. Relevant existing studies from the literature are presented in the related work section (Sect. 6).

Knowledge generated throughout the different experiments presented in this paper is reported in the conclusion section (Sect. 7) and reflected upon in respect to the further development of the methodology in the future research section (Sect. 8).

2   The Need

During architectural computational design processes, usually a parametric model is defined to specify a particular architectural shape space after an intuitive design process has taken place. This parametric model is then evaluated for potential parameters that can provide the flexibility for optimization of the architectural shape to increase the performance of the architectural shape. As a result of this post-rationalization process, the design space for architectural shape is already limited to a small scope of design solutions that reflect the design decisions taken during the design process.

Shape design by grammatical evolution opens up the opportunity for architectural shape to incorporate performance criteria early on in the design process, increasing the impact of the shape generation on the performance potential. Therefore, the design space in this experimental process needs to encode a wide range of design solutions to effectively search for a high performing solution. Shapes generated throughout this evolutionary process can then be used to further inform the design process based on local and site specific criteria.

3   Methodology

The methodology of this research project can be divided in two parts: the grammatical evolution as shape generation process and the representations developed to encode a particular set of shapes that are of experimental interest to the field of architecture. The first will be discussed in this section, before greater detail about the implications of the process is provided in the discussion section. The latter will be described as part of grammatical evolution and in greater detail during the section concerned with the implementation and experiments.

3.1   Grammatical Evolution

Grammatical evolution is an approach to evolutionary computation that introduces genotype-phenotype mapping based on grammatical rules [17]. Search and solution space are often separated. This separation and the translation process allow the search process to iterate all possible genotypical expressions, while ensuring the validity of the grammar output [14]. Through this approach, a particular set of shapes can be generated and uniformity of the phenotypical expression guaranteed based on the translation through the shape grammar.

We present automated shape design by grammatical evolution successively in the next sections based on the components, as they are defined during the establishment of the process, starting from the initialization of the first generation.

3.2   Representations

A set of terminals and functions is provided for different particular design cases based on shape grammar. During this process, a set of terminals and functions is developed for each particular case. The genotype as a result of the generation is a set of integer values, structured through the progression of data throughout the data tree. The representation is tree-based and the root node is used to collect the different shapes in a coherent representation of the solution. These terminal and function sets are reported in the implementation section alongside the different experiments and the resulting sets of shapes that were generated based on these case specific representations.

3.3   Fitness Measure

Decision parameters are defined based on the design criteria for the architectural shapes and a fitness function is developed to allow the constant evaluation of the criteria during the evolutionary process.

3.4   Shape Grammar

During the development of the genotype, a set of grammar rules is applied to develop a phenotypical expression of the architectural shapes that can be evaluated towards the fitness functions. In this research, the objects for evaluation are polygon meshes generated through the application of different algorithms that translate the phenotype consisting of points and/or lines into objects.

Crossover and mutation are powerful operators in GP. When shape grammar is involved, we need to modify these genetic operators so specific grammar restrictions are applied to specify the shape for the design process. During this design process the representation is limited by a coherent grammar structure, while the parameters are flexible to explore the design space during the process. Constraints were integrated into the nodes that generate integer or float values through the definition of domains that allow specific shapes to emerge.

4   Implementation and Experiments

The implementation based on the above methodology and the corresponding experiments are detailed in this section. Note some implementation details are combined with the description of experiment as different parts of experiments involve different implementations.

4.1   Implementation

A variety of tools were used to ensure the accessibility of the process to architectural designers. The main environment used to generate architectural shape was Rhinoceros CAD which supports Rhinoscript, a scripting language for the 3d computer-aided design. It provides a set of specialized methods and functions for the design of complex geometries. Apart from that, a Python-based GP library was developed under Rhino to facilitate grammatical evolution. This gives designers the possibility to modify the process and adjust it towards their own needs during the design process.

Genetic Programming: Our genetic programming implementation utilizes treebased strongly-typed GP. A random initial generation is used as the starting point for the evolutionary process. In the fitness evaluation quantitative properties, of the geometry are used to evaluate individual solutions, in particular shape volume and boundary surface area.

During the evolutionary process, tournament selection (tournament size = 3) is applied as a selection strategy to provide a mating pool (mating pool size = 10) for the crossover operation (crossover rate = 0.6). Besides this selection mechanism, a number of individuals are selected for the modification through a mutation operation (mutation rate = 0.3) to enhance diversity in the population. The best performing individuals of a generation are preserved through elitism (elite rate = 0.1) to ensure a steady advance in performance of the best individuals.

During the experiments a small population of individuals (population count = 500) and a low number of generations (maximal number of generations = 100) are used.

Shape Grammar: After generating the genotype with its associated parameters during grammatical evolution, a phenotypical representation of the design is produced for evaluation. The presented experiments differ in terms of the representations as well as in the translation process from genotype to phenotype. The next section will describe these differences.

4.2   Experiments

A set of experiments has been designed for exploring the feasibility of abstract shapes as initial stage during architectural design processes. As massing is crucial for the design of architectural shapes, auto-generated shape as a starting point for further design opportunities opens a huge potential for reflection on the performance of building envelopes. The pre-optimization of architectural shapes allows a more abstract evaluation of shapes towards performance goals at a stage of the design, when major design decisions are yet to be taken. Therefore this experimental process allows a stronger focus on building performance and furthermore the information of the design process through the adaptation of design features with major impact on the performance of the architectural shape.

Three experiments were undertaken to explore this potential with increasing complexity of the geometry. The experimental process starts with simple shapes for architectural envelopes. The same approach is then extended to the combination of shapes to generate another set of design solutions. In a third experiment, a different shape generation process is introduced to explore the potential application of the process to the generation of complex shapes for facade components.

I. Simple Shapes for Architectural Envelopes: In this experiment, the representation of the architectural envelope is established through the collection of three integer nodes in point nodes to generate three dimensional point clouds. The shape nodes in this case consist of ten points that define the shape of the architectural envelope.

This genotype is then used as input for a convex hull algorithm that translates the point cloud into a polygonal mesh. The fitness function in this case is maximizing the volume of the shape, while minimizing the area of the polygon mesh, leading to compact shapes for the architectural envelope.

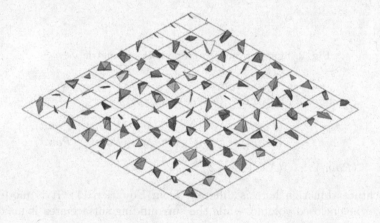

Fig. 1. Shapes from Experiment I - isometric view

$\langle Shape \rangle ::= \langle Point \rangle, \langle Point \rangle, \langle Point \rangle, \langle Point \rangle, \langle Point \rangle,$
$\langle Point \rangle, \langle Point \rangle, \langle Point \rangle, \langle Point \rangle, \langle Point \rangle$

$\langle Point \rangle ::= \langle X \rangle, \langle Y \rangle, \langle Z \rangle$

$$Fitness = Volume/Surface\ Area \qquad (1)$$

These shapes, as shown in Fig. 1, are providing solutions that reduce the surface area of the envelope to minimize thermal loss, while providing a large volume of space for the incorporation of architectural functions as the spatial program gets introduced to the project in the next design stage.

II. Combined Shapes for Architectural Envelopes: In the second experiment, the root node collects two different sets of points that define architectural shapes. Through this process, a new variation of architectural shapes is generated. These shapes are also consisting of ten points each that are defined by a collection of three integer nodes each.

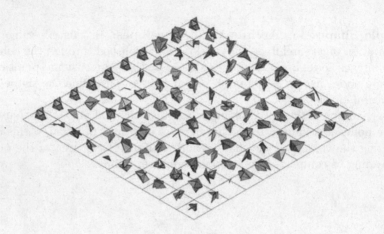

Fig. 2. Shapes from Experiment II - isometric view

$\langle Combined\ Shape \rangle ::= \langle Shape \rangle, \langle Shape \rangle$

$\langle Shape \rangle ::= \langle Point \rangle, \langle Point \rangle, \langle Point \rangle, \langle Point \rangle, \langle Point \rangle,$
$\qquad\qquad \langle Point \rangle, \langle Point \rangle, \langle Point \rangle, \langle Point \rangle, \langle Point \rangle$

$\langle Point \rangle ::= \langle X \rangle, \langle Y \rangle, \langle Z \rangle$

The fitness function here is different from Equation (1). It is modified to minimize the enclosed volume, while the surrounding surface area is maximized during the grammatical evolution.

$$Fitness = Surface\ Area/Volume \qquad\qquad (2)$$

Figure 2 presents the architectural shapes from this experiment. They have different advantages in the context of early design stages. Through the increase in surface area, the interface with the environment is maximized, while the volume for the architectural program is comparatively small. These shapes are feasible for design solutions with a highly differentiated architectural program with specialized spaces. Two application cases are relevant in this context. One is the introduction of collector surfaces for the use of solar radiation with either photovoltaic systems or algae reactors to generate biomass. The second is the use of the envelope for the display of information through the introduction of media facades.

III. Complex Shapes for Facade Components: The third experiment explores the application of grammatical evolution for the generation of facade components. This application case strongly deviates from the other experiments, so that the extend of the design space in z direction is restricted to refer to the two-dimensional character of facade panels. In the tree-structure of the representation, again ten points are collected based on three integer nodes. The domains of these integer nodes are differentiated to incorporate the restrictions in z direction, while the integer nodes for the x and y direction allow values between 0 and 500 mm, the domain set for the integer values of the z direction is between 0 and 100 mm. Another major difference to the previously described experiments is the use of a Delaunay mesh algorithm for the triangulation of the facade panels. Based on the hexagonal geometry, a grammar check is introduced that omits points that are outside of the domain to ensure consistency of the mesh geometry generated. Porosity is introduced to the mesh through the use of a mesh frame algorithm and the resulting meshes are thickened to provide material thickness. Through the smoothing of the shapes an organic aesthetic expression of the geometry is achieved.

$$\langle Shape \rangle ::= \langle Point \rangle, \langle Point \rangle, \langle Point \rangle, \langle Point \rangle, \langle Point \rangle,$$
$$\langle Point \rangle, \langle Point \rangle, \langle Point \rangle, \langle Point \rangle, \langle Point \rangle$$

$$\langle Point \rangle ::= \langle X \rangle, \langle Y \rangle, \langle Z \rangle$$

Fig. 3. Shapes from Experiment III - facade generation

Fitness evaluation for this design case is minimizing the volume of the geometry and therefore the material needed for the production of the facade panel and maximizing the area of the polygon mesh.

$$Fitness = Surface\ Area/Volume \tag{3}$$

The shapes generated in this final experiment that is reported in this paper are presented in Fig. 3. Through their properties, they suggest an application potential for the introduction of thermal mass in the building envelope that exhibits a high rate of heat exchange through the high surface-mass ratio. Another potential application of these geometries is the introduction of evaporative cooling through the material properties, as the fast heating of the surface will enhance these effects.

5   Discussions

In this paper the approaches to shape generation were presented and discussed for their application potential. During the implementation of the grammatical evolution different opportunities and challenges were encountered and will be reflected upon in this section.

5.1   Opportunities and Challenges of Grammatical Evolution

The experiments revealed the potential of grammatical evolution for the generation of abstract case specific architectural shape in three different approaches. The representations developed based on shape, point and integer nodes allow the definition of a design space that is general enough to evaluate possible shapes for a wide range of design cases, while specifying the design space for a particular design situation.

Grammatical evolution proved to be an appropriate methodology for shape design in the context of performance-based architectural design. It allows the generation of a wide range of feasible designs for the optimization of architectural shapes towards a particular set of criteria.

Fig. 4. Post-processed set of combined shapes - perspective view

The challenges are mainly in the implementation level. For example the development of the terminal and function set needs to be traded of against the performance of the architectural shapes generated during the process. Restriction of the integer domains and checking of feasibility of the point clouds in the context of complex shapes for facade components is another challenge.

5.2   Post Processing

Post processing of the generated shapes shows the possible extension of the design space. The use of texture and color in the design process widens the scope for grammatical evolution dramatically. Even if not integrated into the evolutionary process as yet, we want to present initial results for the use of these concepts in the following Fig. 4.

In Fig. 4 the set of combined shapes for architectural envelopes is presented with initial shading to show some of the design potential that is inherent in the shapes generated through the experiments. The use of the paneling process for the use in architectural design processes was tested during the development of the Hoverport as part of an architectural competition for the future of airports. Figure 5 illustrates our entry which is based on the auto-generated facade components described in Experiment III.

5.3   Composition

In this research a variety of simple shapes are generated and the potential of the combination of shapes to generate a variety of architectural envelopes are presented. In general, through the developed process, complex shapes can be generated for a variety of architectural compositions during the architectural design process. This potential for composition needs further exploration to reveal the full capacity of the process.

5.4   Flexible Representation

The flexible representation used in grammatical evolution contributed to the development of the architectural shapes through the provision of a certain design

Fig. 5. Application of facade components in hoverport design

space that could then be systematically extended through the introduction of different shape grammar rules. This flexibility exceeds the current potential used in architectural design processes based on genetic algorithms and conventional search methods.

The presented research established a methodology for the use of grammatical evolution for shape design in architecture and reported on different experiments to generate shapes through the development of case specific representations using shape grammars. Based on the grammatical representation of the architectural shapes, grammatical evolution was introduced based on strongly-typed tree-based genetic programming. This approach allows the generation of a variety of shapes based on flexible representations.

6   Related Work

In this section the background of the research is discussed based on key references and the research is positioned in the context of related work in the field of shape grammar and grammatical evolution. The implementation strategy used for the grammatical evolution is based on the work of Lee, Herawan and Noraziah [11]. As Lees research is focused on the evolution of industrial products, the presented research is set apart through the development of representations for architectural shapes.

The line of research using shape grammar in architectural design originated with Stiny and Mitchell in 1978 [18] through the development of the Palladian grammar. Following on, Koning and Eizenberg (1981) [8] described the Prairie Houses designed by Frank Lloyd Wright, while Woodbury and Burrow (2006) [20] represented the Queen Anne Houses using Shape Grammar. While all three research projects developed an accurate grammatical representation to describe specific architectural designs, they didn't introduce Grammatical Evolution as means to explore the design space and evaluate the architectural performance.

Besides the development of generative design strategies [16] in architectural design during the last decades, a research trajectory exploring the application of evolutionary computational systems in architecture [1,4,6,7] emerged. Genetic encoding of design spaces for particular design cases and strategic exploration of representations present critical components of the related discourse. However, a flexible representation through the use of genetic programming widens the scope of the discourse through the expansion of possible design solutions in particular design spaces.

While Koning and Eizenberg [8] are using a parametric shape grammar to describe the Prairie Houses designed by Frank Lloyd Wright and explore the design space generated through this representation, Lee, Herawan and Noraziah [11] are using a similar representation and extend this approach to an evolutionary framework for product design. The shared approach of using a parametric shape grammar is based on the analysis of the design and extraction of main vocabulary elements, before developing rule schemata. Both researchers are presenting the resulting shapes and compositions in the respective papers and describe their modular setup in great detail.

Lee, Herawan and Noraziah [11] are exploring the application of both genetic algorithm and programming in their framework for grammatical evolution in the context of product development. The parametric representation of designs introduces relationships between the elements of the grammar and allows for the representation of complex product designs. The research presented in this paper, applies a similar methodology to the evolutionary design of architectural shapes. Already, Koning and Eizenberg [8] highlighted the potential for application of shape grammars in architectural design and provide insights on how the design space for Prairie Houses could be extended.

The application of graph grammars, as described by McDermott in 2013 [12], is another option to extend the presented shape design process to another representation that is of interest in the context of architecture. Structural systems are of major importance in the development of architecture and structural frames can be represented as graphs. In his research, McDermott [12] points out that graph grammars reduce complexities faced throughout the implementation of shape grammars. Another interesting aspect of McDermotts work in 2013 [12] is the use of graph grammars with multiple rules, consisting of a selector (for node selection) and an action part (modifying the subgraph). This representation adds more flexibility to the generated design outcomes and allows to encode more complex generative processes. Through the use of graph grammars, arbitrary forms usually encountered with direct representations was reduced.

Both, Lee, Herawan and Noraziah [11] and McDermott [12] are integrating artificial fitness evaluation along with automatic evaluation of numerical values derived from the representation. This approach can be used in further research to allow designers direct impact on the search process. An application case for this method would be the interactive exploration of design spaces for mass-customization in architectural design [2,3,19].

7   Conclusions

In this study we presented a genetic programming based grammar evolution method for automated shape generation. Three shape generation experiments of two types are presented which are architectural envelope design and facade design respectively. Through this study we have shown the applicability of this evolutionary method to facilitate early stage of architectural design. Implications of this intelligent approach are discussed in this paper. We conclude that automated shape generation by grammatical evolution offers a huge potential for the development of performance-based creative systems.

Context and site specifity. At this point of the experiments, the representation of the architectural shapes is independent from context and environment, which would usually impact on the architectural design process. This might also drive the development of a set of fitness functions based on cost, area, volume or other properties of the geometry and the representation.

8   Future Research

During the research process, the presented methodology for the development of representations for design processes and the respective implementation of grammatical evolution will be used as basis for the development of an intelligent framework for interactive decision support in architectural design. In this next steps of the research project, the development of an intelligent design framework for interactive decision support during architectural design processes is approached through an abstract reflection on the process to explore the theoretic potential of the process to provide methodologies for architectural design, before strategically extending the implementation of the developed methodology. It will address the opportunities and limitations presented in the results section of this paper.

Another aspect for the extension of the process is the introduction of multi-criteria evaluation as part of the fitness evaluation to incorporate a variety of fitness measures that are meaningful in the context of architectural design, e.g. structural evaluation, evaluation of environmental parameters, introduction of site restrictions and measures that reflect on the performance of the functional program. Another area of interest in the context of fitness evaluation is the evaluation of the representation itself and especially in the context of more complex design outcomes this approach to fitness evaluation will be of high value to ensure the quality of the representation used during grammatical evolution.

The introduction of new strategies to integrate human interaction into grammatical evolution will be taken forward from there through the development of other methods for artificial selection during the evolutionary process.

References

1. Byrne, J.: Approaches to evolutionary architectural design exploration using grammatical evolution. University College Dublin (2012)
2. Ceccato, C.; Simondetti, A.; Burry, M.C.: Mass-customization in design using evolutionary and parametric methods. In: Proceedings of the 2000 ACADIA Conference (2000)
3. Duarte, J.P.: Towards the mass customization of housing: the grammar of Siza's houses at Malagueira. Environ. Plan. B: Plan. Des. **32**(3), 347–380 (2005)
4. Frazer, J.: An Evolutionary Architecture. Architectural Association, London (1995)
5. Heisserman, J.; Woodbury, R.: Generating languages of solid models. In: SMA 1993 Proceedings on the Second ACM Symposium on Solid Modeling and Applications, pp. 103–112 (1993)
6. Janssen, P.; Kaushik, V.: Evolving lego. Exploring the impact of alternative encodings on the performance of evolutionary algorithms. In: Rethinking Comprehensive Design: Speculative Counterculture, Proceedings of the 19th International Conference on Computer-Aided Architectural Design Research in Asia CAADRIA 2014, pp. 523–532 (2014)
7. Janssen, P.: A design method and computational architecture for generating and evolving building designs. The Hong Kong Polytechnic University (2004)

8. Koning, H., Eizenberg, J.: The language of the prairie. Frank Lloyd Wright's prairie houses. Environ. Plan. B: Plan. Des. **8**(3), 295–323 (1981)
9. Koza, J.R.: Genetic programming. a paradigm for genetically breeding populations of computer programs to solve problems. Stanford University (1990)
10. Langdon, W.B.: Genetic Programming and Data Structures. Genetic Programming + Data Structures = Automatic Programming!. Genetic Programming. Springer, Boston (1998). doi:10.1007/978-1-4615-5731-9
11. Lee, H.C., Herawan, T., Noraziah, A.: Evolutionary grammars based design framework for product innovation. Procedia Technol. **1**, 132–136 (2012). doi:10.1016/j.protcy.2012.02.026
12. McDermott, J.: Graph grammars for evolutionary 3D design. Genet. Program Evolvable Mach. **14**(3), 369–393 (2013). doi:10.1007/s1071001391900
13. Montana, D.J.: Strongly typed genetic programming. Evol. Comput. **3**(2), 199–230 (1995). doi:10.1162/evco.1995.3.2.199
14. O'Neill, M., Ryan, C.: Grammatical evolution. IEEE Trans. Evol. Comput. **5**(4), 349–358 (2001). doi:10.1109/4235.942529
15. Poli, R., Langdon, W.B., McPhee, N.F., Koza, J.R.: A Field Guide to Genetic Programming. Lulu Press, Raleigh (2008). lulu.com
16. Roudavski, A.: Towards morphogenesis in architecture. Int. J. Architect. Comput. **7**(3), 345–374 (2009). doi:10.1260/147807709789621266
17. Ryan, C., Collins, J.J., Neill, M.O.: Grammatical evolution: evolving programs for an arbitrary language. In: Banzhaf, W., Poli, R., Schoenauer, M., Fogarty, T.C. (eds.) EuroGP 1998. LNCS, vol. 1391, pp. 83–96. Springer, Heidelberg (1998). doi:10.1007/BFb0055930
18. Stiny, G., Mitchell, W.J.: The palladian grammar. Environ. Plan. B: Plan. Des. **5**(1), 5–18 (1978)
19. Williams, N., et al.: FabPod: designing with temporal flexibility & relationships to mass-customisation. Autom. Constr. **51**, 124–131 (2015)
20. Woodbury, R.F., Burrow, A.L.: Whither design space? Artif. Intell. Eng. Des. Anal. Manufact. **20**, 63–82 (2006)

Evolutionary Image Transition
Using Random Walks

Aneta Neumann$^{(\boxtimes)}$, Bradley Alexander, and Frank Neumann

Optimisation and Logistics, School of Computer Science,
The University of Adelaide, Adelaide, Australia
aneta.neumann@adelaide.edu.au

Abstract. We present a study demonstrating how random walk algorithms can be used for evolutionary image transition. We design different mutation operators based on uniform and biased random walks and study how their combination with a baseline mutation operator can lead to interesting image transition processes in terms of visual effects and artistic features. Using feature-based analysis we investigate the evolutionary image transition behaviour with respect to different features and evaluate the images constructed during the image transition process.

1 Introduction

Evolutionary algorithms (EAs) have been widely and successfully used in the areas of music and art [1–3]. In this application area the primary aim is to evolve artistic and creative outputs through an evolutionary process [4–7]. The use of evolutionary algorithms for the generation of art has attracted strong research interest. Different representations have been used to create works of greater complexity in 2D and 3D [8–10], and in image animation [11–13]. The great majority of this work relates to the use of evolution to produce a final artistic product in the form of a picture, sculpture or animation.

Another application of evolutionary algorithms to art is the creation of image transitions. Earlier work by Sims [11] described methods for cross-dissolving of images by changes in an expression genotype. Banzhaf [14] used interactive evolution to help determine parameters for image morphing. Furthermore, Karungaru [15] used an evolutionary algorithm to automatically identify features for morphing faces. More recently, deep neural networks have recently been used to create artistic images through the transfer of artistic style from one image to another [16].

Neumann et al. [17] described an image transition process where the key idea is to use the evolutionary process *itself* in an artistic way. The focus of our paper is to study how random walk algorithms can be used in the evolutionary image transition process defined in [17] as mutation operators. We consider the well-studied (1+1) EA, popular random walk algorithms and provide new approach to evolutionary art by using theoretical approaches for evolutionary image transition.

J. Correia et al. (Eds.): EvoMUSART 2017, LNCS 10198, pp. 230–245, 2017.
DOI: 10.1007/978-3-319-55750-2_16

The transition process consists of evolving a given starting image S into a given target image T by random decisions. Considering an error function which assigns to a given current image X the number of pixels where it agrees with T and maximizes this function boils down to the classical ONEMAX problem for which numerous theoretical results on the runtime behaviour of evolutionary algorithms are available [18–20]. An important topic related to the theory of evolutionary algorithms are random walks [21, 22]. We consider random walks on images where each time the walk visits a pixel its value is set to the value of the given target image. By biasing the random walk towards pixels that are similar to the current pixel we can study the effect of such biases which might be more interesting from an artistic perspective. After observing these two basic random processes for image transition, we study how they can be combined to give the evolutionary process additional interesting new properties. We study the effect of running random walks for short periods of time as part of a mutation operator in a (1+1) EA. Furthermore, we consider the effect of combining them with the asymmetric mutation operator for evolutionary image transition introduced in [17]. Our results show that the area of evolutionary image transition based on random walks provides a rich source of artistic possibilities for creating video art. All our approaches are pixel-based and creating videos based on the evolutionary processes show frames corresponding to the images created every few hundred generations[1].

After introducing these different approaches to evolutionary image transition based on random walks, we study their behaviour with respect to different aesthetic features. Feature-based analysis of heuristic search methods has gained increasing interest in recent years [23–25]. In other application areas feature-based analysis is an important method to increase the theoretical understanding algorithm performance and in particularly useful for algorithm selection and configuration [26, 27]. For evolutionary image transition, we study how artistic features behave during the transition process. This allows the measurement of the evolutionary image transition process in a quantitative way and provides a basis to compare our different approaches with respect to artistic measures.

The outline of the paper is as follows. In Sect. 2, we introduce the evolutionary transition process. In Sect. 3, we study how variants of random walks can be used for the image transition process. We examine the use of random walks as part of mutation operators and study their combinations with asymmetric mutation during the evolutionary process in Sect. 4. In Sect. 5, we analyse the different approaches for evolutionary image transition with respect to aesthetic features. Finally, we finish with some concluding remarks.

2   Evolutionary Image Transition

We consider the evolutionary image transition process introduced in [17]. It transforms a given image $S = (S_{ij})$ of size $m \times n$ into a given target image $T = (T_{ij})$ of size $m \times n$. This is done by producing images X for which $X_{ij} \in$

---

[1] Images and videos are available at https://vimeo.com/anetaneumann.

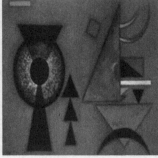

Fig. 1. Starting image X (Yellow-Red-Blue, 1925 by Wassily Kandinsky) and target image T (Soft Hard, 1927 by Wassily Kandinsky) (Color figure online)

Algorithm 1. (1+1) EA for evolutionary image transition

- Let S be the starting image and T be the target image.
- Set X:=S.
- Evaluate $f(X, T)$.
- while (not termination condition)
    - Obtain image Y from X by mutation.
    - Evaluate $f(Y, T)$
    - If $f(Y, T) \geq f(X, T)$, set $X := Y$.

$\{S_{ij}, T_{ij}\}$ holds. Given a starting image $S = (S_{ij})$ a target image $T = (T_{ij})$, and a current image $X = (X_{ij})$, we say that pixel X_{ij} is in state s if $X_{ij} = S_{ij}$, and X_{ij} is in state t if $X_{ij} = T_{ij}$. Our goal is to study different ways of using random walk algorithms for evolutionary image transition.

Throughout this paper, we assume that $S_{ij} \neq T_{ij}$ as pixels with $S_{ij} = T_{ij}$ can not change values and therefore do not have to be considered in the evolutionary process. To illustrate the effect of the different methods presented in this paper, we consider the Yellow-Red-Blue, 1925 by Wassily Kandinsky as the starting image and the T Soft Hard, 1927 by Wassily Kandinsky as the target image (see Fig. 1). In principle, this process can be carried out with any starting and target image. Using artistic images for this has the advantage that artistic properties of images are transformed during the evolutionary image transition process. We will later on study how the different operators used in the algorithms influence artistic appearance in terms of different artistic features.

We use the fitness function for evolutionary image transition used in [17] and measure the fitness of an image X as the number of pixels where X and T agree. This fitness function is equivalent to the ONEMAX problem when interpreting the pixels of S as 0's and the pixels of T as 1's. Hence, the fitness of an image X with respect to the target image T is given by

$$f(X, T) = |\{X_{ij} \in X \mid X_{ij} = T_{ij}\}|.$$

Algorithm 2. Asymmetric mutation

- Obtain Y from X by flipping each pixel X_{ij} of X independently of the others with probability $c_s/(2|X|_S)$ if $X_{ij} = S_{ij}$, and flip X_{ij} with probability $c_t/(2|X|_T)$ if $X_{ij} = T_{ij}$, where $c_s \geq 1$ and $c_t \geq 1$ are constants, we consider $m = n$.

We consider simple variants of the classical (1+1) EA in the context of image transition. The algorithm is using mutation only and accepts an offspring if it is at least as good as its parent according to the fitness function. The approach is given in Algorithm 1. Using this algorithm has the advantage that parents and offspring do not differ too much from the number of pixels. This ensures a smooth process for transitioning the starting image into the target. Furthermore, we can interpret each step of the random walks flipping a visited pixel to the target outlined in Sect. 3 as a mutation step which according to the fitness function is always accepted.

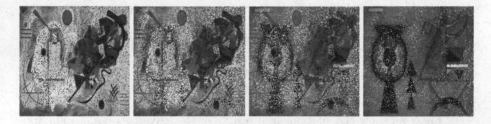

Fig. 2. Image Transition using asymmetric mutation with $c_s = 100$ and $c_t = 50$

As the baseline mutation operator we consider the asymmetric mutation operator which has been studied in the area of runtime analysis for special linear functions [18] as well as the minimum spanning tree problems [28]. Using this mutation operator instead of standard bit mutations for OneMax problems has the advantage that it does not suffer from the coupon collectors effect at the end of the transition process.

We use the generalization of this asymmetric mutation operator already proposed in [17] and shown in Algorithm 2. Let $|X|_T$ be the number of pixels where X and T agree. Similarly, let $|X|_S$ be the number of pixels where X and S agree. Each pixel is starting state s is flipped with probability $c_s/(2|X|_S)$ and each pixel in target state t is flipped with probability $c_t/(2|X|_T)$. The special case of $c_s = c_t = 1$ has been mathematically analyzed with respect to the runtime behaviour on OneMax and other pseudo-Boolean functions.

We set $c_s = 100$ and $c_t = 50$ as in [17]. This allows both a decent speed for the image transition process and enough exchanges of pixels for an interesting evolutionary process. We should mention that obtaining the last pixels of the target image may take a long time compared to the other progress steps when

Algorithm 3. Uniform Random Walk

- Choose the starting pixel $X_{ij} \in X$ uniformly at random.
- Set $X_{ij} := T_{ij}$.
- while (not termination condition)
    - Choose $X_{kl} \in N(X_{ij})$ uniformly at random.
    - Set $i := k$, $j := l$ and $X_{ij} := T_{ij}$.
- Return X.

using large values of c_t. However, for image transition, this only effects steps when there are at most $c_t/2$ source pixels remaining in the image. From a practical perspective, this means that the evolutionary process has almost converged towards the target image and setting the remaining missing target pixels to their target values provides an easy solution.

All experimental results for evolutionary image transition in this paper are shown for the process of moving from the starting image to the target image given in Fig. 1 where the images are of size 200×200 pixels. The algorithms have been implemented in Matlab. In order to visualize the process, we show the images obtained when the evolutionary process reaches 12.5%, 37.5%, 62.5% and 87.5% pixels of target image for the first time. We should mention that all processes except the use of the biased random walk are independent of the starting and target image which implies that the use of other starting and target images would show the same effects in terms of the way that target pixels are displayed during the transition process.

In Fig. 2 we show the experimental results of the asymmetric mutation approach as the baseline. On the first image from left we can see the starting image S with lightly stippling dots in randomly chosen areas of the target image T. Consequently the area of the yellow dimensional abstract face disappears and black abstract figure appears. Meanwhile the background has adopted a dot pattern, where a nuance of dark and light develops steadily. In the last image we barely see the starting image S and the target image T appearing permanently with the background becoming darker blue ton, whereby the stippling effect shown in the middle two frames decreases. The most valuable image in terms of aesthetic and evolutionary creativity emerge at the picture at 62,5% of the evolutionary processes. We can see in this third picture elements of both images compounded with the very special effect that imitate an artistic painting technique.

3   Random Walks for Image Transition

Our evolutionary algorithms for image transition build on random walk algorithms and use them later on as part of a mutation step. We investigate the use of random walk algorithms for image transition which move, at each step, from a current pixel X_{ij} to one pixel in its neighbourhood.

Algorithm 4. Biased Random Walk

- Choose the starting pixel $X_{ij} \in X$ uniformly at random.
- Set $X_{ij} := T_{ij}$.
- while (not termination condition)
    - Choose $X_{kl} \in N(X_{ij})$ according to probabilities $p(X_{kl})$.
    - Set $i := k$, $j := l$ and $X_{ij} := T_{ij}$.
- Return X.

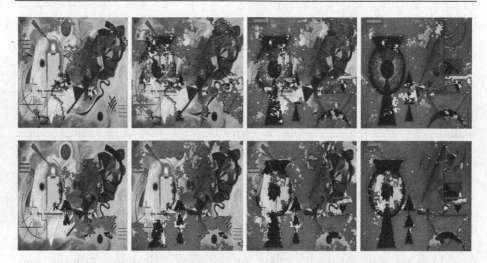

Fig. 3. Image transition for uniform random walk (top) and biased random walk (bottom) (Color figure online)

We define the neighbourhood $N(X_{ij})$ of X_{ij} as

$$N(X_{ij}) = \{X_{(i-1)j}, X_{(i+1)j}, X_{i(j-1)} X_{i(j+1)}\}$$

where we work modulo the dimensions of the image in the case that the values leave the pixel ranges, $i \in \{1, \ldots, m\}$, $j \in \{1, \ldots, n\}$. This implies that from a current pixel, we can move up, down, left, or right. Furthermore, we wrap around when exceeding the boundary of the image.

The classical random walk chooses an element $X_{kl} \in N(X_{ij})$ uniformly at random. We call this the *uniform random walk* in the following. The cover time of the uniform random walk on a $n \times n$ torus is upper bounded by $4n^2(\log n)^2/\pi$ [22] which implies that the expected number of steps of the uniform random walk until the target image is obtained (assuming $m = n$) is upper bounded by $4n^2(\log n)^2/\pi$).

We also consider a *biased random walk* where the probability of choosing the element X_{kl} is dependent on the difference in RGB-values for T_{ij} and T_{kl}. Weighted random walks have been used in a similar way in the context of image segmentation [29]. We denote by T_{ij}^r, $1 \le r \le 3$, the rth RGB value of T_{ij} and define

$$\gamma(X_{kl}) = \max\left\{\sum_{r=1}^{3}|T_{kl}^{r} - T_{ij}^{r}|, 1\right\}$$

In our random walk, we want to prefer X_{kl} if $\gamma(X_{kl})$ is small compared to the other elements in $N(X_{ij})$. In order to compute the probability of moving to a new neighbour we consider $(1/\gamma(X_{kl})) \in [0,1]$ and prefer elements in $N(X_{ij})$ where this value is large.

In the biased random walk, the probability of moving from X_{ij} to an element $X_{kl} \in N(X_{ij})$ is given by

$$p(X_{kl}) = \frac{(1/\gamma(X_{kl}))}{\sum_{X_{st} \in N(X_{ij})}(1/\gamma(X_{st}))}.$$

The biased random walk is dependent on the target image when carrying out mutation or random walk steps and the importance of moving to a pixel with similar color. Introducing the bias in terms of pixels that are similar, the bias can take the evolutionary image transition process to take exponentially long as the walk might encounter effects similar to the gambler's ruin process [30]. For our combined approaches described in the next section, we use the random walks as mutation components which ensures that the evolutionary image transition is carried out efficiently. We will use the biased random walk for evolutionary image transition in Sect. 4.

In Fig. 3 we show the experimental results of the uniform random walk and biased random walk. At the beginning, we can observe the image with the characteristic random walk pathway appearing as a patch in the starting image S. Through the transition process, the clearly recognisable patches on the target image T emerge. In the advanced stages the darker patches from the background of the target image dominate. The effect in animation is that the source image is scratched away in a random fashion to reveal an underlying target image.

The four images of the biased random walk are clearly different to the images of the uniform random walk. During the course of the transition, the difference becomes more prominent, especially in the background where at 87.5% pixels of the target image there is nearly an absolute transition to the target image T. In strong contrast, the darker abstract figure of the images stay nearly untouched, so that we see a layer of the yellow face in starting image S in the centre of the abstract black figure in target image T. In this image the figure itself is also very incomplete with much of the source picture showing through. These effects arise from biased probabilities in the random walk which makes it difficult for the walk to penetrate areas of high contrast to the current pixel location.

4   Combined Approaches

The asymmetric mutation operator and the random walk algorithms have quite different behaviour when applied to image transition. We now study the effect of combining the approaches for evolutionary image transition into order to obtain a more artistic evolutionary process.

4.1 Random Walk Mutation

Firstly, we explore the use of random walks as mutation operators and call this a *random walk mutation*.

The *uniform random walk mutation* selects the position of a pixel X_{ij} uniformly at random and runs the uniform random walk for $t_{\max}$ steps (iterations of the while-loop). We call the resulting algorithm EA-UniformWalk. Similarly, the *biased random walk mutation* selects the position of a pixel X_{ij} uniformly at random and runs the biased random walk for $t_{\max}$ steps. This algorithm is called EA-BiasedWalk. For our experiments, we set $t_{\max} = 100$.

Figure 4 shows the results of the experiments for EA-UniformWalk and EA-BiasedWalk. The transitions produced were significantly different from the previous ones. In both experiments we can see the target image emerging through a series of small patches at first, then steadily changing through a more chaotic phase where elements of the source and target image appear with roughly equal frequency. On the last image of each experiment we can see most details of the target image.

The images from EA-BiasedWalk appear similar to those in EA-UniformWalk in the beginning but differences emerge at the final stages of transition where, in EA-BiasedWalk, elements of the source image still show through in areas of high contrast in the target image, which the biased random walk has difficulty traversing. This mirrors, at a more local scale the effects of bias in the earlier random walk experiments. At a global scale it can be seen that the blue background, which is a low contrast area, is slightly more complete in the final frame of EA-BiasedWalk than the same frame in EA-UniformWalk.

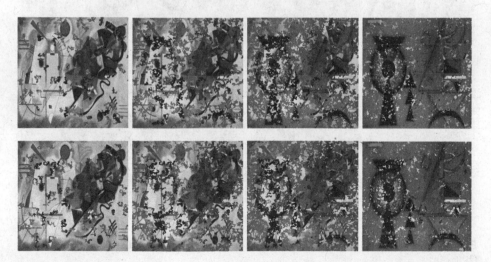

Fig. 4. Image transition for EA-UniformWalk (top) and EA-BiasedWalk (bottom) (Color figure online)

4.2   Combination of Asymmetric and Random Walk Mutation

Furthermore, we explore the combination of the asymmetric mutation operator and random walk mutation. Here, we run the asymmetric mutation operator as described in Algorithm 2 and a random walk mutation every τ generations. We explore two combinations, namely the combination of the asymmetric mutation operator with the uniform random walk mutation (leading to the algorithm EA-AsymUniformWalk) as well as the combination of the asymmetric mutation operator with the biased random walk mutation (leading to Algorithm EA-AsymBiasedWalk). We set $\tau = 1$ and $t_{\max} = 2000$ which means that the process is alternating between asymmetric mutation and random walk mutation where each random walk mutation carries out 2000 steps.

In Fig. 5, we show the results of EA-AsymUniformWalk and EA-AsymBiasedWalk. From a visual perspective both experiments combine the stippled effect of the asymmetric mutation with the patches of the random walk. In EA-AsymBiasedWalk there is a lower tendency for patches generated by random walks to deviate into areas of high contrast. As the experiment progresses, the pixel transitions caused by the asymmetric mutation have a tendency to degrade contrast barriers.

However, even in the final frames there is clearly more background from the target image in EA-AsymBiasedWalk than in EA-AsymUniformWalk. Moreover, there are more remaining patches of the source image near the edges of the base of the abstract figure, creating interesting effects.

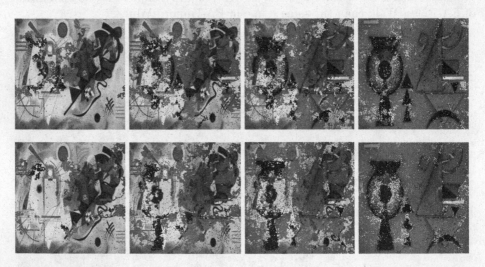

Fig. 5. Image transition for EA-AsymUniformWalk (top) and EA-AsymBiasedWalk (bottom)

5   Feature-Based Analysis

We now analyze the different introduced approaches for evolutionary image transition with respect to some features that measure aesthetic behaviour. Our goal is twofold. First, we analyze how the aesthetic feature values change during the process of transition. Furthermore, we compare the different approaches against each other and show where they differ with respect to the examined features when used for evolutionary image transition. For our investigations, we examine the starting and target image of Fig. 1, the transition of a black starting image into a white target image, and the transition of the starting image Color1 into the target image Color2 as shown in Fig. 6. Taking the last two pairs of images allows us to get additional systematic insights into the process of evolutionary image transition. Note that the images of Fig. 6 are only swapping the colored squares.

Fig. 6. Starting image S (Color1) and target image T (Color2) (Color figure online)

The set of features we use are, in order of appearance, *Benford's Law* [31], *Global Contrast Factor* [32], *Mean Hue*, and *Colorfulness* [33]. We describe each of them in the following.

The *Benford's Law* feature (*Ben*) is a measure of naturalness in an image X. Jolion [31] observed that the sorted histogram of luminosities in natural images followed the shape of Benford's Law distribution of first digits. Here we use the encoding of the Benford's Law feature based on the one used by den Heijer [34].

To calculate $Ben(X)$ we first calculate a nine-bin histogram H_X of the luminosities of X. The bins of H_X are then sorted by frequency and scaled to sum to 1.0. We define

$$Ben(X) = 1 - d_{total}/d_{max}$$

where

$$d_{total} = \sum_{i=1}^{9} H_X(i) - H_{benford}(i)$$

and $H_{benford}$ is a 9-bin histogram, encoding Benford's Law distribution, with the bin frequencies $0.301, 0.176, 0.125, 0.097, 0.079, 0.067, 0.058, 0.051, 0.046$. The value

$$d_{max} = (1 - H_{benford}(1)) + \sum_{i=2}^{9} H_{benford}(i)$$

is the maximum possible value for d_{total} which is the largest possible deviation of H_X from $H_{benford}$.

Global Contrast Factor, GCF is a measure of mean contrast between neighbouring pixels at different image resolutions. To calculate $GCF(X)$ we calculate the local contrast at each pixel at a given resolution r: $lc_r(X_{ij}) = \sum_{X_{kl} \in N(X_{ij})} |lum(X_{kl}) - lum(X_{ij})|$ where $lum(P)$ is the perceptual luminosity of pixel P and $N(X_{ij})$ are the four neighbouring pixels of X_{ij} at resolution r. The mean local contrast at the current resolution is defined $C_r = (\sum_{i=1}^{m} \sum_{j=1}^{n} lc_r(X_{ij}))/(mn)$. From these local contrasts, GCF is calculated as $GCF = \sum_{r=1}^{9} w_r \cdot C_r$.

The pixel resolutions correspond to different *superpixel* sizes of $1, 2, 4, 8, 16, 25, 50, 100$, and 200. Each superpixel is set to the average luminosity of the pixel's it contains. The w_r are empirically derived weights of resolutions from [32] giving highest weight to moderate resolutions.

The *Mean Hue (Hue)* of an image X is

$$Hue(X) = \left(\sum_{i=1}^{m} \sum_{j=1}^{n} h(X_{ij}) \right) /(m \times n)$$

where $h(X_{ij})$ is the hue value for pixel X_{ij} in the range $[0, 1]$. The function *Hue* measures where on average the image X sits on the color spectrum. Because the color spectrum is a circular construct one color, cyan in our case, is mapped to both 1 and 0.

Colorfulness (Color) is a measure of the perceived color of an image. We use Hasler's simplified metric for calculating colorfulness [33]. This measure quantifies spreads and intensities of opponent colors by calculating for the RGB values in each pixel X_{ij} the red-green difference: $rg_{ij} = |R_{ij} - G_{ij}|$, and the yellow-blue difference: $yb_{ij} = |(R_{ij} + G_{ij})/2 - B_{ij}|$. The means: μ_{rg}, μ_{yb} and standard-deviations: σ_{rg}, σ_{yb} for these differences are then combined to form a weighted magnitude estimate for colorfulness for the whole image:

$$Color(X) = \sqrt{\sigma_{rg}^2 + \sigma_{yb}^2} + 0.3\sqrt{\mu_{rg}^2 + \mu_{yb}^2}$$

Figure 7 shows how the features evolve over time during the image transition process. The first column refers to the transition process of the starting and target image given in Fig. 1. The second column shows the transition of a complete black image starting image to a complete white target image, and the third column shows the transition of the color starting image to the color target image of Fig. 6. Each figure shows the results of 10 runs for each algorithm that we have considered for evolutionary image transition.

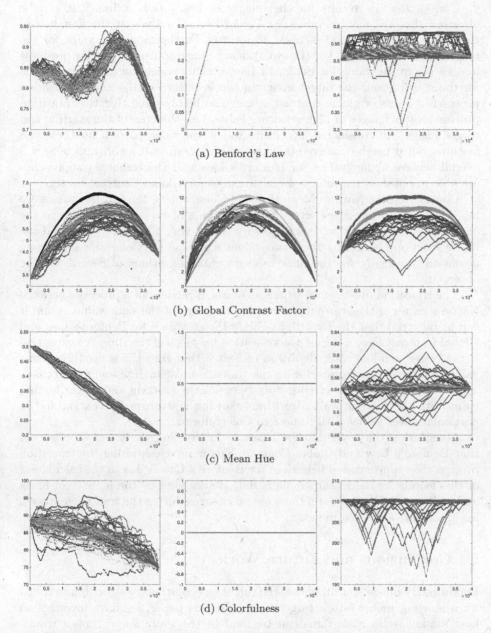

(a) Benford's Law

(b) Global Contrast Factor

(c) Mean Hue

(d) Colorfulness

Fig. 7. Features during transition for images for Asymmmetric Mutation (•), Uniform Random Walk (•), Biased Random Walk (•), EA-UniformWalk (•), EA-BiasedWalk (•), EA-AsymUniformWalk (•) and EA-AsymBiasedWalk (•) for images from Fig. 1 (left), Black-White (middle), Fig. 6 (right). Generation number is shown on the x-axis and features values on the y-axis. (Color figure online)

Considering the results for the images of Fig. 1 (left column), it can be observed that the feature values for Benford's Law reduce at the first half of the transition process and increase afterwards. Furthermore, the value for the target image is quite low, but the evolutionary image transition process produces images where the value for Benford's Law is significantly higher than the one for the starting and the target image in the last third of the image transition process. In terms of global contrast, it can also be observed that the transition process creates images of higher feature value than the ones of the starting and target image. All considered algorithms follow the same pattern for these two features, but it can be observed that the pure random walk algorithms of Sect. 3 overall achieve higher values for Benford's Law and the combined approaches are able to obtain a trajectory of higher values for Global Contrast Factor.

Considering the features Mean Hue and Colorfulness, the features values are following a more direct trajectory from the value of the starting image to the one of the target. For Hue, this trajectory is also very concentrated around the linear function connecting these two values where as for Colorfulness a strong deviation, especially for the pure random walk algorithms of Sect. 3, can be observed.

The transition process for the images of Fig. 6 carries out a process where the features values of the starting and target image are of the same value. Again it can be observed that the algorithms obtain higher values for Benford's Law and Global Contrast Factor during the transition for most of the runs. An exception is the biased random walk algorithm of Sect. 3 that sometimes produces lower values for these two figures during the transition. Mean Hue and Colorfulness again exhibit a more direct trajectory between the starting and target feature value with the random walk algorithms showing a stronger fluctuation and in particular lower values with respect to Colorfulness.

Considering the transition for Black to White images, it can be observed that Benford's Law and Global Contrast Factor increase during the transition process. The concentrated behaviour for Benford's Law is due to the calculation of this feature as the feature value is fully determined by the number of black and white pixel. Furthermore, there are no changes during the transition process for Mean Hue and Colorfulness.

6   Conclusions and Future Work

Evolutionary image transition uses the run of an evolutionary algorithm to transfer a starting image into a target image. In this paper, we have investigated how random walk algorithms can be used in the evolutionary image transition process. We have shown that mutation operators using different ways of incorporating uniform and biased random walks lead to different effects during the transition process. Furthermore, we have studied the impact of the different approaches with respect to different artistic features and observed that the process creates images which significantly differ from the starting and target image with respect to these features.

All our investigations are based on a fitness function that is equivalent to the well-known ONEMAX problem. For future research it would be interesting to study more complex fitness functions and their impact on the artistic behaviour of evolutionary image transition.

Acknowledgement. This work has been supported through Australian Research Council (ARC) grant DP140103400.

References

1. Romero, J., Machado, P. (eds.): The Art of Artificial Evolution: A Handbook on Evolutionary Art and Music. Natural Computing Series. Springer, Heidelberg (2008)
2. Antunes, R.F., Leymarie, F.F., Latham, W.H.: On writing and reading artistic computational ecosystems. Artif. Life **21**(3), 320–331 (2015)
3. Lambert, N., Latham, W.H., Leymarie, F.F.: The emergence and growth of evolutionary art: 1980–1993. In: International Conference on Computer Graphics and Interactive Techniques, SIGGRAPH 2013, Anaheim, CA, USA, July 21–25, 2013, Art Gallery, 367–375. ACM (2013)
4. McCormack, J., d'Inverno, M. (eds.): Computers and Creativity. Springer, Heidelberg (2012)
5. Vinhas, A., Assunção, F., Correia, J., Ekárt, A., Machado, P.: Fitness and novelty in evolutionary art. In: Johnson, C., Ciesielski, V., Correia, J., Machado, P. (eds.) EvoMUSART 2016. LNCS, vol. 9596, pp. 225–240. Springer, Cham (2016). doi:10.1007/978-3-319-31008-4_16
6. al-Rifaie, M.M., Bishop, J.M.: Swarmic paintings and colour attention. In: Machado, P., McDermott, J., Carballal, A. (eds.) EvoMUSART 2013. LNCS, vol. 7834, pp. 97–108. Springer, Heidelberg (2013). doi:10.1007/978-3-642-36955-1_9
7. Greenfield, G.: Avoidance drawings evolved using virtual drawing robots. In: Johnson, C., Carballal, A., Correia, J. (eds.) EvoMUSART 2015. LNCS, vol. 9027, pp. 78–88. Springer, Cham (2015). doi:10.1007/978-3-319-16498-4_8
8. Todd, S., Latham, W.: Evolutionary Art and Computers. Academic Press Inc., Orlando (1994)
9. Greenfield, G., Machado, P.: Ant- and ant-colony-inspired alife visual art. Artif. Life **21**(3), 293–306 (2015)
10. Machado, P., Correia, J.: Semantic aware methods for evolutionary art. In: Arnold, D.V., (ed.) Genetic and Evolutionary Computation Conference, GECCO 2014, Vancouver, BC, Canada, 12–16 July 2014, pp. 301–308. ACM (2014)
11. Sims, K.: Artificial evolution for computer graphics. In: Thomas, J.J., (ed.) Proceedings of the 18th Annual Conference on Computer Graphics and Interactive Techniques, SIGGRAPH 1991, pp. 319–328. ACM (1991)
12. Hart, D.A.: Toward greater artistic control for interactive evolution of images and animation. In: Giacobini, M. (ed.) EvoWorkshops 2007. LNCS, vol. 4448, pp. 527–536. Springer, Heidelberg (2007). doi:10.1007/978-3-540-71805-5_58
13. Trist, K., Ciesielski, V., Barile, P.: An artist's experience in using an evolutionary algorithm to produce an animated artwork. IJART **4**(2), 155–167 (2011)
14. Graf, J., Banzhaf, W.: Interactive evolution of images. In: Evolutionary Programming, pp. 53–65 (1995)

15. Karungaru, S., Fukumi, M., Akamatsu, N., Takuya, A.: Automatic human faces morphing using genetic algorithms based control points selection. Int. J. Innovative Comput. Inf. Control **3**(2), 1–6 (2007)
16. Gatys, L.A., Ecker, A.S., Bethge, M.: Image style transfer using convolutional neural networks. In: Proceedings of the IEEE Conference on Computer Vision and Pattern Recognition, pp. 2414–2423 (2016)
17. Neumann, A., Alexander, B., Neumann, F.: The evolutionary process of image transition in conjunction with box and strip mutation. In: Hirose, A., Ozawa, S., Doya, K., Ikeda, K., Lee, M., Liu, D. (eds.) ICONIP 2016. LNCS, vol. 9949, pp. 261–268. Springer, Cham (2016). doi:10.1007/978-3-319-46675-0_29
18. Jansen, T., Sudholt, D.: Analysis of an asymmetric mutation operator. Evol. Comput. **18**(1), 1–26 (2010)
19. Witt, C.: Tight bounds on the optimization time of a randomized search heuristic on linear functions. Comb. Probab. Comput. **22**(2), 294–318 (2013)
20. Sudholt, D.: A new method for lower bounds on the running time of evolutionary algorithms. IEEE Trans. Evol. Comput. **17**(3), 418–435 (2013)
21. Lovász, L.: Random walks on graphs: A survey. In: Miklós, D., Sós, V.T., Szőnyi, T. (eds.) Combinatorics, Paul Erdős is Eighty, vol. 2, pp. 353–398. János Bolyai Mathematical Society, Budapest (1996)
22. Dembo, A., Peres, Y., Rosen, J., Zeitouni, O.: Cover times for brownian motion and random walks in two dimensions. Ann. Math. **160**(2), 433–464 (2004)
23. Mersmann, O., Preuss, M., Trautmann, H.: Benchmarking evolutionary algorithms: Towards exploratory landscape analysis. In: Schaefer, R., Cotta, C., Kołodziej, J., Rudolph, G. (eds.) PPSN 2010. LNCS, vol. 6238, pp. 73–82. Springer, Heidelberg (2010). doi:10.1007/978-3-642-15844-5_8
24. Mersmann, O., Bischl, B., Trautmann, H., Wagner, M., Bossek, J., Neumann, F.: A novel feature-based approach to characterize algorithm performance for the traveling salesperson problem. Ann. Math. Artif. Intell. **69**(2), 151–182 (2013)
25. Nallaperuma, S., Wagner, M., Neumann, F., Bischl, B., Mersmann, O., Trautmann, H.: A feature-based comparison of local search and the christofides algorithm for the travelling salesperson problem. In: Neumann, F., Jong, K.A.D., (eds.) Foundations of Genetic Algorithms XII, FOGA 2013, Adelaide, SA, Australia, 16–20 January 2013, pp. 147–160. ACM (2013)
26. Nallaperuma, S., Wagner, M., Neumann, F.: Analyzing the effects of instance features and algorithm parameters for max-min ant system and the traveling salesperson problem. Front. Robot. AI **2**, 1–16 (2015)
27. Poursoltan, S., Neumann, F.: A feature-based prediction model of algorithm selection for constrained continuous optimisation. CoRR abs/1602.02862 Conference version appeared in CEC 2016(2016)
28. Neumann, F., Wegener, I.: Randomized local search, evolutionary algorithms, and the minimum spanning tree problem. Theor. Comput. Sci. **378**(1), 32–40 (2007)
29. Grady, L.: Random walks for image segmentation. IEEE Trans. Pattern Anal. Mach. Intell. **28**(11), 1768–1783 (2006)
30. Mitzenmacher, M., Upfal, E.: Probability and Computing: Randomized Algorithms and Probabilistic Analysis. Cambridge University Press, New York (2005)
31. Jolion, J.M.: Images and benford's law. J. Math. Imaging Vis. **14**(1), 73–81 (2001)
32. Matkovic, K., Neumann, L., Neumann, A., Psik, T., Purgathofer, W.: Global contrast factor-a new approach to image contrast. Comput. Aesthetics **2005**, 159–168 (2005)

33. Hasler, D., Suesstrunk, S.E.: Measuring colorfulness in natural images. In: Electronic Imaging 2003, International Society for Optics and Photonics, pp. 87–95 (2003)
34. den Heijer, E., Eiben, A.E.: Investigating aesthetic measures for unsupervised evolutionary art. Swarm Evol. Comput. **16**, 52–68 (2014)

Evaluation Rules for Evolutionary Generation of Drum Patterns in Jazz Solos

Fabian Ostermann(✉), Igor Vatolkin, and Günter Rudolph

Fakultät für Informatik, Technische Universität Dortmund, Dortmund, Germany
{fabian.ostermann,igor.vatolkin,guenter.rudolph}@tu-dortmund.de

Abstract. The learning of improvisation in jazz and other music styles requires years of practice. For music scholars which do not play in a band, technical solutions for automatic generation of accompaniment on home computers are very helpful. They may support the learning process and significantly improve the experience to play with other musicians. However, many up-to-date approaches can not interact with a solo player, generating static or random patterns without a direct musical dialogue between a soloist and accompanying instruments. In this paper, we present a novel system for the generation of drum patterns based on an evolutionary algorithm. As the main extension to existing solutions, we propose a set of musically meaningful jazz-related rules for the real-time validation and adjustment of generated drum patterns. In the evaluation study, musicians agreed that the system can be successfully used for learning of jazz improvisation and that the wide range of parameters helps to adapt the response of the virtual drummer to the needs of individual scholars (Examples of generated music are available at http://sig-ma.de/wp-content/uploads/2017/01/JazzDrumPatterns.zip.).

Keywords: Evolutionary music generation · Rhythm generation · Jazz solo accompaniment

1 Introduction

To play like Miles Davis or John Coltrane is the dream of many amateur and professional jazz students around the world. But that requires many years of experience and practice. The task is so complex, because for jazz improvisation the soloist has to spontaneously invent melodies in his/her mind suitable to background music and play them immediately, matching the solo to the tones played by other members of a band at the same time: from 3–5 musicians in a combo up to 10 and more in a big band. In the end, the successful practice of improvisation comes down to learning by doing.

For learning of improvisation without a band, the accompanying musicians must be replaced by some technical solution. A common approach is to use a prerecorded music track called *playalong* to which the scholar can play. A widely spread collection of playalongs is the *Jamey Aebersold Play-A-Long Series* [1]

© Springer International Publishing AG 2017
J. Correia et al. (Eds.): EvoMUSART 2017, LNCS 10198, pp. 246–261, 2017.
DOI: 10.1007/978-3-319-55750-2_17

that exists since 1967. A currently popular collection can be found online at *learnjazzstandards.com* [2].

In the 1990s, a more versatile way of background music generation was presented by PG Music when releasing the software *Band-in-a-Box* [3]. Its algorithms generate music according to adjustable stylistic rules. As an input, they get a chord progression, time signature, and tempo, therefore enabling the creation of customized playalong tracks with home computers. A currently popular mobile application with similar features is *iReal Pro* [4].

The limits of such systems is that the generated tracks only differ from each other because of rules based on random variables. They can not react to improvised melodies of the soloing musician, which is necessary to learn the interaction in a professional jazz band.

In this paper, we present a novel evolutionary-based approach for the generation of accompaniment with a real-time adaptation to the currently played jazz solo. The evaluation and adaptation are based on the set of rules with regard to desired properties of jazz accompaniment. This provides an opportunity to simulate a band during the practicing of improvisation in order to gain learning effort. The initial implementation is restricted to the generation of drum patterns only, but can be extended to other instruments in the future taking into account melodic and harmonic rules.

In the next section, we first provide a brief overview of related works. Section 3 presents the representation and mutation operators of the proposed evolutionary system for music generation, and its implementation is described in Sect. 4. The rules for an automatic evaluation of the system are listed in Sect. 5. The results of the application and evaluation of the system are discussed in Sect. 6. The conclusions and ideas for future work are provided in Sect. 7.

2   Related Works

Below, we list some works on jazz and drum generation ordered by the impact they had on developing the system presented in Sects. 3 and 4.

First, the *Genetic Jammer* (GenJam) [5] proposed in 1994 by Biles shall be mentioned. It models a virtual jazz student who learns how to improvise by getting feedback from a human mentor whether played melodies are good or bad. This feedback is used to rate melody individuals in an evolutionary algorithm. The special use of jazz harmony theory and fixed rhythmic patterns to create rhythm-based jazz music was inspiring for our system.

Another very creative work on jazz improvising and interactive systems is *Voyager* [6], introduced by jazz trombonist George Lewis in 1986. Lewis calls it a *virtual improvising orchestra* which interactively generates musical responses to analyzed jazz performances as well as own creative musical impulses in a musical dialog with a human jazz soloist. The concept of its performance analysis was fundamentally inspiring for our approach.

Yee-King tries to explore the variety of timbres of percussion instruments with an *evolving drum machine* [7]. He uses evolutionary computing to alter the sound of drums. The concept of the drum machine is adopted in Sect. 3.

The *NEAT drummer* [8] by Hoover and Stanley uses a neural network called *NeuroEvolution of Augmenting Topologies* (NEAT) for evaluating individuals that represent drum patterns. Their focus on balancing structure and innovation of composed drum accompaniment is elementary for creative systems.

The system *CONGA* (Composition in Genetic Approach) [9] by Tokui and Iba tries to generate rhythmic compositions of rhythmical patterns evaluated by humans. It is shown that genre-specific rhythms can be produced on purpose.

Sbeat [10] by Unemi and Nakada does something very similar with compositions consisting of guitar, bass, and drums. The music individuals are interactively evaluated by human. The system is intended to assist beginners in composing music they favorite.

The *GeneticDrummer* [11] by Dostál generates a human-like drum accompaniment for a given score of an instrument such as guitar, bass, or piano by means of evolutionary computing. A drum stroke is not only defined by its time, velocity, and percussion instrument, but also by the playing style, e.g., where to hit the drums and how to hold the stick.

3   Evolutionary System: Representation and Mutation

In our proposal, each solution represents a single musical bar of a *drum pattern* which is a combination of single patterns to be played on individual instruments of the drum set. Drum patterns contain a fixed number of tics and are stored in a two-dimensional array as shown in Fig. 1. For example, a 4/4-bar with eighth notes can be represented by eight tics. The six instruments with individual patterns are bass drum, hihat, snare, ride cymbal, a low tom, and a high tom. A cell of the drum pattern array contains an integer value in range 0–127. This range is adopted from the MIDI[1] specifications. If the value is greater than 0, then it is called an *active cell*. An active cell represents a drum stroke, i.e. a hit with a (wooden) drum stick, with a given strength represented by its loudness value. The higher the value is, the louder the played instrument sounds. A recognizable drum pattern is formed by all its active cells and the corresponding instrument sounds.

Because all patterns will descend from the same user defined initial pattern, we apply no recombination. Only a simple mutation operator is used for evolution. It randomly selects a cell of a drum pattern and sets its value to a random integer number in range 0–127. That is an analogy to how a manual manipulation of a classic drum machine would proceed.

The probability that an active cell will be switched off after the mutation is equal to $\frac{1}{127} < 8‰$. Because the evolutionary process must be able to decrease the density of its patterns for musically reasons, the probability for rests must be increased. Assuming that very silent strokes are not mentioned by a listener or are even disturbing, all cells with a loudness value below 32 are set to 0. This

---

[1] *MIDI* (Musical Instrument Digital Interface) is a technical standard that allows electronic musical instruments, computers, and other related devices to share musical information with one another. For detailed information, see [12].

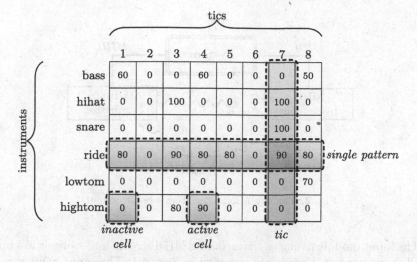

Fig. 1. Drum pattern represented by an array

changes the probability of raising the number of inactive cells in a drum pattern to $\frac{32}{127} \approx 25\%$.

The extent of variation in the patterns should be under control of the user, as sometimes quick changes of the accompaniment are musically desirable. Therefore the variation strength is made adjustable by the number of times the mutation operator is applied to a drum pattern. The parametrization of this mutation strength is carried out indirectly by choosing the number of generated offspring $\mu = 2^k - 1$ with $k \in \mathbb{N}$. Since small changes should be more frequent than large changes, 2^i individuals of the offspring population are mutated $k - i$ times for $i = 0, \ldots, k$.

4   Evolutionary System: Implementation

The environment of the system is sketched in Fig. 2. Human interaction with the system is marked with a ⚥-sign. The $\overset{\circ}{\lambda}_{musician}$ playing the solo is typically identical with the $\overset{\circ}{\lambda}_{user}$ who controls parameters of the system.

In an application scenario, a musician is soloing on an instrument that is able to produce MIDI events. The stream of these events is then evaluated by the system, which also produces an outgoing MIDI event stream for a drum synthesizer. The sound of the synthesizer is interpreted by the musician as accompaniment to the solo. In case there is any audience, the sounds of the musician and the synthesizer are interpreted as a compounded musical performance.

A musically meaningful MIDI drum event stream must be produced in real-time by analyzing a MIDI stream from a human musician's jazz solo. To fulfill that task, the system is split up into three modules named *input module*, *genetic module*, and *playback module*, as visualized in Fig. 3.

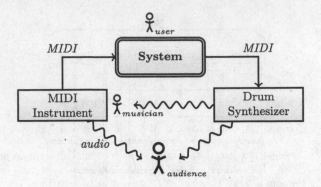

Fig. 2. The evolutionary system's environment

The input module manages the incoming MIDI stream and stores it in a buffer called *input window*. Details are provided in Sect. 4.1. The input window is the basis for the musical analysis carried out by the genetic module. This module generates drum patterns by means of an evolutionary algorithm as described in Sect. 4.2. The generated drum patterns are in turn taken by the playback module to create the output MIDI stream as discussed in Sect. 4.3.

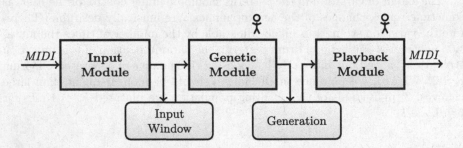

Fig. 3. Modularisation of the evolutionary system

4.1 Input Module

The difference between MIDI events and a musical note representation is that MIDI events are created to describe the actions of a piano player. So, the important actions are key pressing (*NoteOn* event) and releasing (*NoteOff* event). The input module translates these events into a representation of musical notes that will be easier to analyse. After a *NoteOn* event of a certain key is received, it is waited for the *NoteOff* event of that key. Then, a note record is written to the input window with attributes key number, velocity, start time, and length. These attributes are the basis for the analysis carried out by the genetic module.

The problem of waiting for a *NoteOn/NoteOff* pair is the delay when taking the record. Single *NoteOn* events are ignored in the solo analysis. It will be assumed, that in a monophonic jazz phrase the notes are short enough to not delay the analysis gravely. With this practical limitation no presumption of note lengths must be processed.

When using an Audio-MIDI-Converter, nearly every melodic instrument can be used as input instrument. Corrupted MIDI events coming from misinterpretation of the digitized audio signal can be ignored because of the goal to generate creative drum accompaniment. In the following, we will see that if the MIDI input is slightly noised by the use of a converter, this will not distort the musically meaning of the produced drum response.

The way of the MIDI event through the input module is shown in Fig. 4.

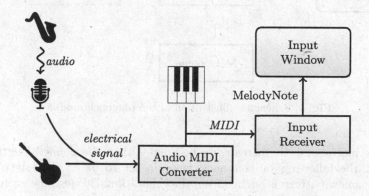

Fig. 4. Schematic illustration of the input module

4.2 Genetic Module

The genetic module is the main component of the system which operates with the help of an evolutionary algorithm. The loop starts with the initialization of the first population. It is filled with copies of user-defined initial drum pattern explained in Sect. 3. The population size is also defined by the user and can be adjusted during a session. Afterwards, the mutation is done as discussed in Sect. 3. Fitness values are assigned to individuals with the help of predefined rules based on musical theory which are explained in detail in Sect. 5. These rules can be weighted with regard to user preferences. Finally, individuals with highest fitness values are selected from the offspring population until the given population size is reached.

The evolutionary loop continues until the user stops the session. After the selection, the loop delays for a user-defined *sleep time* thus influencing the response time of produced accompanying drum patterns. If the sleep time is too short for the calculation of fitness values, the user is warned but the calculation is not interrupted.

4.3   Playback Module

The playback of the current population is separated from its generation by the genetic module. There is no adaptation of rhythmical pulse from the human improvised solo. The created musical timing is based on the principle of a classic drum machine. A metronome sends an increasing tic command to the pattern player which fetches a pattern at random from the current generation and initiates to play all active cells of the tic by sending MIDI events to a drum synthesizer. Figure 5 illustrates this procedure.

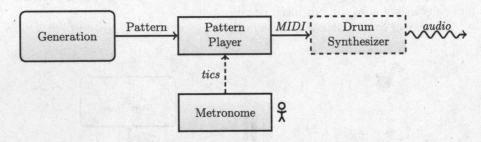

Fig. 5. Schematic illustration of the playback module

If the metronome tic increases to be greater then the defined length of the patterns, the following tic command is again set to 0. At the latest to that moment, a new pattern is fetched from the generation. Because the evolutionary loop can be faster than the playback of an individual pattern, the pattern player checks for a new generation on every tic. If there is a new generation created, a new pattern is immediately taken to be up-to-date with the evolutionary process.

The tempo of playback can be manually controlled by setting the tics-per-minute value (TPM) which defines the time to rest between tic commands as

$$t_\Delta = \frac{60\ 000\ \text{ms}}{TPM}. \tag{1}$$

Note that the patterns grid is not limited to a specific note values level. It can be used for eighth note, 16th note or even triplet patterns by appropriately increasing the TPM value and the pattern grids length.

Additionally, the metronome offers to play the tics as *shuffle* rhythm. Shuffle means a time shift of every second note by a constant factor. It is a typical stylistic feature of swing and funk and is essential for jazz music feel. Mathematically, every second waiting time $t_{\Delta,2}$ is shorter than the prior one $t_{\Delta,1}$. The shuffle factor $f_{shuffle}$ implies the formulas for the waiting times as

$$t_{\Delta,1} = \frac{2 \cdot f_{shuffle} \cdot 60\ 000\ \text{ms}}{TPM} \tag{2}$$

$$\text{and } t_{\Delta,2} = \frac{2 \cdot (1 - f_{shuffle}) \cdot 60\ 000\ \text{ms}}{TPM} \tag{3}$$

$$\text{with } f_{shuffle} \in [0.5, 1] \Rightarrow t_{\Delta,1} \geq t_{\Delta,2}. \tag{4}$$

By adjusting those settings, the user is able to manipulate the playback as desired to support his or her creative process.

5 Weighted Evaluation Rules

The evaluation of individual patterns is based on heuristic features calculated from an excerpt of the input window. The excerpt ignores notes being played too far in the past. The time difference is called *input window size* and can be manually adjusted. The input window size influences the temporal memory length of the virtual drummer. The longer it is, the more musical context can be taken into account. That obviously increases the calculation time. A short memory lets the drummer respond faster, but a too short one cannot bear any meaningful information.

The following features are determined from the input window excerpt:

1. $\#_{notes}$ – total number of notes
2. $\#_{pitches}$ – number of different pitches
3. $\#_{pitches_{12}}$ – number of different pitch classes (max. 12)

Additionally, minimum $MIN(x)$, maximum $MAX(x)$, range $R(x)$, arithmetic average $\bar{x}$, and standard deviation $\sigma(x)$ are calculated over all notes for following note attributes:

1. **length (leng)**: Duration of a note
2. **keynumber (key)**: Mapping of a note to a key of the piano
3. **volume (vol)**: Loudness of a note
4. **interval (int)**: Interval between two consecutive notes
5. **distance (dist)**: Duration between the start time of two consecutive notes
6. **gap**: Duration between end time of a note and start time of the consecutive note

Evaluation rules rate drum patterns with the help of these statistic features concerning their musical compatibility to solo excerpts. Each rule describes an independent musical task the virtual drummer should fulfill. Each drum pattern in the current population is once rated by each rule, and the overall fitness is estimated as the sum of those single ratings.

Before the rules are estimated, some conditions are typically checked. If a condition is negative, the rating of the rule is set to 0. Otherwise, a solo factor f_s and a pattern factor f_p are calculated to describe the intensity of a rule characteristic in a solo excerpt and a drum pattern. f_s and f_p take values between 0 and 1, where 0 corresponds to a low intensity and 1 to a high one. f_s is based on the statistic features only. To calculate f_p, the drum pattern array is analysed. The overall fitness value is a combination of both factors.

The design of the rules is the most subjective property of the whole system, as they define fundamentally the behavior of the virtual drummer. Therefore, the user has the possibility to define weights of rules $w \in [0, 1]$ which are multiplied with the ratings of rule functions. With this procedure, some rules may gain more influence or be completely deactivated.

The three groups of rules (*keep*, *reaction*, and *random*) are explained in the following subsections.

5.1   Keep Rules

The rating of keep rules is based on the distance between a current drum pattern and a user-defined initial pattern. They ensure that the patterns of the population keep a musical link to the initial one. In contrast to reaction rules described in Sect. 5.2, no conditions are checked before the rule estimation. If the solo is pausing and consequently no reaction rule is estimated, the patterns return to the initial pattern. This produces a musical relaxation allowing for the necessary contrast between a low and a high intensity in the musical dialogue between a soloist and a drummer. The keep rules do not calculate a solo factor which means they rate regardless of the solo. Therefore, the fitness is just the value of the pattern factor.

In the following, $P_{i,j}$ is the array of the drum pattern to rate, where $i \in \{0, \ldots, I-1\}$ with the number of tics I, and $j \in \{0, \ldots, J-1\}$ with the number of instruments J. Further, $A_{i,j}$ is the array of the initial pattern.

KeepOriginal Rule. This rule measures the distance between a pattern P and the initial pattern A. The greater the distance is, the lower is the fitness. For every cell, the distance is calculated by the absolute difference between the cell values. For normalization, the difference is divided by the largest possible distance.

$$\textit{Pattern factor} \quad f_p = 1 - \sum_{i=0}^{I-1} \sum_{j=0}^{J-1} \frac{|P_{i,j} - A_{i,j}|}{MAX\{A_{i,j}, 127 - A_{i,j}\}}$$

KeepInstruments Rule. Patterns, which use the same instruments as the initial pattern, are rated higher than others. The idea is that similar instrumentation will lead to similar sound impressions, so that a musical connection between such patterns is recognizable.

$$\textit{Pattern factor} \quad f_p = 1 - \frac{1}{J} \sum_{j=0}^{J-1} XOR\left(\sum_{i=0}^{I-1} P_{i,j}, \sum_{i=0}^{I-1} A_{i,j} \right)$$

where $XOR(x, y) := 1$ if $(x \neq 0 \wedge y = 0) \vee (x = 0 \wedge y \neq 0)$, and 0 otherwise.

KeepTicks Rule. This rule is a supplement of the KeepInstruments rule. The idea is that patterns, which have strokes on the same tics, have a recognizable musical connection. Therefore, patterns with the same tics as the initial pattern are rated higher than others.

Pattern factor $f_p = 1 - \dfrac{1}{I}\sum\limits_{i=0}^{I-1} XOR\left(\sum\limits_{j=0}^{J-1} P_{i,j}, \sum\limits_{j=0}^{J-1} A_{i,j}\right)$

5.2 Reaction Rules

The reaction rules are designed to allow for a musical adaptation of the current solo excerpt to the accompaniment of the virtual drummer. They all have a certain musically meaningful task in the context of the musical dialogue. Therefore, they are only estimated when a certain condition is fulfilled. The fitness of these rules is the arithmetic mean of solo and pattern factor, unless otherwise stated.

Chromatic Rule. The smallest movement possible in Western music corresponds to a semitone. If the solo contains many of such chromatic steps, an accompaniment without rests is desirable. The condition to check is that at least two notes are played and that the smallest interval is a minor second. The solo factor is defined as follows. The smaller the average interval is, the greater is the factor's value. The pattern factor increases with the number of *active tics*. A tic is called active, if one cell of the tic has a value greater than the half of the solo's average volume. If a pattern is without a gap, the pattern factor will be 1.

Condition $\#_{notes} > 1 \wedge Min_{int} \le 1$

Solo factor $f_s = 1/(MAX\{1, \overline{x}_{int}\})$

Pattern factor $f_p = \dfrac{1}{I}\sum\limits_{i=0}^{I-1}\Theta_1\left(\sum\limits_{j=0}^{J-1} h_{0,5}(P_{i,j})\right)$

where $\Theta_n(x) := 1$ if $x \ge n$, and 0 otherwise. Moreover, $h_n(x) := 1$ if $x > n \cdot \overline{x}_{vol}$, and 0 otherwise.

FreeJazz Rule. Common scales in Western music consist of seven pitch classes. If a solo excerpt contains more than seven different pitch classes, uncommon scales are played. This is typical in free jazz. Then, a rather unpredictable accompaniment is desirable. Therefore, the pattern factor is random. The solo factor increases with the number of different pitch classes. Because $\#_{pitches_{12}}$ is maximally 12, the fraction can be maximally equal to $0.8\overline{3}$. Subtracted from 1.08, the solo factor becomes nearly 1.

Condition $\#_{pitches_{12}} > 7$

Solo factor $f_s = 1.08 - 1/(\#_{pitches_{12}})$

Pattern factor $f_s = r \in U[0,1]$ (uniformly distributed random value)

Holdsworth Rule. This rule is inspired by the playing style of the British jazz guitar player Allan Holdsworth. He is famous for using large intervals in his solos. This practice is uncommon for ordinary jazz solos and therefore the virtual drummer shall show a special reaction. If the solo excerpt contains intervals larger than 5 semitones in average, the high fitness values are assigned to patterns, which have accents on different instruments for every two consecutive tics. The greater the interval is, the higher the fitness. A tic has an accent if the corresponding cell value is larger than the current solo's average volume.

Condition	$\#_{notes} > 1 \wedge \overline{x} \geq 5$
Solo factor	$f_s = MIN\{1, \overline{x}_{int}/12\}$
Pattern factor	$f_p = \frac{1}{I}\sum_{i=1}^{I-1} 1 - id\left(acc_P(i-1), acc_P(i)\right)$

where $id(x,y) := 1$ if $x = y \geq 0$, and 0 otherwise. The *accent function* is given by

$$acc_P(x) := \begin{cases} J-1 & \text{if } P_{x,J-1} \geq MAX\{P_{x,0},..,P_{x,J-1}\} \wedge P_{x,J-1} > \overline{x}_{vol} \\ \vdots & \vdots \\ 0 & \text{if } P_{x,0} \geq MAX\{P_{x,0},..,P_{x,J-1}\} \wedge P_{x,0} > \overline{x}_{vol} \\ -1 & \text{otherwise} \end{cases}$$

Legato Rule. Legato indicates that consecutive musical notes are smoothly connected without rests. The accompaniment should avoid rests as well. Legato is detected by the analysis of the gap feature. The smaller the gaps are, the higher is the solo factor. The pattern factor is the same as in the chromatic rule and increases with the number of active tics.

Condition	$\#_{notes} > 1$
Solo factor	$f_s = 1/\sqrt{MAX\{1, \overline{x}_{gap}\}}$
Pattern factor	f_p, see Chromatic rule

Loudness Rule. If the average volume of a pattern is close to the average volume of the current solo excerpt, the fitness value increases. The fitness value is calculated based on the distance between both average volumes.

Condition	$\#_{notes} > 0$
Solo factor	$f_s = \overline{x}_{vol}$
Pattern factor	$f_p = \frac{1}{\#_{active}}\sum_{i=0}^{I-1}\sum_{j=0}^{J-1} P_{i,j}$

where the *number of active cells of P* is $\#_{active} = \sum_{i=0}^{I-1}\sum_{j=0}^{J-1}\Theta_1(P_{i,j})$.

In this particular case, the fitness rule is $f = 1 - \frac{1}{127}|f_s - f_p|$.

Ostinato Rule. An ostinato is a motif or a phrase that persistently repeats. In a solo, the fast repetition of such pattern causes an increase of intensity. This should be maintained in the accompaniment. The condition to check is that in the solo excerpt there must be three times as much notes than different pitches. The solo factor increases with less different pitches and more note repetitions. The pattern factor favors patterns, which have at least two accents at each tic.

Condition $\#_{notes} > 0 \wedge (3 \cdot \#_{notes}) \geq \#_{pitches}$

Solo factor $f_s = 1 - MAX\{1, \#_{pitches} - 3\} / \#_{notes}$

Pattern factor $f_p = \left[\left| I - \sum_{i=0}^{I-1} \Theta_2 \left(\sum_{j=0}^{J-1} h_1\left(P_{i,j}\right)\right)\right| + 1\right]^{-2}$

Pedal Rule. In tonal music, a pedal note is a long sustained note, typically placed below a melody line. When long notes are played, the drum patterns should contain more rests. The longer a note is, the higher the solo factor will be. For notes longer than 5 s, the solo factor is equal to 1. The pattern factor counts the silent tics (inactive tics) and is equal to 1, when active and inactive tics are at the ratio of 2 to 1.

Condition $\#_{notes} > 0$

Solo factor $f_s = MIN\{5000, \overline{x}_{leng}\} / 5000$

Pattern factor $f_p = MIN\{1, \ 3 \cdot \#_{inactiveTics} / (2 \cdot I)\}$

where the number of inactive tics is $\#_{inactiveTics} = I - \sum_{i=0}^{I-1} \Theta_1 \left(\sum_{j=0}^{J-1} P_{i,j}\right)$.

Staccato Rule. Staccato is a form of musical articulation with tones of a shortened duration. Hence, staccato is the opposite to legato, and this technique should influence the accompaniment. A note is detected as played with staccato, if its distance to the consecutive note is maximally half of its length. Additionally, it must by shorter than 380 ms. If it is shorter than 80 ms, the solo factor is equal to 1. Patterns with a strong accent every three tics are preferred by the pattern factor. A strong accent is a tic with a cell value equal or greater than 110, corresponding to the intensity of $\frac{110}{127} \approx 86.6\%$.

Condition $\#_{notes} > 0 \wedge 2 \cdot \overline{x}_{leng} \leq \overline{x}_{dist} \wedge \overline{x}_{leng} \leq 380$

Solo factor $f_s = 1 - MAX(0, \overline{x}_{leng} - 80) / 380$

Pattern factor $f_p = \left[\left| \frac{I}{3} - \sum_{i=0}^{I-1} \Theta_1 \left(\sum_{j=0}^{J-1} \Theta_{110}\left(P_{i,j}\right)\right)\right| + 1\right]^{-1}$

Virtuoso Rule. A virtuoso is an individual who possesses outstanding technical skills, in particular, to play very fast. An eligible accompaniment should try to use as many percussive instruments as possible to produce a versatile sound. The more notes are in the solo, the higher is the solo factor. The pattern factor increases with the number of instruments *used*, i.e. instruments which patterns have at least one cell value greater than the average volume of the solo. If all instruments are already used in the initial pattern, this rule does not make sense. Therefore, it is not recommended to use all instruments in the initial pattern.

Condition $\#_{notes} > 0$

Solo factor $f_s = \left(1 - 1/\#_{notes}\right)^4$

Pattern factor $f_p = \dfrac{\sum_{j=0}^{J-1} AND\left(\sum_{i=0}^{I-1} 1 - \Theta_1(A_{i,j}), \sum_{i=0}^{I-1} h_1(P_{i,j})\right)}{\sum_{j=0}^{J-1} \Theta_1\left(\sum_{i=0}^{I-1} 1 - \Theta_1(A_{i,j})\right)}$

where $AND(x,y) := 1$ if $x \neq 0 \wedge y \neq 0$, and 0 otherwise.

Width Rule. The use of relatively high and low notes in a short period of time causes an increase in intensity, and the accompanying drum patterns should use all instruments to transfer the versatility of sound to the drums. The solo factor is 1, if the range of notes is equal or greater than 36 (3 octaves). The pattern factor increases with the number of non-empty instrument patterns.

Condition $\#_{notes} > 0$

Solo factor $f_s = MIN\{36, R_{note}\}/36$

Pattern factor $f_p = \dfrac{1}{J}\sum_{j=0}^{J-1}\Theta_1\left(\sum_{i=0}^{I-1} P_{i,j}\right)$

5.3   Random Rule

The rating is completely random, and the solo excerpt as well as the drum pattern are ignored completely. The random rule can be used to make the virtual drummer more unpredictable by increasing its weighting. This can be interpreted as a creative impulse in a musical dialogue and is responsible for more variation when the population of patterns becomes too uniform.

Fitness $f = r \in U[0,1]$ (uniformly distributed random value)

6   Evaluation

The success of the system can be measured by active and passive test persons (musicians resp. listeners). In particular, it shall be tested to which extent the

Table 1. List of all parameters which can be adjusted in real time

Parameter	Interval	Number range
Initial pattern	$P_{i,j} \in [0, 127]$	$\mathbb{N}$
Population size	$N_p \in [1, \infty)$	$\mathbb{N}$
Sleep time [ms]	$T_z \in [0, \infty)$	$\mathbb{N}$
Mutation expansion limit	$\mu \in [0, \infty)$	$\mathbb{N}$
Input window size [ms]	$T_w \in [0, \infty)$	$\mathbb{N}$
Weight of a rule	$w \in [0, 1]$	$\mathbb{R}$
Tics-per-minute	$TPM \in [1, \infty)$	$\mathbb{N}$
Shuffle factor	$f_{shuffle} \in [0, 5, 1]$	$\mathbb{R}$

behavior of the virtual drummer can fit the users' expectations, among others by adjustment of system parameters listed in Table 1.

In a user-related study, active test persons were two pianists, two guitarists, a bassist, and a saxophonist with years of experience in jazz bands and at jazz sessions with real drummers. Only the pianists have the benefit of producing entirely correct MIDI events. The string and reed instruments require audio-to-MIDI conversion. Note that the mistakes made at the MIDI conversion are not critical. They do not lead to a malfunction of the system. We will see that a fully predictable behavior of the virtual drummer is even not desirable. Mistakes can be even interpreted by the soloists as musical ideas developed by the drummer.

The test persons were let to play together with the virtual drummer. Afterwards, points of criticism were discussed. An always mentioned issue was that the reactions of the drummer should be as immediate as possible. This can be achieved by the reduction of T_z. For $T_z < 30$ ms, its limit could not be observed any more by the calculation. But $T_z = 30$ ms was rated as sufficient enough by all test persons. Additionally, μ can be increased to produce more versatile offsprings. Therewith, a faster mutation is possible, but the evaluation time increases. To reduce the reaction time, T_w can be manipulated as well. At 100 ms $< T_w < 1000$ ms, the test persons felt a good balance between reaction and detection of solo structures. For $T_w < 100$ ms, the system was not able to recognize meaningful solo excerpts. Many excerpts were empty, leading to an undesired relaxation of the accompaniment. Besides, $T_w \geq T_z$ must always hold to not ignore any notes.

The relaxation of the accompaniment ensured by the keep rules when pausing soloing was reported as a particularly positive experience. Even so, the random rule should always be active to preserve the stagnation of accompaniment.

The test persons have expected an autonomic behavior of the virtual drummer which should provide creative impulses and support a musical dialogue with the soloist. To achieve this only by means of the random rule becomes unsatisfying after a while. Another possibility is to increase N_p, so that also patterns with lower fitness values are played.

Another point of criticism was the rhythmical limitation of the virtual drummer based on the concept of a drum machine. Sometimes, it appears to be very useful in a musical way to place drum strokes between the tics, for example, to play double time fills. It is not possible to do this dynamically because of the input and mutation concept that bases on the drum machine concept to create instant familiarity while interacting with the system. A possible solution is to double the number of tics and to use only even tics for the definition of the initial pattern. Then, the mutation process will add strokes in between. This is described as musically acceptable but not satisfying in comparison with realistic drum fill playing.

The tests always started with a jazz-typical swing drum pattern in medium tempo with $f_{shuffle} = 0.67$. At higher tempos, $f_{shuffle}$ has to be decreased. For tempos above 250 bpm, it was suggested to set $f_{shuffle} = 0.5$. That was confirmed as a reasonable measure by the test persons. Among other tested pattern styles like rock, funk, hiphop, or latin, a particularly favored pattern was the bossa nova. Because of its driving straight eight groove and its shifting single patterns it was often described as thrilling by the test persons. The accentuated use of toms that seems to be typical for the implemented system was described as working perfectly in this style.

An interesting observation was that the test persons often experienced the limits of the system and explored its possibilities. Uneven time signatures, random or intentionally unconventional weighting, unstructured initial patterns, or unusual drum synthesizer sounds were chosen. This led to very experimental and innovative music that was obviously fun to play. The test persons stated that the overall range of possibilities to adjust the system from producing only small variation of initial pattern to a complete chaotic alienation was very satisfying.

Another attempt was to find musical structures of higher levels in the solos. The virtual drummer failed in this task, because of its limited memory and its lack of planning. The use of a high T_w value led to enormous calculation and impractical reaction times.

All active test persons agreed that the system can improve their way of practicing jazz improvisation. They stated that practicing with a virtual accompanying drummer is more interesting, more exciting, more entertaining, and more versatile. Inactive test persons even experienced the music as concert-like if a virtual drummer was supported with human chord instrumentalist as well as a human bassist additionally.

7   Conclusions and Outlook

The aim of this work was to develop a reactive system that is able to create a real-time rhythmical accompaniment for a jazz solo by means of an evolutionary algorithm. The idea derives from the desire to use an interactive playalong track for practising jazz improvisation alone at home. To measure the success of generated drum patterns, a set of evaluation rules was proposed with regard to musical properties of jazz accompaniment.

The system was tested by musicians as well as audience who estimated the musical acceptance of the produced music. It was consistently described as versatile and therefore beneficial for daily practice.

Future work will include improvement of the evaluation rules and their customization. Consideration of weak and strong beats as well as syncopation measuring based on musical theory is planned. Further, it is promising to mix the presented solo and pattern factors to create new rules or to even integrate machine learning approaches like deep neural networks in the rules fitness calculation. Other musically meaningful evolutionary operators can be adopted, e.g., inversions or rotations of patterns. The final attempt will be to create other virtual musicians like pianists, guitarists, or bassists to complete the virtual jazz band and to be able to improvise accompanied by a fully staffed interactive playalong track.

References

1. Aebersold, J.: Volume 1 - How to Play Jazz & Improvise. Jamey Aebersold Jazz (1967) (Jamey Aebersold Play-A-Long Series)
2. Hughes, C.: Learn Jazz Standards (2010). http://www.learnjazzstandards.com/about/. Accessed 2 Nov 2016
3. Gannon, P.: Band-in-a-Box. PG Music Inc., Hamilton (1990)
4. Biolcati, M.: iReal Pro. Technimo LLC, New York (2008)
5. Biles, J.A.: GenJam: a genetic algorithm for generating jazz solos. In: Proceedings of the International Computer Music Conference (ICMC 1994), San Francisco, USA, pp. 131–137. International Computer Association (1994)
6. Lewis, G.E.: Too many notes: computers, complexity and culture in voyager. Leonardo Music J. **10**, 33–39 (2000)
7. Yee-King, M.J.: The evolving drum machine. In: MusicAL 2007 Proceedings (2007). http://cmr.soc.plymouth.ac.uk/Musical2007/proceedings.htm. Accessed 5 Nov 2016
8. Hoover, A.K., Stanley, K.O.: Exploiting functional relationships in musical composition. Connection Sci. **21**(2–3), 227–251 (2009)
9. Tokui, N., Iba, H.: Music Composition with Interactive Evolutionary Computation. Graduate School of Engineering, The University of Tokyo (2001). http://www.generativeart.com/on/cic/2000/ga2000-tokui.htm. Accessed 29 Oct 2016
10. Unemi, T., Nakada, E.: A tool for composing short music pieces by means of breeding. In: Proceedings of the 2001 IEEE Systems, Man and Cybernetics Conference, pp. 3458–3463 (2001)
11. Dostál, M.: Genetic algorithms as a model of musical creativity - on generating of a human-like rhythmic accompaniment. Comput. Inform. **22**, 321–340 (2005)
12. MMA, MIDI Manufacturers Association: General MIDI 1, 2 and Lite Specifications (1991). https://www.midi.org/specifications/category/gm-specifications. Accessed 6 Nov 2016

Assessing Augmented Creativity: Putting a Lovelace Machine for Interactive Title Generation Through a Human Creativity Test

Yasser S. Arenas Rebolledo, Peter van der Putten[✉], and Maarten H. Lamers

Media Technology, Leiden University, Leiden, The Netherlands
arenasgt@gmail.com,
{p.w.h.van.der.putten,m.h.lamers}@liacs.leidenuniv.nl

Abstract. The aim of this study is to find to what extent computers can assist humans in the creative process of writing titles, using psychological tests for creativity that are typically used for humans only. To this end, a computer tool was designed that recommends new titles to users, based on knowledge generated from a pre-built corpus. This paper gives a description of both the development of the system as well as tests applied to the participants, derived from classical psychological tests for human creativity. A total of 89 participants divided in two groups completed two tasks which consisted of generating titles for paintings. One group was allowed to use a template-based system for generating titles, the other group did not use any tools. The results of the experiments show higher creativity scores for the combination of participants augmented by a computational creativity tool.

1 Introduction

Computers are everywhere. At work and at home, people use computer programs in order to draft documents, data sheets and presentations. The programs they use are tested and have a high level of reliability. They have helped professionals become better at their work: they improved their efficiency and accuracy for example.

In 1843, Ada Lovelace already envisioned that computers would assist people with scientific calculations. But she also envisioned that computers would be able to assist humans in creating artwork [14]. And indeed, tools have been developed to assist artists in their creative endeavors. Verbasizer for example is a computer program used by the musician David Bowie to help him in the song writing process [13]. Some tools in the field of computational creativity look to enhance human creativity, while others seek to surpass it, like software capable of generating music in style of Mozart or Bach [4]. Those tools that seek to assist humans, resemble what Lovelace foresaw.

In this paper, we will especially focus on these "Lovelace machines". In early computational intelligence research, autonomous artificial intelligence was contrasted with a more collaborative or even symbiotic relationship between man and machine [10], popularized by Engelbart as intelligence augmentation [5]. By analogy, Lovelace machines can be seen as tools for augmented creativity.

© Springer International Publishing AG 2017
J. Correia et al. (Eds.): EvoMUSART 2017, LNCS 10198, pp. 262–274, 2017.
DOI: 10.1007/978-3-319-55750-2_18

Computational theorists still debate about what can be considered as creative. For example Zhu et al. [17] define creativity as 'the ability to extrapolate beyond existing ideas, rules, patterns, interpretations, etc. and to generate meaningful new ones'. In the same paper parameters are defined for what the output of a creativity tool should look like. Firstly, the item to be measured has to be different from other existing items. If one can model existing items with a statistical model, the new item should be an "outlier". Secondly, the item has to be meaningful. An item that consists of random noise might well be an outlier, but it is not of interest.

A tool that meets both criteria is Titular, a computational creativity tool that suggests novel titles for songs, based on template song title structures learned from a large corpus [11]. A detailed description of a reproduction of this tool can be found in Sect. 2, Materials. The linguistic tool was compared to other title generating systems using Amazon Mechanical Turk. In this test, Titular's automatically generated titles scored relatively well compared to other systems on the first criterion of the output for a creativity tool. However, this experiment did not evaluate to what extent the tool actually enhanced human creativity. In an additional test, participants in a songwriter contest experimented with Titular as a support in writing and naming songs. This experiment generated some positive anecdotal evidence; the actual creative output of the participants was not assessed however [11]. Although the combined results of the tests look promising, they leave space to assess to what extent the system actually enhances human creativity.

As such, the interest of this research is not to measure the creative output of the tool nor human alone but to quantify the creativity of the combination of human and tool, leveraging generally accepted psychological tests for assessing human creativity. The evaluation of the creativity of the participants was made using one of the methods described by J.P. Guilford [8]. In Guilford's work participants were given a plot story and then were asked to provide original titles for the plot story. This approach matches with the purpose of the tool itself. What type of work (song, story or painting) is used in the test does not influence the creative thinking of the participants, as shown in a psychological study on creativity where participants were asked to give titles for paintings instead of titles for a story [3]. The score of the titles generated by the participants was calculated according to Torrance's test for creativity using two of the four scales: fluency and originality [12].

This paper proposes an experiment to evaluate the creativity of two groups of participants to generate titles for two paintings, leveraging standard psychological tests for human creativity. One group was allowed to use the tool and the other was not. Afterwards, the creative performance of each of the individuals was evaluated and the average creative performance of the group that had the tool at its disposal was compared to the creativity of the group that did not. For the purpose of the experiments we also implemented our own version of Titular, called The Title Generator, and extended it from a system to generate titles 'one off' (artificial creativity) to a system with interactive capabilities to assist a user's creative process (augmented creativity). The results showed that the group of participants using the tool demonstrated a significant increase in both fluency and originality. In addition this also proves that a creativity tool developed for one modality (coming up with new songs) can be useful across other modalities (naming a painting).

Note that our aim was not necessarily to create a tool that was better than other tools, as discussed the goal was more to provide an explorative example of how an augmented creativity system can be tested in a structured manner. Even a relatively simple experiment like this immediately raises interesting questions and discussion points. Should the creativity system be judged purely by its output? If that were true, does it matter to differentiate between different drivers of creativity, such as 'true' creativity versus motivation or engagement? Is it useful to run tests without checking for meaningfulness or quality of the ideas? In the experiments we will make some simplifying assumptions and justify these, and in the discussion we will further revisit some of these interesting questions. Again, here we do not claim to have the right or the best answers, or even to answer or even raise all important questions. What we do want to argue is that by running formal evaluation experiments, embedded in human psychology research, questions like these can be researched in a structured manner.

The remainder of the paper is structured as follows. The next section explains how The Title Generator was constructed building from the original idea from Titular, and then extended to turn it into an interactive tool. Section 3 introduces the experimental set up, results of the experiments can be found in Sect. 4, and the paper ends with the discussion and conclusion in Sects. 5 and 6.

2   Materials

The first part of this section describes how the Title Generator was built that was used in the experiments. The second part provides details on the addition of interactive extensions to the program, which were not available in Titular, the system it was based on.

2.1   Title Generator

The Title Generator was built according to a similar methodology used for Titular, an algorithm that can automatically generate song titles, based on knowledge extracted from a song title corpus [11]. The Title Generator tool uses a template based approach [15] to learn the syntactical structures required to create titles. For this study, two functionalities were added to give users more control over the tool and the output; these functionalities will be justified later in this paper. As the source code of the original Titular program could not be modified, which was necessary to add the new functions, a new version of the program was developed from scratch. In the following paragraphs, the steps to develop the Title Generator will be described.

The dataset used to construct The Title Generator was The Million Song Data Set [1]. This data set consists of 1,000,000 titles of songs of different genres, artists and languages. First, the dataset was filtered for English titles and duplicate titles and words that were not part of the titles of the tracks (such as 'featuring', 'live', 'mixed by', etc.) were eliminated. What remained was a final list of 138,257 tracks that served as a base to extract the templates and the words to create new titles.

strangers in the night → *('NNS', 'in',*
'the', 'NN')
wicked woman → *('JJ', 'NN')*

VBG, *"['flattering',   'forgiving',*
'tuning', 'songwriting', 'bloodsucking',
'shapeshifting', 'spouting', 'arching',
'putrefying',   'waiting',   'spearing',
'deceiving',',   'asleeping',   'nutting',
'starring',   'ainging',   'dampening',
'examining', ... 'bugging', ...]"

Fig. 1. Titles converted to templates using POS-tags.

Fig. 2. Word associated with POS-tag **VBG**.

Subsequently, the templates of the titles were obtained by replacing every word in the real titles in the list with their respective Part-Of-Speech tag (POS-tag) using the Natural Language Toolkit (NLTK) [2]. A full list of all the tags is available in the NLTK documentation. Examples of the tag definitions used are noun-plural (NNS), noun-singular (NN) and adjective (JJ), see Fig. 1. It is important to mention that not all the words were replaced: words such as conjunctions, determiners, prepositions and pronouns remain the same. As a result of this process, a new list with 45,844 templates was generated, including a registration of how often each template occurs. The list of possible templates is a lot shorter than the original track list as many templates follow the same structure.

As a next step, the words in the titles were grouped according to their corresponding POS-tag. This resulted in a list of 20 POS-tags that each are associated with different types of nouns, verbs, adverbs and adjectives. Figure 2 shows a few words associated with the POS-tag VBG (verb, gerund/present participle). The words and their relative frequency were stored in a different file which contains 28,252 different words. Words with a higher occurrence have a higher chance to be selected by the Title Generator. The same is the case for the list of templates mentioned earlier.

Finally, to produce new context free grammar titles [9], the Title Generator selects five templates from the templates list and replaces each POS-tag in the templates with a word associated with that tag from the word list. Figure 3 shows a few examples of the generated titles.

('NN', 'these', 'NNS')
→ *make these robotics*
('we', 'VBD', 'the', 'NN', 'NN')
→ *we loved the Alcatraz man*
('if', 'I', 'could', 'VB', 'the', 'NNS')
→ *if I could heal the spirits*
('like', 'NN', 'in', 'the', 'NNS')
→ *like isle in the hands*

('like', 'NN', 'in', 'the', 'NNS')
->*like turn in the plants*
->*like fox in the thieves*
->*like water in the clichés*
->*like spuce in the years*

Fig. 3. New titles generated by the program replacing POS-tags.

Fig. 4. Different words fill a template chosen by the user.

love VBP
→ *little love*
['RB', 'VPB']
→ *the firebird that you love*   faces in the rhythm
['the', 'NN', 'that', 'you', 'VBP']   ('NNS', 'in', 'the', 'NN')
→ *you love me to steal*
['you', 'VBP', 'me', 'to', 'VB']   → sings in the rhythm
→ *I love a life*   → moves in the rhythm
['i', 'VBP', 'a', 'NN']   → shepherds in the rhythm
→ *mending love*   → things in the rhythm
['VBG', 'VBP']   → stones in the rhythm

Fig. 5. A selected word displayed in different   **Fig. 6.** Extension that allows user to keep a part
templates   of the generated title while the rest is replaced.

2.2 Interactive Extensions to the Title Generator

Three new functionalities were added to the Title Generator that are not included in Titular to give the user more control over the output of the program and to provide the user with more meaningful and familiar titles. The added functionalities allow the user to keep words and obtain different suggestions for a given template, using these functionalities recursively until the desired output is achieved.

The first of the functionalities (see Fig. 4) provides the user with the option to click on a template of his or her liking. After selecting the template, different words fill the template. This allows the user to explore all the possibilities of a given template while keeping a structure that the user can distinguish and interpret easily. The second functionality (illustrated by Fig. 5) gives the user the possibility to click on a word in the suggested titles in order to obtain new templates using that word. This option aims to inspire users to come up with more title structures as well as to discover new associations with other words. The third and last functionality added to the Title Generator is the ability to change or maintain certain words in the suggested template while keeping the same template (see Fig. 6). This allows the user to obtain titles with partial changes, only giving new suggestions for a specific part of the template. The objective of these extensions was to make the tool more interactive so that the output of the program would approximate the preference of the user more.

3 Experiments

This study aims to assess to what extent the Title Generator enhances human creativity. Therefore, an experiment was conducted to evaluate whether the tool helps users to be more creative in terms of quantity and originality.

3.1 Writing Tasks

In essence, the Title Generator should enhance the creative writing capacity of its users, regardless of modality of the narrative. In order to test whether it actually does so, the tool was incorporated in a writing task which is based on examples from studies that measure human creativity in other domains than songwriting. Most notably a study in which participants had to write down any number of titles for a short story that was given to them [6] and a study where they were asked to give one title to each one of four images [3]. For the purpose of this research, the set-up of the two studies was combined.

The experiment was conducted with a total of 89 individuals who were asked to write down titles for two paintings. In total, 39 of the participants had the Title Generator at their disposal, the rest did not. Tests were taken one by one and to minimize the influence of what time it was we alternated between using the tool and not using the tool for every new participant. The paintings that were selected had to give the participants of the study some guidance, but should also allow them to come up with their own creative interpretations of the image. Therefore, two paintings were chosen that are figurative (they represent clear objects and people) but painted in an expressionist style. Furthermore, in order to prevent participants having seen the paintings before or associating their titles with the artist rather than with the painting, a painter was selected who is relatively unknown in the country where the experiment was run. The paintings that are used are both from the hand of the Ecuadorian painter Oswaldo Guayasamin. The first painting, "Pareja en Silencio" (in English: couple in silence), portrays a couple hugging each other (Fig. 7). The second one, "El Guitarrista" (the guitar player), portrays a guitar player (Fig. 8). So this is an experimental evaluation for augmented creativity, where the tool was created on one modality (song texts) and tested on another (naming pictures).

Fig. 7. Painting used in task 1

Fig. 8. Painting used in task 2

Participants were provided with a sheet of A4 paper containing a reduced size but color image of "Pareja en Silencio" (task 1) on the front side and "El Guitarrista" (task 2) on the back side. Both images were followed by the instruction "please list all the possible titles you can think of for this painting" and contained a number of lines for the participants to write down the titles. The exercise was accompanied by the verbal instructions: "take all the time you need and when you are ready please return the sheet to the interviewer".

Actually, it was a conscious decision not to set any time limits and let participants use all the time they need. One of the reasons was that we didn't want to give users of the tool any 'unfair' advantage in terms of simply being able to complete a task quicker. Also we wanted to investigate whether the tool actually helped users to remain motivated to keep working on the task, which would have been harder to measure with a fixed time limit.

One may argue that creating more titles is a function of spending more time, but likewise one may argue that the tool motivates people to work longer on the task, by providing more support, stimulation and reducing any potential fear hampering fluency. Would we really be testing the creativity of the individual? And would be testing 'true' creativity? The point is that in our research we are interested in the creativity of the symbiotic system of man and machine, our goal is not to assess the creativity of the individual. Also the notion of 'true' creativity is questionable, or very much intertwined with other factors driving output at least, such as motivation, engagement and removing creative blockers such as fear. By analogy, creativity is more transpiration than inspiration, and throughout history artists have used a variety of artificial tools to let creativity flow, from incorporating randomness to drugs.

3.2 Group 1: Control

The objective of the first study was to establish a baseline of 'human' creativity. Hence, the participants in this group did not have the Title Generator to their disposal nor any other tools. This group served as a control group for the experiment. Their results were compared to those of group 2.

3.3 Group 2: Augmented Creativity

To measure the difference in creativity with the control group, the participants of group 2 were allowed to use the Title Generator as inspiration to write down titles for the two images. Users were allowed to explore the capabilities of the Title Generator to produce titles according to their preferences. Participants were permitted to use the generated titles in two possible ways. Either make an exact copy of the title given by the computer program or modify the given title by adding or replacing words of their own inspiration. To immediately verify the usefulness of the tool after the task, participants were asked to provide feedback about the Title Generator, asking whether the Title Generator helped coming up with more respectively better titles, plus an open question on in what ways the tool could help.

3.4   Evaluation of Creative Performance

We purposely want to align with the most generally accepted tests for human creativity and the Torrance Tests of Creative Thinking are a key example of seminal work in human creativity testing [12]. It builds on the foundational work on creativity and divergent thinking by Guilford [7, 8]. The key dimensions in these tests are fluency, flexibility, originality and elaboration. For the purpose of our study we focus on two dimensions, fluency and originality [12].

In the real world fluency may be associated with a range of skills beyond productivity such as eloquence, apparent effortlessness in idea generation, ability to generate many ideas under very specific constraints, ability to generate a continuous flow of ideas over time etc. However in the context of Torrance, fluency is restricted to a very specific metric: the total number of ideas generated. In the context of our task we simply operationalized this to the number of titles each participant gave for each painting. The participants received one point for every given title.

Torrance also states that the ideas should be interpretable, meaningful and relevant. However in our experiments we explicitly decided not to take this into account. Part of the reason not to do this is that the decision whether a title for a painting is meaningful is very subjective; there shouldn't even have to be a connection between what is figuratively displayed and the title. Of course given its subjective nature, you can then have panels of 'experts' evaluate the titles, similar to the use of Mechanical Turk in the original Titular experiment [11], but for modern art it will be questionable to define a specific audience as the experts here – the general audience would be as good as artists or art critics. Practical constraints were obviously also a reason, but given that for both experimental conditions we did not check for meaningfulness or relevance, in our view insight can still be derived from the experiments. In this case it was important not to give participants any detailed information on how their results were going to be evaluated – specifically that we were not going to score for title meaningfulness or quality.

The originality was measured by establishing how often the title was mentioned by the other participants of the same group and for the same painting. If a title was given by less than 5% of the participants, a score of 1 was given to the title. If the title was given by less or equal than 1% of the participants 2 points were given for each title. A score of 0 was given if more or equal than 5% of the participants came up with the same title. To obtain a clean measure of originality for each participant, the points obtained were divided by the total fluency of each participant. Again, it was considered whether the scores for originality should be influenced by their quality too, as an important criterion of creativity is that output should be meaningful. However, evaluating quality is a very subjective task and requires a separate research to assess whether the titles present high or low quality; hence, it was decided not to include it in this work.

4   Results

This section provides an overview of the results focusing on fluency and originality.

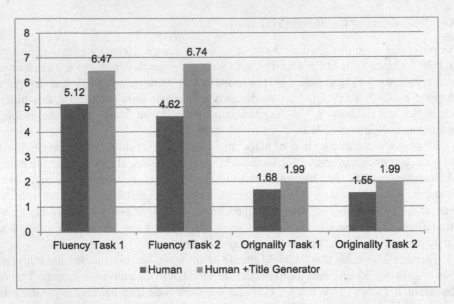

Fig. 9. Flexibility and originality results for the two conditions and tasks

4.1 Fluency

As can be seen from the experimental results in Fig. 9, using The Title Generator improved results in terms of fluency (number of titles) for both tasks. These results are also proven to be statistical significant at the 0.05 level through a T test, $p < 0.044$ for Task 1 and $p < 0.003$ for Task 2.

4.2 Originality

The originality is based on how often each title was mentioned for the same painting by other participants in the group. Also for this measure using The Title Generator improves results (Fig. 9). Again these results are statistically significant at the 0.05 level, with $p < 1.79E\text{-}08$ (Task 1) and $p < 9.49E\text{-}09$ (Task 2).

The most common responses given by participants for the various tasks and experimental conditions are shown in Table 1. For the human only group, in both tasks the high occurrence of the same titles reduces originality, as can also be seen in Fig. 9. Table 1 shows that for the group using The Title Generator there is almost no repetition of titles in both tasks. In both studies, only one title was mentioned by more than one participant.

Table 1. Results of survey questions for group that used The Title Generator

Task 1				Task 2			
Human		Human + Title Generator		Human		Human + Title Generator	
Title	%	Title	%	Title	%	Title	%
love	18%	the hug	6%	music	6%	guitar man	6%
embrace	14%	glorious love	1%	the guitar	14%	far from home	1%
big hands	10%	wolves of night	1%	love for music	6%	for whom the life is silent	1%
the kiss	10%	there is no day like you	1%	lonely	6%	blues of the outcast	1%
the hug	10%	love the cool existance	1%	guitar hero	6%	wild repetition	1%

As explained in Sect. 3.4, participants that used The Title Generator were also asked to provide feedback after they completed both tests. Table 2 shows the results of this survey.

Table 2. Percentage of participants mentioning a specific title for the various tasks and conditions

Helped produce more titles	82%
Helped produce better titles	59%
Used words from generator	66%
Used syntax from generator	35%
Other: gave me inspiration	41%

5 Discussion

The participants assisted by the creativity tool had an increase in the number of their responses. Their fluency is 26% greater for task 1 and 46% for task 2. This is also reflected by the feedback: 82% of the participants in group 2 indicated that the Title Generator helped them to come up with more titles. The data suggests that the Title Generator also helped the participants to come up with more original titles: the group that used the Title Generator was on average 18% and 28% more original for task 1 and task 2 respectively than those that did not use the tool. The survey shows that 58% of the participants using the Title Generator found that the tool helped them to think of better titles. Although there was a significant difference between the group that used the Title Generator and the one that did not for both fluency and originality, the tool seemed especially helpful to enhance the participants' fluency.

As for how the Title Generator assisted the participants: those that said that the tool helped them to come up with more or better titles indicated they used words displayed by the generator or used part of the titles. Under the possibility to indicate other reasons (open question), a large number of participants (41%) indicated that the tool somehow inspired them (it helped them think out of the box, made them be more creative, etc.). The participants were not given a fixed time to finish both tasks as it was deemed important that the participants would concentrate on the creative process, rather than focusing on finishing on time. Whilst not formally measured, it was observed that the participants of the second group took more time to finish the task; the fact that the tasks entertained them longer seems to support their claim that they did not run out of inspiration as soon as the participants in the first group. This observation could also be attributed to the 'novelty bias': participants might have spent more time on the task because of the novelty of the tool, and therefore showed higher scores on fluency. However, if the tool helps people to stay engaged in a creative task that is a merit in itself as we were not interested in efficiency, but purely in the output of the task. More experiments are needed to test which of the two assumptions holds true. Note also to repeat, we are not claiming that humans become more creative using the tool. We simply state that the symbiotic combination of both human and tool create better results – again in our experiments we are mainly interested in the output.

The majority of the titles generated by the first group can be seen as more literal descriptions of the painting. The titles refer to the figures on the painting ("big hands", "music") or to concepts that were commonly seen as being related to the painting ("love", "lonely"). The group that had to their disposal the Title Generator came up with more figurative titles. Next to more literal descriptions such as "the hug", free interpretations such as "wolves of night" were generated by the second group; instead of the more common concept of "lonely", the more freely associated "far from home" was used for example. Nonetheless, the most 'obvious' titles for both paintings ("the hug" for the first painting and "the guitar" for the second painting) were still the most mentioned responses in group 2 as well. This seems to indicate that the participants did not just merely copy the suggestions of the Title Generator but also relied on their own creativity. Also just from informal inspection it was interesting to see that for most of the titles at least some kind of association with the painting can be found, even if remote. Clearly participants made an effort to come up with 'meaningful' song titles.

A creative dimension that we considered but didn't include because of scope constraints and subjectivity in evaluation is elaboration. A very crude way to measure this in this task is simply the number of words used in the title. Whilst not formally measured on all titles, Table 1 already seems to indicate that the group using The Title Generator provides more elaborate answers, so this could be an interesting direction for follow up analysis or research.

6   Conclusion and Future Work

Ada Lovelace wrote about the possible collaboration between humans and technology, and how both science and art can benefit from working with computers. The aim of this research was to assess whether The Title Generator, a tool developed to assist artists, actually enhances creative output. An adapted tool, the Title Generator, was used in experiments to answer the research question. The results of the experiments seem to suggest that in this case, indeed, the collaboration between computer and man enhances human creativity. The results showed that participants in a creative writing test present more fluency and originality when they use the Title Generator. Four t-tests furthermore showed that the difference between the group that used the Title Generator and the one that did not, was significant.

The difference in fluency was most significant between the two groups. Participants indicated that both the words as well as the templates of the tool gave them inspiration to think of new titles. Although less significant, the group that used the tool was also clearly most original. With respect to the originality, the participants that used the Title Generator were able to come up with more figurative titles for the paintings, as opposed to the more literal descriptions of the participants in the other group. This seems to suggest that the Title Generator helped them to think more out of the box and generate titles that went beyond the obvious. The aspect of generating (meaningful) outliers that users are able to interpret is an important characteristic for creativity tools.

For this study new adaptations were made to the original Titular program in order to give the user more control over the program, and enhance interaction. In further studies a measurement of the usability of these extensions could be an interesting subject of research. Furthermore, those that used the tool and completed a survey on if and how it helped them were asked for possible areas of improvement. Several participants suggested that a possibility to obtain titles based on a seed-word or other personalized data as an input to the program could be a valuable extension. Such an addition raises interesting questions for follow up research; for example, what will be the difference in creativity using a personalized vs a non-personalized title generator?

References

1. Bertin-Mahieux, T., Ellis, D.P., Whitman, B., Lamere, P.: The million song dataset. In: ISMIR 2011: Proceedings of the 12th International Society for Music Information Retrieval Conference (2011)
2. Bird, S.: NLTK: the natural language toolkit. In: Proceedings of the COLING/ACL on Interactive Presentation Sessions. Association for Computational Linguistics (2006)
3. Chen, C.: Boundless creativity: evidence for the domain generality of individual differences in creativity. J. Creative Behav. **40**, 179–199 (2006)
4. Cope, D.: Experiments in music intelligence. In: Proceedings of the International Computer Music Conference. Computer Music Assn., San Francisco (1987)
5. Engelbart, D.: Augmenting Human Intellect: A Conceptual Framework. SRI Summary Report AFOSR-3223, October 1962
6. Eisenberger, R., Rhoades, L.: Incremental effects of reward on creativity. J. Pers. Soc. Psychol. **81**, 728–741 (2001)
7. Guilford, J.P.: Creativity. Am. Psychol. **5**(9), 444–454 (1950)
8. Guilford, J.P.: The Nature of Human Intelligence. McGraw-Hill, New York (1967)
9. Lari, K., Young, S.J.: The estimation of stochastic context-free grammars using the inside-outside algorithm. In: Computer Speech and Language (1990)
10. Licklider, J.C.R.: Man computer symbiosis. IRE Transactions on Human Factors in Electronics, HFE-1, 4–11 (1960)
11. Settles, B., 2010. Computational creativity tools for songwriters. In: Proceedings of the NAACL HLT 2010 Second Workshop on Computational Approaches to Linguistic Creativity. Association for Computational Linguistics

12. Torrance, E.P.: The Torrance Tests of Creative Thinking-Norms-Technical Manual Research Edition-Verbal Tests, Forms A and B-Figural Tests, Forms A and B. Personnel Press, Princeton (1966)
13. Thompson, D.: Hallo Spaceboy: The Rebirth of David Bowie. ECW Press, Toronto (2007)
14. Toole, B.: Poetical science. Byron J. **15**, 55–65 (1987)
15. Van Deemter, K., Krahmer, E., Theune, M.: Real versus template-based natural language generation: a false opposition? In: Computational Linguistics (2005)
16. Wallach, M.A., Kogan, N.: Modes of thinking in young children: a study of the creativity-intelligence distinction Holt. Rinehart & Winston, New York (1965)
17. Zhu, X., Xu, Z., Khot, T.: How creative is your writing? In: Proceedings of the Workshop on Computational Approaches to Linguistic Creativity (CALC) (2009)

Play it Again: Evolved Audio Effects and Synthesizer Programming

Benjamin D. Smith[✉]

Indiana University-Purdue University-Indianapolis, Indianapolis, USA
bds6@iupui.edu

Abstract. Automatic programming of sound synthesizers and audio devices to match a given, desired sound is examined and a Genetic Algorithm (GA) that functions independent of specific synthesis techniques is proposed. Most work in this area has focused on one synthesis model or synthesizer, designing the GA and tuning the operator parameters to obtain optimal results. The scope of such inquiries has been limited by available computing power, however current software (Ableton Live, herein) and commercially available hardware is shown to quickly find accurate solutions, promising a practical application for music creators. Both software synthesizers and audio effects processors are examined, showing a wide range of performance times (from seconds to hours) and solution accuracy, based on particularities of the target devices. Random oscillators, phase synchronizing, and filters over empty frequency ranges are identified as primary challenges for GA based optimization.

Keywords: Sound synthesis · Machine learning · Adaptive genetic algorithms · Audio effects

1 Introduction

Programming modern professional grade software synthesizers to discover or recreate desirable sounds requires extensive expertise and a time intensive process. Popular commercial products, such as Massive, FM8, Sylenth, and Ableton Live's native devices offer the user many hundreds of parameters resulting in countless potential combinations. Tuning an audio effects device or plug-in (such as compression, reverb, distortion, etc.), many offering similar complexity and possibilities, presents the user with the same challenge. As a partial solution most software synthesizers have extensive libraries of parameter presets to help novice users, and there is a distinct market for additional libraries targeting specific aesthetics. However exploring these libraries (which may contain thousands of presets) to locate a desirable sound remains a daunting process. If the user has a specific target sound in mind, the recreation process may be time prohibitive and highly disruptive to the creative work flow of composing and producing.

Automating the process of programming synthesizer and audio effects parameters to reproduce a target sound saves studio time and allows creators to

© Springer International Publishing AG 2017
J. Correia et al. (Eds.): EvoMUSART 2017, LNCS 10198, pp. 275–288, 2017.
DOI: 10.1007/978-3-319-55750-2_19

focus on the creative process. Ideally this computational process will function independent of the specifics of individual software synths or other audio effects processors, allowing users to work with their favorite devices, and it should function regardless of the origin of the target sound, programming the synth or audio effects processor to reproduce the target sound as accurately as possible (given the device's capabilities and limits).

This work examines the use of Genetic Algorithms (GA) to develop presets for devices hosted in Ableton Live with the goal of being instrument independent, and aiming to flexibly target any VST or Live native instrument or device. While other research has proven largely successful in supporting individual instruments [9,14,16], this work's primary challenge is designing a flexible optimization algorithm capable of handling the variety and number of parameters that may be available (simple FM synthesis models, or effects devices, often have a few parameters while commercial implementations, such as Native Instruments Massive and FM8, have over a thousand parameters), without using privileged knowledge about the synthesis model in question or specific parameter assignments.

After background context, this paper presents the GA design and considerations in Sect. 3. The implementation is presented in Sect. 4, followed by evaluations in Sect. 5. Conclusions and future work are found in Sect. 6.

2 Background

The objective of reverse engineering synthesis parameters has been examined by a number of researchers, looking at different synthesis techniques and synthesizers, going back as far as the early 1990s [2–4,13]. Most of these successful approaches employ evolutionary computing and GAs to automatically recreate specific sounds. The most popular synthesis technique seems to be FM synthesis [2–5,7,9,13,14,16], perhaps due to the diversity of possible sounds while minimizing system complexity, enabling effective computation within hardware and CPU constraints.

More recent work has targeted specific FM software synths [16] and modular software synths [9,14]. The former case looks at relatively simple software synthesizers, comparing the GA's ability to match timbre with that of human programmers, finding that the computer finds more accurate results in significantly less time. In the later cases the OP-1 synthesizer is employed, involving a much more complex process, due to the modular and non-deterministic aspects of this synthesizer.

Employing spectral analysis to inform fitness determination is used from the first examples, although the early cases [2,4] employ fixed filter bank analysis techniques, moving to FFT analysis of the harmonics in a synthesized signal. Full FFT analysis is employed in recent work [14], along with amplitude envelope characterization for higher GA discrimination. An alternative [10] is to leverage Discrete Fourier Transforms, filling critical frequency bands in a custom perceptual model which aims to reflect human audition characteristics.

Other work [8,16] employs Mel-Frequency Cepstrum Coefficient (MFCC) analysis to capture timbrel information and inform fitness decisions. MFCCs are standard in natural language processing and are shown to work effectively in capturing both spectral and amplitude aspects of time varying sounds. This also has the advantage of providing a degree of fundamental frequency independence, where MFCC data will be close for similar spectra regardless of pitch (proving analogous to human perception which can identify an instrument independently of which notes are being sounded).

Attempting to design the whole synthesizer based on a given target sound is also being examined [1,8], but presents a much more complicated domain compared to programming a given synthesizer. Optimizing the synthesis model or discovering completely new synthesis algorithms to accurately reproduce a given target sound is showing promise but has significant research ahead.

3   Design

3.1   Genetic Model

This work assumes the reader is familiar with the basics of GAs and presents the specifics of this particular design only. Selection, crossover, and mutation operators are all employed and presented in the following sections. Discussion of adaptive modifications and selection pressure considerations follow.

Each chromosome is a vector of device parameters, represented as floating point numbers, which are non time varying (i.e. the device parameters are set and remain fixed during sound generation and fitness calculation). Each individual is applied to the target device and the resulting audio is analyzed for GA selection and reproduction.

Selection. Timbre matching is the primary problem being faced and the fitness function takes a standard approach [8,16], using the first 13 cepstra of a windowed MFCC analysis. This has the advantage of representing the spectral energy in a consistent and comparable way, capturing both envelope (amplitude) and tone generator (spectral weighting) characteristics.

Fitness is based on the sum squared error (Err) between the target and candidate MFCC data (Eq. 1, where t is the time sequence target MFCC data, X is the individual solution MFCC data, W is the number of windows). Window and hop sizes from 10 to 90 ms were empirically tested and found to have negligible impact on solution quality (data in this paper uses a 20 ms window with 10 ms overlap, which is common to many examples in the literature). Normalization is not applied at the audio level as this would preclude optimization for volume, gain, and dynamics parameters. In this case matching based on loudness and amplitude envelopes is desirable.

$$Err(t,x) = \frac{\sum_{i=1}^{W} \sqrt{(\sum_{j=1}^{N_x}(t_{i,j} - x_{i,j})^2}}{W} \tag{1}$$

Unlike FFT or simple spectral-mean based fitness functions, measuring Euclidean distance between cepstra is problematic due to the widely varying ranges of each order (i.e. the first cepstra has a range of $[-96,96]$ and the twelfth might be $[-0.001,0.001]$). Unaddressed this imbalance incidentally weights the orders, diminishing the potential for higher order cepstra to contribute to the selection process. To provide balance to the MFCC vectors the standard deviation of each cepstra over the entire dataset is tracked and used for normalization. While this analysis and fitness evaluation has proven adequate, and is based on findings in other sources, a detailed comparison of fitness variants (using other spectral analysis methods) and additions would be required to prove the most efficient and effective solution.

Roulette-wheel selection is employed, and in-breeding (where an individual is chosen to reproduce with itself) is prevented. A small percentage (4–6%) of *elite* individuals are carried through to the next generation unchanged (individuals with the highest fitness are allowed to survive, in addition to participating in breeding the next generation).

Crossover. Three crossover models were applied and examined for this system. Since the chromosomal elements (the device parameters) are independent, their cardinal order is probably coincidental and relationships between neighbor elements may or may not be significant (depending on the software designer's choices in parameter ordering). This would seem to indicate Uniform Crossover as optimal, however alternatively, the target device's parameters may be grouped in a meaningful way (such as in groups of filter bank parameters, oscillator parameters, amplitude envelope parameters all appearing sequentially) and using single point, or dual point crossover may be more effective in maintaining these relationships. These three crossover models (single-, dual-, and uniform crossover) were all examined and the results were inconclusive: all three arrived at equally satisfactory winning individuals, and any variances in efficiency could not be attributed to the crossover model alone.

Mutation. Several mutation models were tested for impact on efficiency. Uniform replacement and scaled Gaussian addition with both range clamping and range fold-over were implemented and tested. Gaussian addition (using a normal distribution with center $= 0$ and standard deviation $= 1.0$), clamped to $[0, 1]$ has the effect of increased probability for parameters to sit at the edges of their range, with diminished exploration of the middle ground. Uniform replacement functions better, but this model has trouble, after locating a near optimal solution, of resolving the final small adjustments to find the ideal match. Gaussian addition where the parameter is folded over the range $[0, 1]$ in ping-pong fashion appears to provide the advantages of both the previous two models, although empirically conclusive data has yet to be produced. The lack of efficiency in making the final adjustments may further indicate the need for an adaptive Gaussian distribution to encourage mutations of finer granularity as convergence is being approached.

Termination. Convergance of the population is either taken as a solution making a > 99.99% fitness match, or observing a < 0.001 change in the magnitude of the winning fitness over 200 generations. Both conditions are necessary due to the unknown particularities of the target device. Many digital synthesizers and audio effects employ random oscillators and noise generators, which will render a 100% match virtually impossible to achieve. Further, while the effectiveness of recreating sounds that originated on the same synth or effect is an initial concern, attempting to get a best approximation of target sounds from other sources (such as analog synths, or non-synthetic audio) will have further convergence difficulties.

3.2 Adaptive Genetic Algorithm

Due to the potential range of audio processes that may be targeted, determining fixed crossover and mutation coefficients is problematic. The examination of a single synthesizer or synthesis technique can empirically test these rates and set them for optimal convergence. However, in order to be independent of a specific synthesis engine, and support any number of parameters in the model (typically 6 to 200), an adaptive crossover and mutation algorithm is employed [6,12].

The objective of the adaptive model is to maintain adequate population diversity, preventing local minima solutions, and improve movement through exploration and exploitation phases. Here, population diversity is calculated simply using the maximum F_{max}, mean $\bar{F}$, and individual parent fitness F' during an evaluation stage [12]. The adaptive probabilities of crossover P_c and mutation P_m are:

$$P_c = K_c \left(\frac{F_{max} - F'}{F_{max} - \bar{F}} \right) \tag{2}$$

$$P_m = K_m \left(\frac{F_{max} - F'}{F_{max} - \bar{F}} \right) \tag{3}$$

Both probabilities are constrained in the range $[0,1]$ and constants are set to allow maximum exploration $K_c = 1, K_m = 0.5$ when population diversity declines.

In practice this adaptive approach results in overpopulation of copies of the best solutions, due to $P_m = 0$ when $F' = F_{max}$, preventing effective solution space exploration. This is partly solved by imposing a minimum P_m to disturb the winning solutions [12], forcing a range of $[0.005, 0.5]$ on the mutation coefficient. Observations showed crossover being applied to around 10% of the parents in each generation, again limiting exploration and exploitation. Using the mean of both parent's fitness values (to encourage crossover of the most fit individuals) brings this rate up to $\approx 30\%$

$$F'' = \frac{F_1' + F_2'}{2} \tag{4}$$

$$P_c = K_c \left(\frac{F_{max} - F''}{F_{max} - \bar{F}} \right) \tag{5}$$

3.3   Selection Pressure

The squared error fitness function frequently fails to make progress towards convergence, based on observational data. This appears to be due to inadequate selection pressure (i.e. the gradient across the population is too flat). Thus the following selection pressure S constant, and coefficient P_s are employed to calculate the fitness of each individual Fit_x:

$$P_s = S\left(\frac{\bar{F}}{F_{max}}\right) \qquad (6)$$

$$Fit_x = \left(\frac{1}{Err_x + 1}\right)^{P_s} \qquad (7)$$

The impact of various values of S is examined in the Evaluation Sect. 5, below.

4   Implementation

The system is implemented and tested as a combination of Ableton Live sets using built-in software instruments and audio effects devices, Max for Live devices, and a Max external encompassing the GA model. Due to the nature of the software employed the model has to run in real-time (in Live). First each individual chromosome is used to configure a software instrument or processor, MIDI notes (if the target is a synthesizer) or an audio clip is played (if the target is an audio processor), and the output is recorded. An MFCC analysis is performed and the data is transmitted back to the GA for storage and comparison as fitness features (see above).

   A primary challenge facing this implementation is the timing priority afforded by Ableton Live. Due to the asynchronous connection between Max and Live the initiation of audio sample playback can occur with latencies between $0 - 75ms$. This results in some unknown amount of erroneous artifact audio at the beginning of each MFCC analysis. The successful implementation of this project required delaying sample playback to align with the signal vector rate and the hop point of the MFCC analysis (to remove leading zeros in the audio recording), and then a least-squares comparison between each chromosome's MFCC features and the target across up to 20 time offsets to locate the most probable alignment.

   The evaluation commenced by randomizing the target device's parameters and playing 2 s of notes (for synthesizers) or one of 8 pre-selected audio clips (through audio effects). The audio clips were selected to encompass a wide range of dynamic spectral content (including drum loops, complex synth tones, environmental noise, etc.). In addition to the real-time constraints, parameter transmission between Max and Live is a primary limiting factor resulting in epoch times between 5 and 50 s, depending on population and chromosome size. Reaching adequate epoch counts can thus be time prohibitive (5000 evolutions could take $\approx$70 h) with desirable population sizes (of 600–1000 for larger chromosomes) [11,15], and thus smaller devices and population sizes are employed in this initial exploration.

5   Evaluation

Evaluation of the proposed system is carried out in the following tests, examining selection pressure values and performance over different synthesizers and audio devices with different sound sources. Data for a synthesizer, a distortion effect, a three-band equalizer, a phaser, and a reverb unit is presented and discussed in Sects. 5.1 and 5.2 below. Identifying the highest degree of accuracy as well as the efficiency of the optimization process (in terms of population, epochs, and total execution time to convergence) is of primary concern.

The primary test for synthesizer and audio device programming is perceptual, proving the sonic accuracy of the arrived at solution(s). This proof can be approached with both empirical and perceptual measures. MFCC analysis is accepted as an accurate reflection of human perception and is used extensively in natural language processing for this reason, thus statistics of the fitness values are presented as a characterization of perceptual accuracy (the alternative of an extensive study with human subjects, is beyond the scope of this initial work).

The accuracy of a reverse engineered solution, in the literature, is typically shown by taking a distance measure between the converged device parameters (chromosome) and the target parameters. However, this measure must be qualified for several reasons. In a constrained space (where each dimension is confined to range [0,1]) the maximum distance is only achievable when the target is at one of the limits. If the target is in the center, the possible error (distance from target) is greatly diminished and thus even very bad solutions will appear to have low error rates. In the metrics presented herein this is accounted for by first computing the maximum possible error using the target parameter vector (of length N with parameters p):

$$E_{max} = \sqrt{\sum_{i=1}^{N}((|0.5 - p_i| + 0.5)^2)} \tag{8}$$

The parameter mismatch error is calculated as a percentage of the maximum possible for a given target.

Secondly, some parameters in a synthesizer or audio device may be part of dependent chains, such as a bypass switch which negates the effect of dependent parameters. Based on the state of the switch it is impossible for the GA to ascertain the state of the bypassed parameters. In extreme cases the target may be completely silent, allowing many parameter sets to achieve 100% fitness but have a high parameter mismatch.

Multiple solutions may be possible, presenting a third consideration. For example, many synths have both macro and micro tuning parameters (half steps and cents) which provide different methods of obtaining the same transposition, multiple gain stages which can obtain equal output levels, or many ways of obtaining the same wave form through different oscillators. These alternative solutions achieve a sonic match but evaluate badly when comparing the parameters (see Figs. 2, 3, 4 and 5).

In order to make this system generalized and abstracted from the specifics of the target audio device all parameters are represented as floating points in range [0,1]. These are then scaled appropriately when applied to the device (most commonly to MIDI control range [0,127]). However, many audio device parameters are treated as switches or selectors with incremental steps (such as bypass switches, filter type or wave form selectors, etc.). In these cases any value within the quantized range is treated equally, i.e. [0,0.5) is "off" and [0.5,1] is "on." If the GA arrives at the correct setting it may still show as being inaccurate due to this incremental quantization.

Finally, a special subset of problems faces audio processing devices wherein the source audio may not reveal the impact of the effect. For example, a filter impacting a range of frequencies that are not present in the source (such as a very low filter on high material), or dynamic effects on a static source.

While this system has been tested with numerous synths and audio devices, an evaluation of three devices is presented here: a 33 parameter synthesizer, a 20 parameter phaser effect, and a 7 parameter distortion.

5.1   Synthesizer: Electric

The Electric synthesizer is a native device in Ableton Live, intended to reproduce a wide range of electric keyboard and piano sounds. It offers 33 parameters (see Figs. 3 and 5). For evaluation purposes all target parameters were fully randomized for each run, and a 4 note pattern (Cs in 3 octaves) with different velocities for each note is employed. Two example executions (maximum and mean fitness for each generation) are shown in the following figure (Fig. 1).

As expected, the adaptive crossover and mutation rates ensure population diversity throughout the runs (as reflected by the mean and maximum fitness

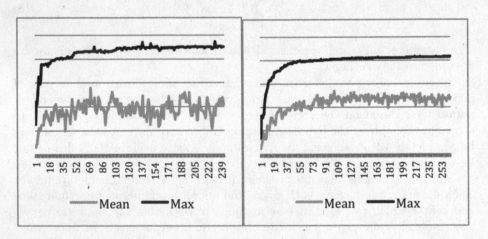

Fig. 1. Maximum and Mean fitness over epochs, for start of two sample runs (scale of [0,1]). Diversity is reflected by gap between mean fitness and maximum fitness.

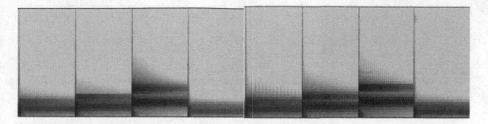

Fig. 2. Sonogram (frequency over time) of electric target (left) and best fit solution (right).

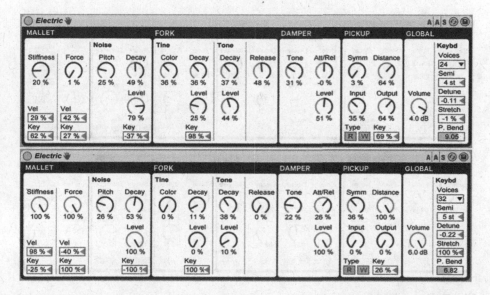

Fig. 3. Parameters for target (top) and best fit solution (bottom), resulting in spectra shown in Fig. 2.

across all epochs, Fig. 1). The sonograms for the first target and solution show a high degree of accuracy (Fig. 2), with a parameter mismatch magnitude of 2.113 (91.1% accuracy in parameter settings). However the solution arrived at, while perceptually very similar, is different from the target (see Fig. 3, Note: Mallet parameters appear to be compensated by Fork parameters in solution).

A second selected example (Figs. 4 and 5) finds a closer parameter solution with a mismatch magnitude of 1.823373 (93.14% accuracy in parameter settings). Note in both example cases the solution is played at a different pitch (see Semi and Detune parameters), yet the timbrel match is very close.

5.2 Audio Effects

Two audio effects are examined below comprising a distortion and a phaser.

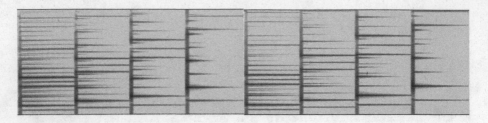

Fig. 4. Sonogram (frequency over time) of electric target (left) and best fit solution (right).

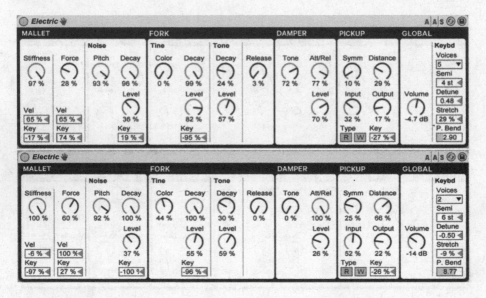

Fig. 5. Parameters for target (top) and best fit solution (bottom), resulting in spectra shown in Fig. 4.

Distortion. The Live native Overdrive implements a typical distortion effect and has 7 parameters. This system arrives at a solution in an average of 57.47 epochs (N = 109, 12 failures to converge), using a population of 128, comparing with a single input sound file. Unlike the Electric synthesizer there do not appear to be any alternative solutions: the maximum fit solutions are nearly perfect matches with the target parameter set. The failures are from extreme frequency ranges versus the sonic content of the source material (e.g. the distortion filter frequency $< 60\,hz$ on a middle C4 note).

Comparison of time-to-converge for different selection pressure coefficient values is shown in Fig. 6. Based on this data values for S around 6 appear optimal, for this device. Both higher and lower values take longer to converge, in both worst-case and average scenarios.

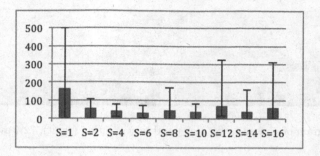

Fig. 6. Epochs to converge for values of S on the Overdrive/Distortion device. Gray bar is mean, error bars indicate maximum and minimum epochs-to-converge ($N = 109$). $S = 6$ is the best mean and maximum point.

Phaser. The Phaser device in Live is a series of all-pass filters designed to create phase shifts in the frequency spectrum of a sound. It includes an attack/release envelope in addition to a low frequency oscillator (LFO) to drive the stereotypical phase sweep effects. This device poses a challenge to the GA due to the synchronization of the LFO. The same chromosome evaluated many times produces different fitness values depending on the starting phase of the LFO (which is purely a function of time since launch of Live). However, the GA arrives at sonicly accurate results, yet they statistically compare poorly.

Fig. 7. Sonogram of example unprocessed original sound sample (synthesizer tone) used in phaser device tests.

An example is shown in Fig. 7 (the original, unphased source sample), where the highly processed target (Fig. 8, left) is answered by a best-fit solution (Fig. 8, right). Visually, and aurally, the solution appears to be a match, however it is only 72% accurate in MFCC fitness and has an 8.64% parameter mismatch (Fig. 9). Due to the inability of the GA to arrive at a 100% match a comparison is made across values of S (selection pressure coefficient) using the fitness of the best final solution (Fig. 10). The peak is around $S = 8$, close to the empirical evidence from the Distortion data, above.

6   Improvements

Examining convergence failures informs further improvements to the system, leading to the identification of several contexts that provide additional

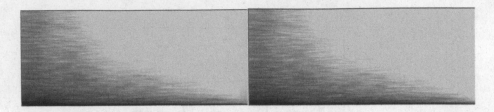

Fig. 8. Sonogram of phased target (left) and best solution (right), 72% average MFCC match.

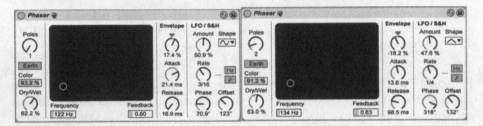

Fig. 9. Phaser device parameters for target (left) and best solution (right), 8.64% average parameter mismatch.

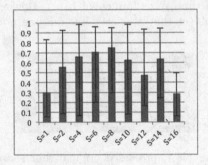

Fig. 10. Final fitness for Phaser with different selection pressure coefficient values. Gray bar indicates mean, error bars show maximum and minimum final fitness. (N = 46). The best mean performance is seen with S = 8.

challenges. The following conditions either cause convergence on an inaccurate solution, or require extreme numbers of generations to resolve.

(1) When an audio effect (or synthesizer parameter) is minimally present, such as a dry/wet parameter that is < 4%, the original sound is very nearly unaffected and while the GA appears to identify the presence of some audio manipulation it does not have enough impact to properly discriminate an accurate solution.

(2) As previously noted, random oscillators or noise generators pose particular problems due to the non-determinism of their output. The GA continues to search, exploring and going over old areas in an attempt to locate a stable

match. In these cases even the best match may evaluate poorly in consecutive epochs depending on the state of the random generators. Other audio analysis models could characterize the overall timbre, augmenting the MFCC analysis and allowing the GA to accept these noisy conditions.

(3) Oscillators, particularly low frequency effects, that are not phase locked pose a specific challenge. In cases where the phase is reset on a key trigger the GA is successful, but if the oscillator runs freely there is a high probability that even the best solution will compare unfavorably with the original. For many native devices Live provides the option for oscillators and generators to be synchronized to the master clock and configured in units of beats rather than hertz. Because the GA is evaluating continuously, regardless of the master clock, these generators will produce phase mismatch errors and disrupt identification of the optimal solution. Synchronizing to the master clock is generally undesirable since this would require delaying chromosome audition, greatly increasing operation time per epoch. However, this may be implemented as an optional operating mode based on user preference.

(4) Effects that operate on frequency ranges not present in the source material prevent convergence on an optimal solution. This was the primary cause of failures with Distortion and EQ effects. This is a challenge for the randomized parameter testing method employed herein, but does not pose an issue for actual application.

7   Conclusions

This work shows a high degree of success in being able to recreate or reverse engineer parameter presets, with performance varying based on specifics of the target audio process. In the best conditions an audio effect can be fit in less than one minute. However, the limits and idiosyncrasies of the system remain to be further investigated, beyond anecdotal observation. Certain conditions prevent an optimal solution from being discovered, but the particularities of those circumstances remain to be fully identified and addressed.

We have shown the ability to match randomly generated configurations or presets, however human musical preferences are rarely random, and may exhibit tendencies that either simplify or extend the challenge to a GA based solution. Further examination of performance versus human generated presets and sound samples is required to determine efficacy in more realistic applications. Another common approach and interest in GA based synthesizer programming is to employ targets originating from other sources (either other synthesizers or natural, sampled sounds) [8]. Recreating presets from samples of analog synthesizers, or acoustic instruments, can pose additional challenges and remains to be examined in the context of this system.

The potential for creators and producers to employ this system in creative musical and sound work is distinct and may provide both time saving and transformational benefits. Rather than spending hours programming synths and audio devices this system can quickly provide a ready starting point for tweaking. Further, it may be able to provide new sounds based on matching non-synthetic sounds (such as vocalizations), or by using targets as guidance where one is looking to identify sounds that combine elements from multiple sources in one result. The optimal solutions between samples with contrasting spectral signatures has the distinct potential to discover unique and compelling sounds.

References

1. Garcia, R.: Growing sound synthesizers using evolutionary methods. In: Proceedings ALMMA 2001: Artificial Life Models for Musical Applications Workshop (ECAL 2001) (2001)
2. Horner, A.: Double-modulator FM matching of instrument tones. Comput. Music J. **20**(2), 57–71 (1996)
3. Horner, A.: Nested modulator and feedback FM matching of instrument tones. IEEE Trans. Speech Audio Process. **6**(4), 398–409 (1998)
4. Horner, A., Beauchamp, J., Haken, L.: Machine tongues XVI: Genetic algorithms and their application to FM matching synthesis. Comput. Music J. **17**(4), 17–29 (1993)
5. Johnson, A., Phillips, I.: Sound Resynthesis with a Genetic Algorithm. Imperial College London (2011)
6. Karafotias, G., Hoogendoorn, M., Eiben, A.E.: Parameter control in evolutionary algorithms: Trends and challenges. IEEE Trans. Evol. Comput. **19**(2), 167–187 (2015)
7. Lai, Y., Liu, D.T., Jeng, S.K., Liu, Y.C.: Automated optimization of parameters for FM sound synthesis with genetic algorithms. In: Proceedings of the International Workshop on Computer Music and Audio Technology. Citeseer (2006)
8. Macret, M., Pasquier, P.: Automatic design of sound synthesizers as pure data patches using coevolutionary mixed-typed cartesian genetic programming. In: Proceedings of the 2014 Annual Conference on Genetic and Evolutionary Computation, pp. 309–316. ACM (2014)
9. Macret, M.M.J.: Automatic Tuning of the Op-1 Synthesizer Using a Multi-Objective Genetic Algorithm. Doctoral dissertation, Simon Fraiser University, Vancouver, CN (2013)
10. Riionheimo, J., Välimäki, V.: Parameter estimation of a plucked string synthesis model using a genetic algorithm with perceptual fitness calculation. EURASIP J. Adv. Signal Process. 8(1–15) (2003)
11. Rylander, S.G.: Optimal population size and the genetic algorithm. Population 100(400) (2002)
12. Srinivas, M., Patnaik, L.M.: Adaptive probabilities of crossover and mutation in genetic algorithms. IEEE Trans. Syst. Man Cybern. **24**(4), 656–667 (1994)
13. Tan, B., Lim, S.: Automated parameter optimization of double frequency modulation synthesis using the genetic annealing algorithm. J. Audio Eng. Soc. **44**(1/2), 3–15 (1996)
14. Tatar, K., Macret, M., Pasquier, P.: Automatic synthesizer preset generation with presetgen. J. New Music Res. **45**(2), 124–144 (2016)
15. Weise, T., Wu, Y., Chiong, R., Tang, K., Lässig, J.: Global versus local search: The impact of population sizes on evolutionary algorithm performance. J. Global Optim. **66**(3), 1–24 (2016)
16. Yee-King, M., Roth, M.: Synthbot: An unsupervised software synthesizer programmer. In: Proceedings of the International Computer Music Conference, Ireland (2008)

Fashion Design Aid System with Application of Interactive Genetic Algorithms

Nazanin Alsadat Tabatabaei Anaraki[✉]

College of Architecture, Texas Tech University, Lubbock, Texas, USA
Nazanin.tabatabaeianaraki@ttu.edu

Abstract. These days, consumers can make their choice from a wide variety of clothes provided in the market; however, some prefer to have their clothes custom-made. Since most of these consumers are not professional designers, they contact a designer to help them with the process. This approach, however, is not efficient in terms of time and cost and it does not reflect the consumer's personal taste as much as desired. This study proposes a design system using Interactive Genetic Algorithm (IGA) to overcome these problems. IGA differs from traditional Genetic Algorithm (GA) by leaving the fitness function to the personal preference of the user. The proposed system uses user's taste as a fitness value to create a large number of design options, and it is based on an encoding scheme either describing a dress as a whole or as a two-part piece of clothing. The system is designed in the Rhinoceros 3D software, using python, which provides good speed and interface options. The assessment experiments with several subjects indicated that the proposed system is effective.

Keywords: Fashion design · Interactive genetic algorithm · Artificial evolution · Human-computer interface

1 Introduction

Fashion design is an industry closely connected to our lives and it is one of the most important representations of human civilization. People used to either make their own clothes or buy them from small local producers. However, the Industrial Revolution enabled the shift to mass production. In order for the clothing industry to compete with other industries, it should utilize modern technologies.

Since most consumers are not professional designers, those who desire custom-made clothing contact a designer to help them with the process. This approach, however, is not efficient in terms of time and cost, and it often does not reflect the consumer's personal taste as much as desired.

This paper proposes a design system using the Interactive Genetic Algorithm (IGA) to overcome these problems. I classified dress in two categories: whole dresses and dresses with upper and lower parts. The system is designed in the Rhinoceros 3D software, using python #. The user will find the design that appeals to her the most by exploring this evolutionary environment and evaluating the designs in each step until the satisfactory result is achieved.

© Springer International Publishing AG 2017
J. Correia et al. (Eds.): EvoMUSART 2017, LNCS 10198, pp. 289–303, 2017.
DOI: 10.1007/978-3-319-55750-2_20

The remainder of the paper is organized as follows: in Sect. 2, I introduce the background and current methods used in fashion design as well as the concept of the applications of IGA. Section 3 gives the overview of the system. The system implementation and user interface are given in Sect. 4. The effectiveness of the proposed methodology is analyzed by experiments in Sect. 5. Finally, conclusions are summarized in Sect. 6.

2   Background

2.1   Fashion Design Aid System

Today, advances in information technology along with the globalization of the world's economy have made a major change in the production process of many industries. The fashion industry, which is closely connected to our lives, has also shifted from small local production to mass production and then to make-to-order [1].

The fashion industry's main concern is to respond to clients' demands rapidly but at minimal cost [2, 3]. To achieve this goal, computer-aided fashion design can be used to enhance the efficiency of product development. Although there are different systems available for fashion design, such as Adobe Illustrator and AutoCAD, available for fashion design, they are usually for professionals only and hard for non-professionals to use.

Appearance style is a metaphor for identity: through this kind of personal interpretation of or resistance to fashion, individuals announce who they are and who they hope to become [4]. Hence, a fashion design system that helps an ordinary person design her appealing fashion is very useful.

But what is "the most appealing fashion design"? People have different opinions on this matter, and the image they have in mind is often ambiguous and fuzzy. Therefore, it is difficult to design human evaluation functions. To reflect such human sensibility, this study proposes using Interactive Genetic Algorithm (IGA). IGA differs from the conventional Genetic Algorithms (GA) in its evaluation of the fitness function [5]. In conventional GA, the objective function is numerically defined, while in IGA, the fitness function is replaced by a human user. IGA "interacts" with users and incorporates the emotion and preference of the user in the process of evolution; therefore, the user's subjective evaluation determines the fitness of the solution [6]. IGA has been applied to optimization [7, 8], designing the layout and lighting of rooms [9], fashion [11], web sites [10], data mining [11, 12], and music composition [13], all according to the users' preferences.

2.2   Genetic Algorithm and Interactive Genetic Algorithm

Interactive genetic algorithms are a subset of genetic algorithms. Genetic algorithms, first explored by John Holland, are used to solve optimization and search problems by emulating principles of biological evolution [14]. GAs set variables up as genes and combine them to make designs. GAs then iteratively evaluate designs against a fitness test and select the best designs from a group to be used to make the next generation, these designs are subject to mutation and combination of traits, similar to biological evolution [15]. Instead of improving a single design, as seen in Fig. 1, a population of solutions is examined to find the best solutions, which are then recombined or mutated to generate a new population of (better) solutions. The general GA process is as follows [16]:

- Step 1: Initialize the population of chromosomes.
- Step 2: Calculate the fitness for each individual in the population using a fitness function.
- Step 3: Reproduce individuals to form a new population according to each individual's fitness.
- Step 4: Perform crossover and mutation on the population.
- Step 5: Go to step (2) until a particular condition is satisfied.

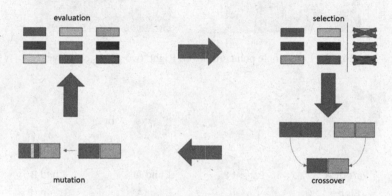

Fig. 1. GA process

Here is the pseudocode [17]:

```
BEGIN
Create a random set of initial solutions
  LOOP:
    Choose a subset of good solutions according to some
     ''fitness measure''.
    Perform recombination on randomly chosen pairs of so
     lutions.
    Perform mutation on randomly chosen solutions.
  UNTIL (population is stable)

END
```

To use GA for a problem, we should be able to parameterize the problem. Each parameter is considered a gene that can be represented through binary numbers. A combination of several genes creates a chromosome. Each singular representation of the chromosome within a population is termed a genotype [18]. An individual's fitness is calculated through decoding its genotype. In order to create new individuals for the next population, two individuals are submitted to crossover. To perform a single point crossover, as shown in Fig. 2, a crossover point is selected in each parents' genotype, then all data beyond that point is swapped between the two parents. The two-point crossover occurs by swapping everything between two selected points in parents' genotype. Figure 3 shows a visual representation of the crossover operation.

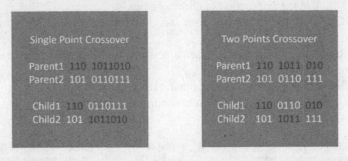

Fig. 2. Left: single point crossover. Right: two points crossover

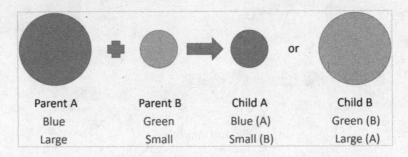

Fig. 3. Visual representation of crossover operation

To introduce more diversity, and potentially find drastically new solutions, mutation is utilized [18] (Fig. 5). Mutation occurs according to a user-definable mutation probability that is typically set to low. In the mutation process, as seen in Fig. 4, one point of the individual's genotype is modified.

Fig. 4. Mutation operation

IGAs are similar to GAs, except the fitness function of the GA is replaced by a user evaluation of solution options. Human subject selects the most successful designs that will be used to create the next generation of designs. Interactive Genetic Algorithms were originally proposed by Dawkins [19] and led to works within computer art community. One example of evolutionary art is "Galapagos" exhibit by Karl Sims which allowed for discovery of interesting figures in a population of three-dimensional virtual organisms [20].

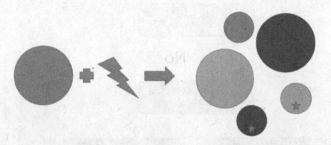

Fig. 5. Visual representation of mutation operation. In this example, those offspring marked with a star are those that required more than one mutation to produce.

3   System Design

Figure 6 shows the overview of the entire system. The proposed system has a database of partial dress elements, considering the dress as a whole or as a two-part piece of clothing. To create the initial population, the system selects random elements and combines them into 9 dress designs. User gives fitness value to each of the displayed designs. The system uses the roulette-wheel parent selection to choose pairs of designs for divergent and convergent recombination. After applying crossover and mutation, the results are displayed again and this loop continues until a desired design is found. As a summary, here is a pseudocode description of the algorithm:

```
BEGIN
Create a random set of initial designs.
  LOOP:
    Decode the solutions.
    Display each of the nine design solutions.
    Ask the user to assign scores to the current designs
      in the population.
    Use the roulette-wheel parent selection to choose
      pairs of designs for divergent and convergent recom-
      bination.
    Perform mutation on randomly chosen solutions.
  UNTIL (a desired design is found)
Store the 3D file, which was chosen by the user.

  END
```

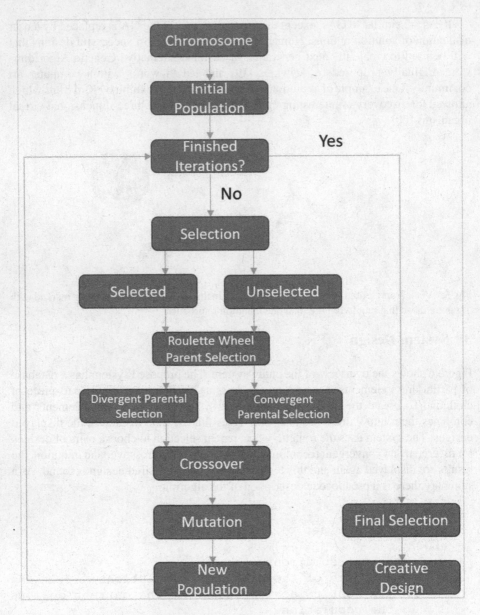

Fig. 6. Flow diagram showing the system overview

First, a dress is classified into either a whole piece (Fig. 7) or a two-piece object (Fig. 8). The two-piece consists of two parts: neck and body, skirt and waistline. Next, they are encoded with extra 16 bytes, which allows for continuous growing of each part during the design process (Fig. 9).

Each part is created by applying the generation rules on primitive solids, inspired by Latham and Todd's Form grow [21]. Primitive solids include sphere, box, ellipsoid,

Fig. 7. Dress as a whole piece

Fig. 8. Dress as a two-piece object

Fig. 9. Continuous growth of a design

Fig. 10. A truncated star-shaped five-sided pyramid to modify

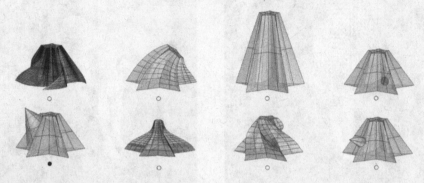

Fig. 11. The effect of the application of transform rules on the object in Fig. 10

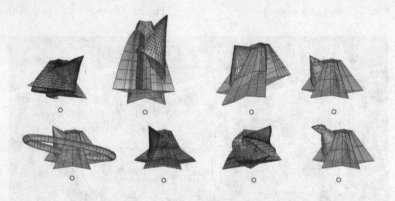

Fig. 12. The effect of transform rules on the object selected in Fig. 11

torus, cylinder, cone, pyramid and plane. The grow rules consist of horn, fractal, ribcage in addition to transform rules such as stack, twist, bend, grow, and branch. Figures 10, 11 and 12 show the effect of transform rules and crossover on a truncated star-shaped five-sided pyramid.

3.1 Algorithm Mechanics

Some researchers have proposed IGA systems in which users must rank each design within each set of a generation, while others have argued that such ranking causes user fatigue [18, 22–24]. Instead, they propose that users should identify their most and least preferred among a set because these are easier to identify than a whole ranking. This consequently reduces fatigue and provides a wealth of information [18]. For this research, 9 designs are displayed in each generation and the user will select his/her three favorite designs in each generation and rank them. However, in the last generation the user will only select one final favorite design.

For the mating process roulette wheel selection (Fig. 13) is implemented in the algorithm. The three selected designs are given large percentage of the roulette wheel based on the scores allocated to them, while the other designs are given small and equal wheel percentages. A sector of the wheel is chosen at random. The probability of selection being weighted by the score allocated to that design, the design corresponding to that sector is then used as one of the parents for the new design. In this algorithm, the probability of user-selected designs to be a parent is high, while other designs also have a chance of being parents. This crossover scheme allows for a larger design diversity.

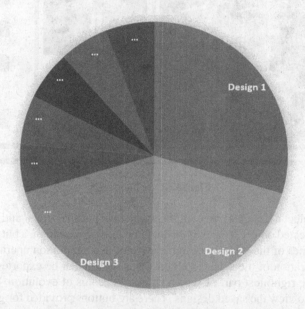

Fig. 13. Roulette wheel selection

Mutation increases diversity in the population and potentially exposes a more preferred design [18]. Since this research allows for continuous growing of each part during the design process, simple binary mutation methods are not useful enough. Thus, we introduced a mutation operator that allows local search in a preferred area, helping the design change in one dimension. This mutation operator functions gradually on the design options, but it also serves to change the binary structure of the individuals in a larger way than bit-space mutation operators do [18].

4   System Implementation

The system is designed in the Rhinoceros software because of the good speed and interface options. Figure 14 shows the user interface of the system. The system displays current population composed of nine designs (Fig. 15) in one screen. There are ranking buttons next to each individual model. To decrease user fatigue, only 3 designs should be ranked. The selected designs have a higher probability to be used for recombination in the next generation. An example of simple 1-point crossover is given in Fig. 16.

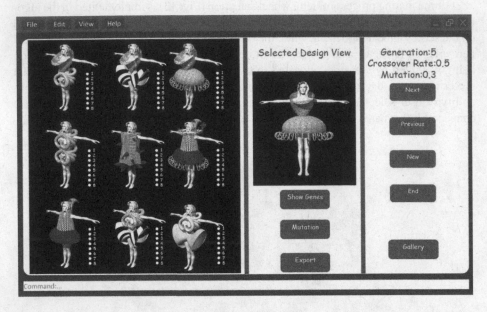

Fig. 14. User interface

User can select a design to view in the middle part of the screen and can rotate and zoom in the selected object. To get more detail on a selected design, a button is provided to show the genes of the selected design. There is also a mutation operator that allows for local exploration of the selected object. This design can be exported and saved to the gallery. The rightmost part of screen shows the status of evolution. There is a link to the gallery, to view the saved designs. There are buttons provided for generating next population and to going back to the previous population.

IGAs can help enhance creativity because they can operate in both divergent and convergent ways [18]. We have implemented both operations in order to find new and interesting design concepts while retaining the characteristics that the user already liked. To model convergence, we gave a high probability to the designs selected by the user to become parents for the population of the new generation. Then, we applied a low mutation rate on the created designs. This process allows users to converge toward a design that is appealing to them.

Mistuned GAs may cause drastic diverge from their intended goal which increases in each generation. Thus, high mutation rates should be avoided in order to prevent random mutation during the process. In this system, by using roulette wheel selection we managed to explore the design environment without diverging drastically from the main goal, while expecting unique inspiring designs.

Fig. 15. A population composed of nine designs

Fig. 16. Crossover

5   Experimental Overview

To examine the performance of the system, we tested it in both convergent and divergent process. In the convergent process, we allocated 75% to the three designs selected by the user according to the rank given by him/her. Then, we divided the remaining 25% of the wheel between the remaining 6 designs. For the divergent process, we allocated 75% of the wheel between the 6 unselected designs while dividing the rest between the selected designs. The selected designs had a high probability to be selected in the convergent process, while in the divergent process the unselected designs had a higher likelihood. This system allows for the implementation of both these processes simultaneously. Our tests showed that the first 4–6 generations were considered divergent, and after that the generations tend to be more convergent.

We examined user satisfaction with our system by testing a group of 15 experiment subjects. In the first part of this examination, we asked the subjects to design a costume they would wear to a costume party, using the proposed IGA system. They had to find their favorite design in less than 15 generations and rank it from 0 to 10. In the second part of the experiment, we asked 5 subjects to carefully choose 20 designs between 100 randomly selected designs. Then, we asked 15 subjects to find and rank their favorite design between those 20 selected designs presented to them pair to pair. The results showed that the users gave a higher rank to the designs created by our IGA system (avg. 8.8 compared with avg. 6.9). The experiments showed a promising result on the application of IGA for a fashion design aid system for non-professionals.

An important aspect of human computer interaction is the idea that a computer system should allow the user to make use of their increased knowledge about the system as they use it more frequently [25]. As users gain more experience with the system, they grow some understanding of how different parameters affect the design. To use this knowledge, we propose an option in the interface to adjust values of those parameters, which affect not the absolute values of those parameters but how much the algorithm explores those regions of parameter space, i.e. the extent to which corresponding regions of the genome are mutated [17]. In addition, an option to adjust the exploration rate could be also useful. This option could provide the user with control over the mutation and crossover rate.

In the first version of this system, no local mutation operator was provided. However, by testing the system by 15 experiment subjects, we were able to gain constructive opinions. The main request was adding a new operator to allow local exploration on a selected design, helping it change in one dimension e.g. texture change of an element (Fig. 17). The result of the experiments showed the proposed additional operator to be effective. However, some users tend to use this operator too often, which did not allow the system to be as explorative as it is intended to be. For further exploration of this system, we propose focusing on the limitation of the use of this operator to prevent this situation.

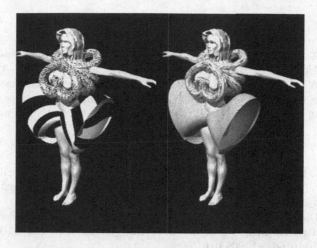

Fig. 17. Mutation in texture operated by the mutation function.

6 Conclusion

This study proposed a fashion design aid system for non-professionals by using IGA. The proposed system, which is designed in the Unity game engine, uses user's taste as a fitness value to create a large number of design options. The experimental tests showed that users are significantly satisfied with the system; therefore using IGA for a fashion design aid system is a proper choice.

We tested the performance of our system in both convergent and divergent ways. The results showed the system to be effective in both ways. These two concepts can help enhance creativity. With an appropriate implementation of both these operations, we can allow for reaching to an appealing design, while exploring the unique interesting designs in our system environment.

To improve our system, we propose to add an option in the interface to adjust the parameters that affect the design. In addition, an option to adjust the exploration rate could be also useful. This option can provide the user with control over the mutation and crossover rate.

The experimental results suggested for adding a mutation operator allowing local exploration of a selected design. To further improve this study, we propose to focus on limitation of this operator, to prevent the system from being one-dimensional and not being able to explore the design environment.

Acknowledgements. I would like to express my deep gratitude to Professor Upe Flueckiger and Dr. Nelson Rushton for their patient guidance and enthusiastic encouragement during the development of this research work.

References

1. M'Hallah, R., Bouziri, A.: Heuristics for the combined cut order planning two-dimensional layout problem in the apparel industry. Int. Trans. Oper. Res. **23**(1–2), 321–353 (2016)
2. Guo, Z.X., Wong, W.K., Leung, S.Y.S., Fan, J.T., Chan, S.F.: Mathematical model and genetic optimization for the job shop scheduling problem in a mixed and multi-product assembly environment: a case study based on the apparel industry. Comput. Ind. Eng. **50**, 202–219 (2006)
3. Rose, D., Shier, D.: Cut scheduling in the apparel industry. Comput. Oper. Res. **24**, 3209–3228 (2007)
4. Kaiser, S.B.: Fashion and Cultural Studies. Berg, London (2012). English edition
5. Gonsalves, T., Kawai A.: Fourth International conference on Computer Science & Information Technology, pp. 169–174 (2014)
6. Hu, Z.-H., Ding, Y.-S., Zhang, W.-B., Yan, Q.: An interactive co-evolutionary CAD system for garment pattern design. Comput. Aided Des. 40(12), 1094–1104 (2008). doi:http://dx.doi.org/10.1016/j.cad.2008.10.010
7. Sakawa, M., Yauchi, K.: Interactive decision making for multiobjective nonconvex programming problems with fuzzy numbers through coevolutionary genetic algorithms. Fuzzy Sets Syst. **114**(1), 151–165 (2000)
8. Sakawa, M., Nishizaki, I.: Interactive fuzzy programming for two-level nonconvex programming problems with fuzzy parameters through genetic algorithms. Fuzzy Sets Syst. **127**(2), 185–197 (2002)
9. Fukada, Y., Sato, K., Mitsukura, Y., Fukumi, M.: The room design system of individual preference with IGA. In: International Conference on Control, Automation and Systems, Seoul, Korea (2007)
10. Oliver, A., Monmarche, N., Venturini, G.: Interactive design of web sites with a genetic algorithm. In: Proceedings of the IADIS International Conference WWW/Internet, Lisbon, Portugal, pp. 355–362 (2002)
11. Kim, H.-S., Cho, S.-B.: Application of interactive genetic algorithm to fashion design. Eng. Appl. Artif. Intell. **13**(6), 635–644 (2000)
12. Gong, D.-W., Hao, G.-S., Zhou, Y., Sun, X.-Y.: Interactive genetic algorithms with multi-population adaptive hierarchy and their application in fashion design. Appl. Math. Comput. **185**(2), 1098–1108 (2007)
13. Tokui, N., Iba, H.: Music composition with interactive evolutionary computation. In: Proceedings of the Generative Art 2000, International Conference on generative Art, Milan, Italy (2000)
14. Holland, J.: Adaptation in Natural and Artificial System. The University of Michigan Press, Ann Arbor (1975)
15. Nathan-Roberts, D.: Using Interactive Genetic Algorithms to Support Aesthetic Ergonomic Design. Dissertation, University of Michigan. Ann Arbor, Michigan: ProQuest/UMI (in-press)
16. Eberhart, R., Simpson, P., Dobbins, R.: Computational Intelligence PC Tools. Waite Group Press, Corte Madera (1996)
17. Johnson, C.B.: Exploring the sound-space of synthesis algorithms using interactive genetic algorithms. In: Patrizio, A., Wiggins, G.A., Pain, H. (eds.) Proceedings of the AISB 1999 Symposium on Artificial Intelligence and Musical Creativity. Brighton: Society for the Study of Artificial Intelligence and Simulation of Behaviour (1999)
18. Kelly, J.C.: Interactive genetic algorithms for shape preference assessment in engineering design. ProQuest (2008)

19. Dawkins, R.: The Blind Watchmaker: Why the Evidence of Evolution Reveals a Universe Without Design. Norton, New York (1996)
20. Frauenfelder, M.: Do-it-yourself darwin. Wired **6**(10), 164 (1998)
21. Todd, S., Latham, W.: Evolutionary art and Computers. Academic Press, Orlando (1994)
22. Buonanno, M.A., Mavris, D.N.: Small supersonic transport concept evaluation using interactive evolutionary algorithms. In: Collection of Technical Papers – AIAA 4th Aviation Technology, Integration, and Operations Forum, ATIO, vol. 1, pp. 411–426, 20–23 September 2004
23. Cho, S.B.: Towards creative evolutionary systems with interactive genetic algorithm. Appl. Intell.: Int. J. Artif. Intell. Neural Netw. Complex Probl. Solving Technol. 16(2), 129–38 (2002)
24. Kamalian, R., Zhang, Y., Takagi, H., Agogino, A.M.: Reduced human fatigue interactive evolutionary computation for micromachine design. In: Proceedings of 2005 International Conference on Machine Learning and Cybernetics, vol. 9 (2005)
25. Preece, J., Rogers, Y., Sharp, H., Benyon, D., Holland, S., Carey, T.: Human-Computer Interaction. AddisonWesley, Essex (1994)

Generalisation Performance of Western Instrument Recognition Models in Polyphonic Mixtures with Ethnic Samples

Igor Vatolkin[⊠]

Department of Computer Science, TU Dortmund, Dortmund, Germany
igor.vatolkin@tu-dortmund.de

Abstract. Instrument recognition in polyphonic audio recordings is a very complex task. Most research studies until now were focussed on the recognition of Western instruments in Western classical and popular music, but also an increasing number of recent works addressed the classification of ethnic/world recordings. However, such studies are typically restricted to one kind of music and do not measure the bias of "Western" effect, i.e., the danger of overfitting towards Western music when the classification models are optimised only for such tracks. In this paper, we analyse the performance of several instrument classification models which are trained and optimised on polyphonic mixtures of Western instruments, but independently validated on mixtures created with randomly added ethnic samples. The conducted experiments include evolutionary multi-objective feature selection from a large set of audio signal descriptors and the estimation of individual feature relevance.

Keywords: Music instrument recognition · Evolutionary multi-objective feature selection · Generalisation performance of classification models

1 Introduction

Music instrument recognition belongs to the most challenging applications among music classification tasks. Started more than a decade ago, many efforts were spent to develop acoustic features which describe timbral properties of sound and build models which successfully predict instruments in the audio signal. [4,10] belong to the earliest studies on automatic instrument recognition of individual tones. In [4], four classes of instruments, and in [10] six classes and 30 individual instruments were categorised. In particular, the recognition of instrument groups was very successful with accuracy rates over 90% [10].

However, instrument recognition becomes a very hard problem if two or more instruments are recorded together. One solution is to apply source separation algorithms, such as non-negative matrix factorisation [18]. Another approach is based on the estimation of the predominant fundamental frequency [8]. Complex signal processing methods like source separation and music transcription can be

© Springer International Publishing AG 2017
J. Correia et al. (Eds.): EvoMUSART 2017, LNCS 10198, pp. 304–320, 2017.
DOI: 10.1007/978-3-319-55750-2_21

also avoided by means of statistical pattern recognition [12]. With an increasing number of available audio signal descriptors, feature selection may help to select the most relevant characteristics for each category. Multi-label feature selection was applied for the prediction of instruments in mixtures of two samples [29], and a hierarchical taxonomy of labels was used in [11].

Generally, the choice of data is a crucial aspect of algorithm evaluation and optimisation. Almost all MIR systems are restricted to a certain kind of music data without a sufficient variability [31]. An excellent example of overfitting towards genres is described in [6]. Here, several previously designed systems on emotion recognition for classical, film, popular, and mixed music (9 data sets) are compared, and it is observed that the performance of regression models drops significantly when switching from one to another validation set. Another issue is the "album" effect in genre recognition, where classification models rather learn properties of albums and not genres [34].

The works mentioned at the beginning of this section and a predominant share of internationally available and peer-reviewed studies deal only with Western classical and popular music. In recent years, however, an increasing attention was paid to the analysis of world music, consider for example a list of studies from the CompMusic project[1]. To name a few works, [1] proposed a system for the recognition of Persian scales, the transcription of Turkish music was addressed in [3], and rhythm identification in Turkish and Indian music in [30]. However, there is a lack of studies on automatic recognition of ethnic instruments, and the published studies have often a very limited scope and are not supported with statistical analysis. In [13], three Indian instruments (harmonium, tabla, and santoor) are classified, however for a rather small set of 60 monophonic recordings. [15] describes a study on the recognition of 10 Indian instruments, either as individual samples or mixtures of two instruments, and [21] on the categorisation of 9 Pakistani instruments in monophonic recordings into 3 families. Several years ago, [19] stated that "at the level of musical instruments there is practically nothing done", and this situation is barely changing.

There exist only a few studies which compare music characteristics between Western and world music, e.g., for several music categorisation tasks using Western, Latin American, and African music [22], or on the properties of musical form [2]. However, to our knowledge, there are no investigations which directly measure the classification quality of instrument recognition models optimised for the best performance on Western music and validated on world music and vice versa. Such experiments, together with a deeper analysis of commonalities and differences between both music worlds may help to create classification and regression models of better general performance and to support application scenarios also beyond instrument classification, such as music recommendation, transcription, mood and genre recognition, etc.

Transferred to the recognition of instruments, the basic idea for our study was to examine how well models trained and optimised for the recognition of Western instruments in Western music recordings would perform when applied in

---

[1] http://compmusic.upf.edu/publications, accessed on 15.11.2016.

recordings with ethnic instruments. If a theoretically expected significant drop of performance would be confirmed, this would indicate the needs to include such instruments in training and test sets. The most realistic testbed would comprise polyphonic audio pieces played with both Western and ethnic instruments. However, the lack of such databases with precise ground truth and the complexities arising with different playing styles and different properties of attack/decay/release/sustain-envelope hinder the creation of such a data set. As a counterexample, the classification of individual tones is too simple and less relevant for a real-world classification scenario. Therefore, our intention for this initial study was to identify instruments in polyphonic mixtures of two and more simultaneously played sources.

In the next section, we will describe data sets and setup of algorithms for the recognition of Western instruments in polyphonic mixtures. Then, the results of the conducted study are analysed. We conclude with the most important findings and provide an outlook for the next research steps.

2   Setup and Methods

2.1   Data Sets and Categorisation Tasks

Tables 1 and 2 list instruments selected for the study. The samples of Western instruments are compiled from MUMS collection [7], RWC collection [14], and University of Iowa collection[2]. The samples of ethnic instruments are taken from Ethno World 5 Professional & Voices collection[3]. The instruments are assigned to four groups according to the manual of the Ethno World collection: bowed, strings, key, and woodwind/brass. For ethnic instruments, the origins of their creation are listed if possible. The range indicates minimum and maximum available pitches in scientific pitch notation.

Four data sets each consisting of 3000 polyphonic mixtures were generated:

Table 1. Western instruments.

Instrument	Category	Range
Cello	Bowed	C2–G♯5
Viola	Bowed	C3–D6
Violin	Bowed	G3–F♯6
Piano	Key	C2–F♯6
Acoustic guitar	Stringed	E2–E5
Electric guitar	Stringed	E2–E5
Flute	Woodw./brass	B3–F♯6
Trumpet	Woodw./brass	E3–G6

---

[2] http://theremin.music.uiowa.edu/MIS.html, accessed on 15.11.2016.

[3] http://www.bestservice.de/en/ethno_world_5_professional__voices.html, accessed on 15.11.2016.

Table 2. Ethnic instruments.

Instrument	Category	Origin	Range
Dilruba	Bowed	India	C3–G5
Egyptian fiddle	Bowed	Egypt	E2–E5
Erhu	Bowed	China	A3–C6
Jinghu opera violin	Bowed	China	A4–A6
Morin khuur violin	Bowed	Mongolia	D3–C6
Hohner melodica	Key	Germany	A3–C6
Scale changer harmonium	Key	India	G1–C5
Balalaika	Stringed	Russia	A2–F5
Bandura	Stringed	Ukraine	C2–G6
Banjolin	Stringed	different	E3–A5
Banjo framus	Stringed	n.a	A2–D6
Bouzouki	Stringed	Greece	C3–B5
Ceylon guitar	Stringed	Ceylon	C2–C4
Cümbüs	Stringed	Turkey	C2–C5
Domra	Stringed	Russia	C3–B5
Kantele	Stringed	Finland	C1–C6
Oud	Stringed	Persia	F2–F5
Sitar	Stringed	India	C3–G5
Tampura	Stringed	India	F2–A♯3
Tanbur	Stringed	Middle East	F♯1–C5
Saz	Stringed	Turkey	E2–E5
Ukulele	Stringed	Hawaii	A3–B5
Bawu	Woodw./brass	China	B3–F5
Dung dkar conch trumpet	Woodw./brass	Himalayas	E3–A5
Fujara	Woodw./brass	Slovakia	C2–C7
Pan flute	Woodw./brass	different	F2–D♯5
Pinkillo	Woodw./brass	S. America	C5–C7
Pivana	Woodw./brass	Corsica	C5–C7
Shakuhachi	Woodw./brass	Japan	A3–E6

- WI: Intervals with Western instrument samples
- WC: Chords with Western instrument samples
- MI: Intervals with one Western/one ethnic sample
- MC: Chords with at least one Western and at least one ethnic sample

The procedure of mixing is similar to our previous study on the recognition of instrument families [33]. For an interval, the following steps are done based on random decisions with equal probability:

- Draw two instruments (Western instruments for WI data set and one Western/one ethnic for MI data set)
- Draw an interval (minor third, major third, fourth, fifth, major sixth, or minor seventh)
- Draw the first pitch from available pitches of the 1st instrument; repeat the drawing of the first pitch if the second pitch is not available for the second instrument
- Mix both pitches to produce an interval

For a chord, the steps comprise:

- Draw a key (from 12 halftones)
- Draw a mode (major or minor)
- Draw a tonal function (tonic, subdominant, dominant, or submediant)
- Draw four instruments (Western instruments for WC data set and at least one Western/one ethnic for MC data set)
- Draw the first pitch from available pitches of the 1st instrument; repeat if the pitches are not available for the 2nd - 4th instrument
- Mix four tones for a subdominant, otherwise three

The overall number of categorisation tasks is equal to 16 (recognition of 8 Western instruments either in intervals or in chords).

2.2 Features

Table 3 provides a summary of features extracted from mono audio signal downsampled to 22,050 Hz. As proposed in [32], they are grouped into timbral, rhythmic, and pitch characteristics. For characteristics of timbre the information about the source domain is additionally provided. Note that some descriptors with less importance for instrument recognition – such as chroma vector – are intentionally integrated in this large feature set, because we (1) expect that feature selection and classification will identify and remove noisy features and (2) combinations of individually less relevant features with others, also less relevant ones, may increase the relevancy, e.g., when some timbral characteristics of lower pitches (identified by means of fundamental frequency) would characterise an instrument (examples of features which are individually irrelevant and relevant in their combination are provided in [16, p. 10]). The features are extracted with various tools including MIR Toolbox [20], jAudio [24], Chroma Toolbox [27], own implementations in Matlab, etc., and are integrated together after [33].

The selection of feature extraction frames with regard to attack-decay-sustain-release envelope may increase the performance of instrument recognition. For example, experiments in [28] underlined the significance of the attack stage, and in [23] it was showed that non-harmonic components of timbral sounds contain relevant timbral information. Some sources for noisy components such as violin bow or piano key strike may have a larger impact on timbre at the beginning of the tone. Therefore, to pay better attention to the tone progress

Table 3. Audio signal features. No.: overall number of dimensions after feature processing (see the text).

Group/domain	Examples	No.
Timbre/time	Zero-crossing rate, linear prediction coefficients, low-energy	39
Timbre/spectrum	Spectral centroid, bandwidth, skewness, kurtosis, flatness	138
Timbre/cepstrum	MFCCs (various implementations), delta MFCCs	138
Timbre/phase	Angles and distances in phase domain [25]	6
Timbre/Bark	Magnitudes in the Bark scale	69
Timbre/ERB	Zero-crossing rate for ERB bands	90
Rhythm	Periodicity peaks, sum of correlated components	12
Pitch	Fundamental frequency, inharmonicity, chroma energy normalized statistics [26]	303

before and after the identified onset (peak in the onset curve), for each dimension of short-framed features three different vectors are built: features extracted from the middle of the attack interval, the onset frame, and the middle of the release interval. This means that, e.g., 20 original dimensions of MFCC vector lead to 60-dimensional vector used in classification models (1st MFCC dimension from the attack interval, 2nd MFCC dimension from the attack interval,..., 1st MFCC dimension from the onset frame, etc.). The third column in the table lists numbers of final dimensions after processing.

2.3 Classification and Evaluation

The classification is done using random forest with the default number of 100 trees, as it is a fast algorithm which is less sensitive to parameters compared to, e.g., support vector machines. Random forest [17] has proved a good generalisation performance in our previous experiments [33] and has usually lower classification errors than a single decision tree being an ensemble classifier.

As a measure of classification quality, the relative classification error is estimated which is balanced across positive and negative examples, i.e.:

$$e = \frac{1}{2} \left(\frac{FN}{TP + FN} + \frac{FP}{TN + FP} \right), \tag{1}$$

where TP denotes true positives (number of mixtures which contain an instrument to recognise and were properly classified), TN true negatives (mixtures without an instrument which were correctly classified), FP false positives (mixtures with a predicted instrument which however do not contain it) and FN false negatives (mixtures with an instrument which were predicted as not containing it).

2.4 Feature Selection

To select the best features for each classification task, we apply evolutionary multi-objective feature selection for the minimisation of classification error (cf. Eq. 1) and the relative number of selected features [33]. In terms of multiobjective comparison of solutions, feature set Φ_1 *dominates* feature set Φ_2 (denoted with $\boldsymbol{f}(\Phi_1) \prec \boldsymbol{f}(\Phi_2)$) when it is not worse with regard to all d evaluation criteria (objectives) $f_1, ..., f_d$ and better for at least one criterion f_k:

$$\boldsymbol{f}(\Phi_1) \prec \boldsymbol{f}(\Phi_2) \text{ when } \forall i \in \{1, ..., d\} : f_i(\Phi_1) \leq f_i(\Phi_2) \text{ and}$$
$$\exists k \in \{1, ..., d\} : f_k(\Phi_1) < f_k(\Phi_2). \tag{2}$$

Here, the objective functions are minimised; functions to be maximised can be adapted for the minimisation. The goal is to find *non-dominated* fronts with feature sets which are not dominated by any other feature set.

A non-dominated front with D solutions (feature sets) $\Phi_1, ..., \Phi_D$ is evaluated by the estimation of dominated hypervolume or $\mathcal{S}$-metric [35]:

$$h(\boldsymbol{f}(\Phi_1), ..., \boldsymbol{f}(\Phi_D)) = vol\left(\bigcup_{i=1}^{D} [\Phi_i, \boldsymbol{r}]\right), \tag{3}$$

where $\boldsymbol{r}$ is the reference point (usually worst possible solution[4]), and $vol(\cdot)$ is the united volume of all hypercubes between $\boldsymbol{r}$ and $\boldsymbol{f}(\Phi_1), ..., \boldsymbol{f}(\Phi_D)$.

SMS-EMOA [9] is applied as the multi-objective evolutionary algorithm. To support the search in different regions of decision space favouring smaller feature sets, the expected initial feature set sizes are set to 1%, 5%, and 10% of all features. For each initial feature set size and each classification task, 20 statistical repetitions of experiments are done. As a compromise between sufficient optimisation time and requirements on computing resources in our grid system, the number of generations was set to 1200, and the population size to 40 solutions.

As the second feature selection strategy, we adapt the scheme of minimum redundancy - maximum relevance (MRMR) [5] using Pearson correlation coefficient. The idea is to search for features which have a strong correlation with a classification target (i.e., the instrument to recognise) and are as least correlated as possible to already selected features. Let $\boldsymbol{x}_1, ... \boldsymbol{x}_F$ be F individual feature dimensions for N observations (music intervals or chords) and $\boldsymbol{y}$ the vector with binary ground truth (classification target).

First, the correlation between each feature and a target is estimated:

$$r_{\boldsymbol{x}_i, \boldsymbol{y}} = \frac{\sum_{k=1}^{N}(x_{i,k} - \bar{x}_i)(y_k - \bar{y})}{\sqrt{\sum_{k=1}^{N}(x_{i,k} - \bar{x}_i)^2 \sum_{k=1}^{N}(y_k - \bar{y})^2}}, \tag{4}$$

where $\bar{x}_i$ denotes the mean value of $\boldsymbol{x}_i$ and $\bar{y}$ the mean value of $\boldsymbol{y}$, respectively.

---

[4] In this study, the reference point is (1,1): a theoretical solution which uses all features and leads to the classification error $e = 1$.

Then, the inter-correlation between all features r_{x_i, x_j} ($i, j \in \{1, ..., F\}$) is estimated by a replacement of y with x_j in 4.

The first selected feature is the one with the largest absolute $r_{x_i, y}$. Now, the set of selected features is extended one by one with features which are most correlated with the target and least correlated with already selected ones. Let $s = 1, ..., S$ denote the selected features and $c = 1, ..., C$ the remaining candidate features. The relevance criterion V_c for a candidate feature c is given by:

$$V_c = |r_{x_c, y}|, \tag{5}$$

and the redundancy criterion by:

$$W_c = \frac{1}{S} \sum_{s=1}^{S} |r_{x_c, x_s}|. \tag{6}$$

Note that we do not distinguish between positive and negative correlation estimating absolute values for each correlation coefficient.

At each step, the feature with the maximum value of $V_c - W_c$ is added. Usually, this process continues until a given number of features is selected. In our setting, we add features until all of them are selected, to compare the change of classification quality on Western mixtures to the change of quality of the same feature sets for mixtures with ethnic instruments.

3   Experiments

3.1   Nomenclature

The target of the experiments is first to find optimal feature sets for the recognition of Western instruments in Western mixtures with the help of feature selection as described in Sect. 2.4 and then validate the performance of corresponding classification models on mixtures with Western and ethnic instruments. The optimisation of feature selection is done with 5-fold cross-validation on data sets WI and WC. This means that 4/5 of data instances (training set) are used for the training of the model, and its performance is validated with a balanced relative error e (Eq. 1) on the remaining 1/5 of data instances (validation set). After 5 folds, the mean error is estimated. As the target of this procedure is to select the best optimised feature set for Western mixtures, we refer to this error as the *optimisation error* e_W. The *test error* e_M is the error of models trained on Western mixtures but tested on data sets with ethnic samples. The following variables are used:

- $\widehat{\varPhi}_W$: Feature set with the smallest optimisation error for WI or WC
- $|\widehat{\varPhi}_W|$: Corresponding number of features
- $e_W(\widehat{\varPhi}_W)$: Corresponding optimisation error
- $e_M(\widehat{\varPhi}_W)$: Corresponding test error for MI or MC
- $e_M(\widehat{\varPhi}_M)$: Best test error estimated for MI or MC

- $\widehat{h}_W$: Hypervolume of the non-dominated front for WI or WC
- $h_M(\widehat{h}_W)$: Corresponding hypervolume for MI or MC using solutions from the non-dominated front for WI or WC
- $\widehat{h}_M$: Hypervolume of the non-dominated front for MI or MC

3.2 Evolutionary Feature Selection

Figure 1 plots two examples of non-dominated fronts after evolutionary multi-objective feature selection for recognition of trumpet (left subfigure) and acoustic guitar (right subfigure) in chords. The smallest $e_W(\widehat{\Phi}_W)$ is 0.1585 for trumpet using 49 features. For the next non-dominated feature set, the error increases to 0.1587 (46 features), 0.1598 (42 features), etc. As we can observe, the classification performance of these sets decreases significantly when the same models are applied for the recognition of instruments in mixtures with ethnic samples. For the three above mentioned feature sets $e_M(\widehat{\Phi}_W) = 0.2483, 0.2505$, and 0.2716 which shows the poor generalisation performance of models trained and optimised with only Western instrument samples. The gap between $e_W(\widehat{\Phi}_W)$ and $e_M(\widehat{\Phi}_W)$ is even larger for the recognition of acoustic guitar in chords (Fig. 1, right subfigure).

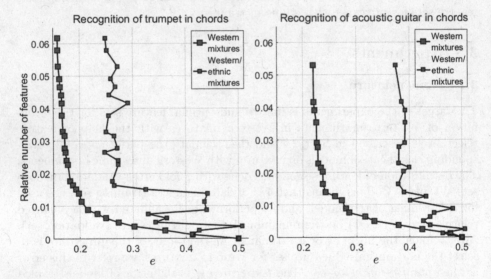

Fig. 1. Non-dominated fronts after optimisation of instrument recognition in Western mixtures (large rectangles) and the same solutions validated on Western/ethnic mixtures (small rectangles).

Table 4 provides a summary of experiment results. For example, the set with the smallest optimisation error $e_W(\widehat{\Phi}_W) = 0.1815$ for the recognition of acoustic guitar in intervals contains 64 features (first line). When the model trained with

Table 4. Errors and hypervolumes after evolutionary multi-objective feature selection.

Category			Feature sets with smallest errors			Hypervolumes		
Instrument	Mixture	$|\widehat{\Phi}_W|$	$e_W(\widehat{\Phi}_W)$	$e_M(\widehat{\Phi}_W)$	$e_M(\widehat{\Phi}_M)$	$\widehat{h}_W$	$h_M(\widehat{h}_W)$	$\widehat{h}_M$
Ac. guitar	Intervals	64	0.1815	0.3272	0.2677	0.8159	0.6714	0.7304
Ac. guitar	Chords	42	0.2194	0.3732	0.3471	0.7785	0.6257	0.6521
Cello	Intervals	31	0.1672	0.4721	0.3226	0.8306	0.6430	0.6763
Cello	Chords	52	0.2571	0.4463	0.3616	0.7405	0.5532	0.6362
El. guitar	Intervals	63	0.0417	0.4282	0.0672	0.9550	0.8297	0.9290
El. guitar	Chords	58	0.0740	0.2534	0.1239	0.9225	0.7544	0.8726
Flute	Intervals	42	0.2818	0.3199	0.2580	0.7167	0.6992	0.7397
Flute	Chords	53	0.3282	0.3926	0.3306	0.6697	0.6526	0.6669
Piano	Intervals	61	0.1576	0.2652	0.1726	0.8401	0.7519	0.8249
Piano	Chords	64	0.1672	0.2241	0.2021	0.8300	0.7731	0.7954
Trumpet	Intervals	51	0.1294	0.2046	0.1709	0.8683	0.8112	0.8271
Trumpet	Chords	49	0.1585	0.2483	0.2236	0.8392	0.7516	0.7741
Viola	Intervals	50	0.2796	0.4761	0.3603	0.7179	0.5811	0.6382
Viola	Chords	84	0.3243	0.4622	0.3909	0.6738	0.5674	0.6080
Violin	Intervals	42	0.3036	0.2821	0.2708	0.6941	0.7143	0.7261
Violin	Chords	41	0.3530	0.3665	0.3294	0.6453	0.6415	0.6669

these features is used for the recognition of acoustic guitar in data set MI, the error drops to $e_M(\widehat{\Phi}_W) = 0.3272$. The best test error $e_M(\widehat{\Phi}_M) = 0.2677$. Respectively, hypervolume on the non-dominated front drops from $\widehat{h}_W = 0.8159$ to $h_M(\widehat{h}_W) = 0.6714$, and the hypervolume of the non-dominated front for test set $\widehat{h}_M = 0.7304$. In the following, we will discuss several observations after the statistical analysis of the optimised feature sets with regard to generalisation performance.

Comparison of Optimisation and Test Performance: The generalisation ability of random forest models can be measured when they are optimised on Western mixtures and tested on mixtures with ethnic samples. First, let the null hypothesis be that errors $e_W(\widehat{\Phi}_W)$ and $e_M(\widehat{\Phi}_W)$ have the same distribution. It is rejected by means of paired Wilcoxon test[5] with p = 6.4304e-04 (columns 4 and 5 of Table 4 are compared). Second, we prove the null hypothesis that the hypervolumes $\widehat{h}_W$ and $h_M(\widehat{h}_W)$ have the same distribution (columns 7 and 8). The hypothesis is rejected with p = 9.3509e-04. This stands for the following statistical observation: *models trained for the recognition of Western instruments in intervals/chords and optimised with evolutionary multi-objective feature selection have a significantly decreased generalisation performance with regard to smallest*

---

[5] For all applied tests in this paper, we use a standard value of 5% for the significance level.

classification error as well as hypervolume when applied for mixtures with ethnic samples.[6]

Comparison of Test Performances: Even if optimisation errors of the best feature sets are generally smaller than the corresponding test errors, it may be questioned whether the best feature set for Western mixtures would be also the best for Western/ethnic mixtures. $e_M(\widehat{\Phi}_M)$ (column 6) is always lower than $e_M(\widehat{\Phi}_W)$ (column 5). This means that if models are optimised only on Western mixtures, the best model is not necessarily equal to the model with the best generalisation ability. The null hypothesis that both errors have the same distribution is rejected with p = 4.3778e-04. Similarly, $\widehat{h}_M$ is significantly larger than $h_M(\widehat{h}_W)$ with p = 4.3778e-04 (columns 8 and 9). This leads to the following observation: *models with lowest optimisation errors found after multi-objective evolutionary feature selection do not correspond to models with lowest test errors, and non-dominated fronts after the optimisation of feature selection in Western mixtures have significantly lower hypervolume than non-dominated fronts built from feature sets validated on mixtures of Western and ethnic samples.*

3.3 MRMR Feature Selection

In the second experiment, MRMR feature selection is applied, starting with an empty feature set and increasing the number of features adding each time a feature with the smallest correlation to the already selected ones and the largest correlation to the category. Figure 2 plots the progress of e_W (thick line) and e_M (thin line) during MRMR selection for the recognition of flute in music intervals. e_W strongly drops at the beginning (using too few features is obviously not sufficient) and starts to increase slowly after some optimal number of features is achieved. Too many features increase the danger of overfitting, so that some of them are identified as relevant for the training set but are less relevant for the validation set. The general trend of e_M progress is similar to e_W (an initial decrease followed by a slow increase), but is characterised by higher errors and other local optima. The left half of Table 5 (columns 4–7) provides a summary of experiment results with MRMR feature selection.

Comparison of Optimisation and Test Errors: As in the experiment from the previous section, let the null hypothesis be that $e_W(\widehat{\Phi}_W)$ and $e_M(\widehat{\Phi}_W)$ (columns 4 and 5) have the same distribution. It is rejected by means of paired Wilcoxon test with p = 4.3778e-04. This stands for the observation that *models trained for the recognition of Western instruments in intervals and chords and optimised with the help of MRMR feature selection have a decreased generalisation performance when applied for mixtures with ethnic samples.*

Comparison of Test Errors: The null hypothesis that $e_M(\widehat{\Phi}_W)$ and $e_M(\widehat{\Phi}_M)$ (columns 5 and 6) have the same distribution is rejected with p = 4.3778e-04.

---

[6] The statistical observations are shortened for simplicity reasons and should be interpreted with certain restrictions. Obviously, they hold only for tested instruments, mixtures, features, feature processing, and feature selection method.

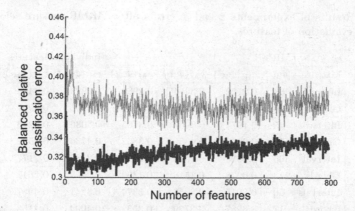

Fig. 2. Progress of optimisation (thick blue line) and test (thin red line) errors during MRMR feature selection for the recognition of flute in music intervals. (Color figure online)

This means that *models with lowest optimisation errors found after MRMR feature selection typically do not correspond to models with lowest test errors.*

Comparison to Evolutionary Feature Selection: Let us now compare the smallest errors found with the help of MRMR and evolutionary multi-objective feature selection. First, let the null hypothesis be that both optimisation errors $e_W(\widehat{\Phi}_W)$ have the same distributions (4th columns in Tables 4 and 5). It is rejected with p = 4.3778e-04. The largest relative advantage of evolutionary feature selection is for electric guitar/intervals ($e_W(\widehat{\Phi}_W)$ increases from 0.0417 to 0.0623 with MRMR), and the smallest for flute/chords (0.3283 to 0.3287). For the comparison of generalisation performance of sets $\widehat{\Phi}_W$, let the null hypothesis be that test errors $e_M(\widehat{\Phi}_W)$ have the same distributions (5th columns in Tables 4 and 5). It is not rejected with p = 0.7564. Comparing individual values of $e_W(\widehat{\Phi}_W)$, we observe that only for 9 of 16 categories $e_W(\widehat{\Phi}_W)$ is smaller (better) after evolutionary feature selection. The results of both tests let us suggest that *the significant advantage of evolutionary multi-objective feature selection at finding the feature set with the smallest error compared to MRMR turns to a possible drawback of overfitting when the generalisation ability is measured by the performance on mixtures with added ethnic instruments.*

3.4 Analysis of Individual Features

In the final experiment, we do not apply feature selection at all but validate the features individually. The models – using all features from 795 individually one after each other – are trained again on Western data sets. Similarly to the first experiment, the 5-fold cross-validation error is estimated and the performance is measured on Western/ethnic mixtures. We select the "best" feature ($|\widehat{\Phi}_W| = 1$) which recognises an instrument with the lowest classification error, and measure

Table 5. Results of experiments 2 and 3: errors after MRMR feature selection and individual evaluation of features.

Category		MRMR				Individual features		
Instrument	Mixture	$|\widehat{\varPhi}_W|$	$e_W(\widehat{\varPhi}_W)$	$e_M(\widehat{\varPhi}_W)$	$e_M(\widehat{\varPhi}_M)$	$e_W(\widehat{\varPhi}_W)$	$e_M(\widehat{\varPhi}_W)$	$e_M(\widehat{\varPhi}_M)$
Ac. guitar	Intervals	63	0.2102	0.3454	0.2989	0.2923	0.3492	0.3197
Ac. guitar	Chords	136	0.2420	0.3933	0.3677	0.2995	0.3927	0.3837
Cello	Intervals	45	0.1977	0.3907	0.3705	0.4036	0.4213	0.3976
Cello	Chords	59	0.2840	0.4336	0.3862	0.4436	0.4522	0.4355
El. guitar	Intervals	305	0.0623	0.1928	0.1044	0.2182	0.2177	0.2177
El. guitar	Chords	213	0.0978	0.1516	0.1369	0.2523	0.2617	0.2617
Flute	Intervals	50	0.3015	0.3738	0.3374	0.3241	0.4987	0.3519
Flute	Chords	17	0.3287	0.3729	0.3539	0.4094	0.4158	0.4059
Piano	Intervals	172	0.1864	0.2812	0.2245	0.3864	0.3638	0.3638
Piano	Chords	191	0.1990	0.2348	0.2164	0.3258	0.2849	0.2849
Trumpet	Intervals	35	0.1311	0.2138	0.1900	0.2701	0.3073	0.3050
Trumpet	Chords	28	0.1608	0.2372	0.2282	0.2784	0.2958	0.2958
Viola	Intervals	426	0.3284	0.4841	0.4051	0.4523	0.3858	0.3858
Viola	Chords	274	0.3411	0.4404	0.4081	0.4417	0.4760	0.4385
Violin	Intervals	15	0.3298	0.3836	0.3346	0.4492	0.4307	0.4146
Violin	Chords	64	0.3633	0.3777	0.3441	0.4387	0.4355	0.4355

the performance of models created with this feature also on mixtures with ethnic samples. The right half of Table 5 lists the results (columns 7–9).

Comparison of Best Optimisation and Corresponding Test Errors: For individual features, we may expect that the difference between $e_W(\widehat{\varPhi}_W)$ and $e_M(\widehat{\varPhi}_W)$ will be lower than for sets optimised with feature selection towards the best performance for Western mixtures. Further, we would expect higher errors on mixtures with ethnic instruments. The null hypothesis ($e_W(\widehat{\varPhi}_W)$ and $e_M(\widehat{\varPhi}_W)$ have the same distribution) is rejected with p = 6.4304e-04. *Individual features which perform best for identification of Western instruments in Western mixtures perform worse on Western/ethnic mixtures.*

Comparison of Best Optimisation and Best Test Errors: Here, the null hypothesis is that $e_M(\widehat{\varPhi}_W)$ and $e_M(\widehat{\varPhi}_M)$ have the same distribution. It is rejected with p = 1.2207e-04. *The individually best features for recognition of Western instruments in Western mixtures are not the same as the individually best features for the recognition of Western instruments in mixtures with ethnic samples.*

Table 6 lists individually best features for the recognition of Western instruments in Western (column 2) and Western/ethnic mixtures (column 3). Only for the classification task trumpet/intervals the same feature was individually best for both data sets.

Table 6. Individually best features for recognition of Western instruments in intervals and chords. Att: features from middles of attack intervals; Ons: features from onset frames; Rel: features from middles of release intervals; RMS: root mean square; ZCR: zero crossing rate.

Category		Western mixtures	Western/ethnic mixtures
Ac. guitar	Intervals	Att(MFCC 2)	Att(Angles in phase domain)
Ac. guitar	Chords	Att(MFCC 2)	Att(Angles in phase domain)
Cello	Intervals	Ons(RMS for ERB band 2)	Ons(ZCR for ERB band 1)
Cello	Chords	Att(MFCC 4)	Att(Bark scale magnitude 19)
El. guitar	Intervals	Rel(RMS for ERB band 8)	Rel(MFCC 3)
El. guitar	Chords	Rel(RMS for ERB band 8)	Rel(Bark scale magnitude 18)
Flute	Intervals	Ons(RMS for ERB band 2)	Rel(RMS for ERB band 2)
Flute	Chords	Ons(MFCC 3)	Att(RMS for ERB band 2)
Piano	Intervals	Rel(RMS for ERB band 9)	Rel(Bark scale magnitude 18)
Piano	Chords	Rel(Bark scale magnitude 19)	Att(Low energy)
Trumpet	Intervals	Rel(MFCC 2)	Rel(MFCC 2)
Trumpet	Chords	Rel(Spectral extent)	Rel(Spectral extent)
Viola	Intervals	Ons(RMS for ERB band 2)	Att(MFCC 5)
Viola	Chords	Ons(Low energy)	Att(Low energy)
Violin	Intervals	Att(RMS for ERB band 2)	Rel(RMS for ERB band 2)
Violin	Chords	Att(y-axis intercept)	Att(Bark scale magnitude 2)

4   Conclusions

In this study, we measured the generalisation performance of Western instrument recognition models when applied to polyphonic mixtures with ethnic samples. As optimisation method, evolutionary multi-objective and MRMR feature selection were used. Classification errors increased significantly for feature sets optimised on Western mixtures, and these sets were not the best for the recognition of the same instruments in Western/ethnic mixtures. Evolutionary feature selection seems to be a more powerful method for the identification of best feature sets, but is more sensitive to over-optimisation on the other side. In the last experiment with models created with individual features, the features with the highest relevance for Western and Western/ethnic mixtures were almost always not the same.

Being probably the first study which compares instrument recognition models between two musical "worlds", many research issues remain to be resolved. First of all, we should try to understand why the models optimised on Western mixtures fail when ethnic instruments are added. This can help us to build new signal and timbral descriptors with a higher generalisation ability. Because a clear distinction between "Western" and "ethnic" timbre is not always possible, such mid-level features will increase the objectivity and interpretability of instrument recognition models. Second, more different parameters must be taken into account: other methods for feature aggregation, other classification algorithms, other feature selection and optimisation methods. Third, the complete setup can be reversed, with the goal to recognise ethnic instruments.

We hope that with our study the interest will grow to establish new measures of robustness and generalisation performance in music classification tasks, and that the research community will be aware of the danger of over-optimisation towards specific data sets.

References

1. Abdoli, S.: Iranian traditional music dastgah classification. In: Proceedings of the 12th International Society for Music Information Retrieval Conference (ISMIR), pp. 275–280 (2011)
2. Agarwal, P., Karnick, H., Raj, B.: A comparative study of Indian and western music forms. In: Proceedings of the 14th International Society for Music Information Retrieval Conference (ISMIR), pp. 29–34 (2013)
3. Benetos, E., Holzapfel, A.: Automatic transcription of Turkish makam music. In: Proceedings of the 14th International Society for Music Information Retrieval Conference (ISMIR), pp. 355–360 (2013)
4. Brown, J.C., Houix, O., Mcadams, S.: Feature dependence in the automatic identification of musical woodwind instruments. J. Acoust. Soc. Am. **109**(3), 1064–1072 (2001)
5. Ding, C.H.Q., Peng, H.: Minimum redundancy feature selection from microarray gene expression data. J. Bioinform. Comput. Biol. **3**(2), 185–205 (2005)
6. Eerola, T.: Are the emotions expressed in music genre-specific? An audio-based evaluation of datasets spanning classical, film, pop and mixed genres. J. New Music Res. **40**(3), 349–366 (2011)
7. Eerola, T., Ferrer, R.: Instrument library (MUMS) revised. Music Percept. **25**(3), 253–255 (2008)
8. Eggink, J., Brown, G.J.: Instrument recognition in accompanied sonatas and concertos. In: Proceedings of the IEEE International Conference on Acoustics, Speech, and Signal Processing (ICASSP), pp. 217–220 (2004)
9. Emmerich, M., Beume, N., Naujoks, B.: An EMO algorithm using the hypervolume measure as selection criterion. In: Coello Coello, C.A., Hernández Aguirre, A., Zitzler, E. (eds.) EMO 2005. LNCS, vol. 3410, pp. 62–76. Springer, Heidelberg (2005). doi:10.1007/978-3-540-31880-4_5
10. Eronen, A.J., Klapuri, A.: Musical instrument recognition using cepstral coefficients and temporal features. In: Proceedings of the IEEE International Conference on Acoustics, Speech, and Signal Processing (ICASSP), pp. 753–756 (2000)
11. Essid, S., Richard, G., David, B.: Instrument recognition in polyphonic music based on automatic taxonomies. IEEE Trans. Audio Speech Lang. Process. **14**(1), 68–80 (2006)
12. Fuhrmann, F.: Automatic musical instrument recognition from polyphonic music audio signals. Ph.D. thesis, Universitat Pompeu Fabra (2012)
13. Gaikwad, S., Chitre, A.V., Dandawate, Y.H.: Classification of Indian classical instruments using spectral and principal component analysis based cepstrum features. In: Proceedings of the 2014 International Conference on Electronic Systems, Signal Processing and Computing Technologies (ICESC), pp. 276–279 (2014)

14. Goto, M., Hashiguchi, H., Nishimura, T., Oka, R.: RWC music database: Music genre database and musical instrument sound database. In: Proceedings of the 4th International Conference on Music Information Retrieval (ISMIR), pp. 229–230 (2003)
15. Gunasekaran, S., Revathy, K.: Fractal dimension analysis of audio signals for Indian musical instrument recognition. In: Proceedings of the International Conference on Audio, Language and Image Processing (ICALIP), pp. 257–261 (2008)
16. Guyon, I., Nikravesh, M., Gunn, S., Zadeh, L.A.: Feature Extraction: Foundations and Applications. Springer, Heidelberg (2006)
17. Hastie, T., Tibshirani, R., Friedman, J.: The Elements of Statistical Learning. Springer, New York (2009)
18. Heittola, T., Klapuri, A., Virtanen, T.: Musical instrument recognition in poly-phonic audio using source-filter model for sound separation. In: Proceedings of the 10th International Society for Music Information Retrieval Conference (ISMIR), pp. 327–332 (2009)
19. Koduri, G.K., Miron, M., Serrà, J., Serra, X.: Computational approaches for the understanding of melody in carnatic music. In: Proceedings of the 12th International Society for Music Information Retrieval Conference (ISMIR), pp. 263–268 (2011)
20. Lartillot, O., Toiviainen, P.: MIR in Matlab (II): A toolbox for musical feature extraction from audio. In: Proceedings of the 8th International Conference on Music Information Retrieval (ISMIR), pp. 127–130 (2007)
21. Lashari, S.A., Ibrahim, R., Senan, N.: Soft set theory for automatic classification of traditional pakistani musical instruments sounds. In: Proceedings of the International Conference on Computer Information Science (ICCIS), pp. 94–99 (2012)
22. Lidy, T., Silla Jr., C.N., Cornelis, O., Gouyon, F., Rauber, A., Kaestner, C.A.A., Koerich, A.L.: On the suitability of state-of-the-art music information retrieval methods for analyzing, categorizing and accessing non-Western and ethnic music collections. Signal Process. **90**(4), 1032–1048 (2010)
23. Livshin, A., Rodet, X.: The significance of the non-harmonic "noise" versis the harmonic series for musical instrument recognition. In: Proceedings of the 7th International Conference on Music Information Retrieval (ISMIR), pp. 95–100 (2006)
24. McEnnis, D., McKay, C., Fujinaga, I.: jAudio: Additions and improvements. In: Proceedings of the 7th International Conference on Music Information Retrieval (ISMIR), pp. 385–386 (2006)
25. Mierswa, I., Morik, K.: Automatic feature extraction for classifying audio data. Mach. Learn. J. **58**(2–3), 127–149 (2005)
26. Müller, M.: Information Retrieval for Music and Motion. Springer, Heidelberg (2007)
27. Müller, M., Ewert, S.: Chroma toolbox: MATLAB implementations for extracting variants of chroma-based audio features. In: Proceedings of the 12th International Society for Music Information Retrieval Conference (ISMIR), pp. 215–220 (2011)
28. Newton, M., Smith, L.: A neurally inspired musical instrument classification system based upon the sound onset. J. Acoust. Soc. Am. **131**(6), 4785–4798 (2012)
29. Sandrock, T.: Multi-label feature selection with application to musical instrument recognition. Ph.D. thesis, Stellenbosch University (2013)
30. Srinivasamurthy, A., Holzapfel, A., Serra, X.: In search of automatic rhythm analysis methods for Turkish and Indian art music. J. New Music Res. **43**, 94–114 (2014)
31. Sturm, B.: Evaluating music emotion recognition: Lessons from music genre recognition? In: Proceedings of the IEEE International Conference on Multimedia and Expo (ICME), pp. 1–6 (2013)

32. Tzanetakis, G., Cook, P.: Musical genre classification of audio signals. IEEE Trans. Speech Audio Process. **10**(5), 293–302 (2002)
33. Vatolkin, I., Preuß, M., Rudolph, G., Eichhoff, M., Weihs, C.: Multi-objective evolutionary feature selection for instrument recognition in polyphonic audio mixtures. Soft Comput. Fusion Found. Methodologies Appl. **16**(12), 2027–2047 (2012)
34. Vatolkin, I., Rudolph, G., Weihs, C.: Evaluation of album effect for feature selection in music genre recognition. In: Proceedings of the 16th International Society for Music Information Retrieval Conference (ISMIR), pp. 169–175 (2015)
35. Zitzler, E., Thiele, L.: Multiobjective optimization using evolutionary algorithms - A comparative case study. In: Proceedings of the 5th International Conference on Parallel Problem Solving from Nature (PPSN), pp. 292–304 (1998)

Exploring the *Exactitudes* Portrait Series with Restricted Boltzmann Machines

Sam D. Verkoelen, Maarten H. Lamers[✉], and Peter van der Putten

Media Technology Group, Leiden Institute of Advanced Computer Science,
Leiden University, Leiden, The Netherlands
sdverkoelen@gmail.com,
{m.h.lamers,p.w.h.van.der.putten}@liacs.leidenuniv.nl

Abstract. In this paper we explore the use of deep neural networks to analyze semi-structured series of artworks. We train stacked Restricted Boltzmann Machines on the Exactitudes collection of photo series, and use this to understand the relationship between works and series, uncover underlying features and dimensions, and generate new images. The projection of the series on the two major decorrelated features (PCA on top of Boltzmann features) results in a visualization that clearly reflects the semi structured nature of the photos series, although the original features provide better classification results when assigning photographs to series. This work provides a useful case example of understanding structure that is uncovered by deep neural networks, as well as a tool to analyze the underlying structure of a collection of visual artworks, as a very first step towards a robot curator.

1 Introduction

Exactitudes is an ongoing photography project by artists Ari Versluis and Ellie Uyttenbroek that started in 1993 and aims to document subcultures omnipresent in society ([14] of which a 6th ed. appeared in 2014). A sample of each subculture is captured in a series of twelve photographic portraits, all systematically taken from the same angle, with a neutral background and similar body position (Fig. 1). At present the project contains 154 series of twelve color photographs each, taken in the period 1994–2014.

The artists themselves appointed in an interview [9] that it "should almost be a scientific anthropological record of people's attempts to distinguish themselves from others by assuming a group identity". Although the portraits were intentionally presented by Versluis and Uyttenbroek within an arts context, they hold potential value for academics. In a study of sensory experience and affect in relation to denim clothing [1] the author compares different ways of wearing jeans, using Exactitudes as one of the resources. Other studies have discussed Exactitudes in the context of design [5], design teaching [12], fashion [6, 11], and societal changes in consumption, activities and household organization [2].

J. Correia et al. (Eds.): EvoMUSART 2017, LNCS 10198, pp. 321–337, 2017.
DOI: 10.1007/978-3-319-55750-2_22

Fig. 1. Four example *Exactitudes* series [14]. Left to right: 4 (*Bimbos*), 41 (*Surfistas*), 100 (*Cocktails & Dreams*), 154 (*United Americans*).

Our work aims to explore the potential of applying deep learning neural networks to highly structured sets of visual artworks, and uses the Exactitudes collection of photographs as an example exploration. Primarily, we are interested in uncovering shared visual structure and variations thereon in sets of visual artworks, in relation to meta-features of these works. For example, do variations among a collection of visual artworks co-vary over time, or do sub-classes within a larger collection of visual artworks correlate with combinations of visual features and meta-variables?

Clearly, the Exactitudes collection of photographs is visually highly structured. Within each series (twelve photographs depicting a subculture) the subjects wear clothing that identifies their sharing of a group identity, and the body position and stance are uniformly fixed. Moreover, visual identification of gender and (often) race and age are fixed within series. Across different series clothing, gender, race and body position/stance vary. However, overall position of the subject's body within the frame appears stable with few exceptions: most series depict subjects as full upper body, starting between hip and knee height, but five series deviate from this. Ultimately, all portraits are of equal horizontal and vertical dimensions and all backgrounds are uniformly grey/white, lacking visual features.

Our approach adheres to a common computational intelligence paradigm of viewing individual images, i.c. portraits, as separate points in a high-dimensional "pixel space". The many dimensions of this space each express a pixel position (from a digital representation) of the portraits, and span the possible values that each pixel can have. Through non-linear dimensionality reduction [3, 8], we attempt to describe the individual portraits in much fewer dimensions, whilst as much as possible maintaining their underlying topographic relations in high-dimensional pixel space. The newly constructed dimensions in effect are non-linear combinations of the original pixel space dimensions, combined such that shared (non-linear) co-relations between the original dimensions in the data are encoded in fewer shared new dimensions.

A mapping of vectors in pixel space (i.e. images) to vectors in low-dimensional feature space, and v.v., are constructed within the resulting encoder. In effect, detailed information concerning shared visual structures in the portraits is captured inside the encoder tool, whereas individual portrait variations thereon are expressed as vectors spanning the lower-dimensional feature space. A deep neural network, in the form of stacked Restricted Boltzmann Machines [8], makes up our encoder.

In an exploratory manner and under the paradigm of discovery science [4], this study attempts to uncover as much information as possible about shared visual features among the Exactitudes portraits. It does so by scrutinizing the workings of the encoder, whilst treating it as a "black box" input-output processing device (c.f. [3]). In particular, we interpolate and extrapolate portraits by decoding positions in feature space to which none of the original Exactitudes portraits was precisely mapped. This method will expose both shared visual structures captured within the encoder, in particular their mutual non-linear dependencies and the individual variations thereon captured in portrait encodings. Subsequently, the individual portrait encodings will be viewed in relation to their meta-variables, such as series membership, so as to extract meaning from the encoding.

Through this process, we create knowledge on the application of deep learning neural networks to highly structured sets of visual artworks. Potential applications of this knowledge include in-depth discovery and understanding of structure in visual collections, automated curatorship of visual collections, automated authorship identification through outlier detection in lower-dimensional representations of visual works, and generative construction of new within-class visual works. Undoubtedly other potential applications exist.

In the remainder of this paper, we describe the methods applied, focusing more on the data set and the construction of our encoder tool. Subsequently, a subset from a larger series of small experiments [13] is described, along with their outcomes.[1] Finally, results and implications are reflected upon in a wider context of applying computational intelligence to visual arts.

2   Turning Exactitudes into Data

Digital portrait files were gathered with permission from the artists from their dedicated Exactitudes website (www.exactitudes.com). Structured filenames allowed for simple web-scraping techniques to collect all 1848 image files (154 series of 12 portraits), each a 600×600 pixel JPEG compressed RGB image with no alpha channel.

Five series of which the overall body framing of the portraits deviated significantly from the remaining 149 series were removed from further consideration (1, 7, 16, 18, and 77 using original numbering). Despite the uniform nature of the data, some further pre-processing was undertaken as to facilitate learning of the data by a deep learning neural network. Digital portraits were resized to 50×50 pixels and converted to 8-bit greyscale. Although this step is irreversible, it allows our experiments to be executed within a feasible timespan. Subsequently, pixel intensities were scaled to range [0,1] since our neural network model requires input values within this range.

Sparse representations of data have a number of benefits for energy based learning models as Restricted Boltzmann Machines [10], such as increased likelihood of correct classification. Sparsity i.c. means that the input vector representations consist mainly of

---

[1] Generated experimental images, including and extending beyond those in this paper, are available for download at high resolution from https://samverkoelen.com/evomusart17/.

zero values, with a limited number of non-zero values. A histogram of the greyscale pixel values shows that our data is not sparse. To magnify sparsity of the data, pixel values $x \in [0, 1]$ were inverted after applying a sigmoidal contrast enhancing function in which gain g was set to 40 and cutoff c to 0.25, both values determined experimentally:

$$invcontrast(x) = 1 - \left(\frac{1}{1 + e^{g(c-x)}} \right)$$

This inverting contrast enhancement function creates sparse binary-like data vectors and can be reversed without any loss of information through applying $invcontrast^{-1}(x)$. What remains of our data is a sparse image space ($\mathbb{R}^{2500}$) containing 1788 representations of Exactitudes portraits.

3 Deep Learning Stacked Restricted Boltzmann Machines

We apply auto-encoding neural network models [3, 8] for non-linearly reducing the dimensionality of our data. Specifically, our model consists of stacked Restricted Boltzmann Machines (RBMs), as proposed by Hinton [8]. Multiple RBMs are stacked on top of each other, where the feature layer (a.k.a. hidden layer) of a lower RBM provides inputs to the visible layer of the next-higher RBM. Moving upward, the number of feature layer units in each subsequent RBM is gradually reduced, and through an unsupervised contrastive divergence learning procedure this "deep" neural network stack is trained layer by layer.

Through pilot experimentation applying different numbers and sizes of RBM layers, it was assessed that five stacked RBMs were suitable for our data. In five steps, they transform the sparse image data from $\mathbb{R}^{2500}$ to feature vectors in $\mathbb{R}^{4000}$, $\mathbb{R}^{2000}$, $\mathbb{R}^{1000}$, $\mathbb{R}^{500}$ and finally $\mathbb{R}^{50}$. The initial "broadening" of the input vectors from $\mathbb{R}^{2500}$ to $\mathbb{R}^{4000}$ is required to capture the subtle non-binary character of the input data (with values approaching 0 and 1, but not limited to 0 and 1) in a larger subsequent binary layer [8]. The process of propagating data "upward" through the RBM stack transforming it from $\mathbb{R}^{2500}$ to $\mathbb{R}^{50}$ is referred to as "encoding". The resulting feature space is referred to as "RBM^{50}".

This architecture was selected based on visual assessment of the images' reconstruction quality for different architecture variations. Image reconstructions are created by propagating encoded RBM^{50} vectors "downward" through the RBM stack, expanding to $\mathbb{R}^{2500}$ and subsequently pixel-wise applying $invcontrast^{-1}(x)$, a process henceforth referred to as "decoding". Since the mentioned reduction of Exactitudes portrait sizes and color information is irreversible, decoded (a.k.a. reconstructed) images are 50×50 pixels and in greyscale tone.

Figure 2 illustrates a selection of input images and their reconstructions. A loss of visual information after RBM^{50} encoding/decoding of images is apparent and expected. Nonetheless, the human shape, posture, body position and luminosity of clothing are retained.

Fig. 2. Two selected portraits. Left to right: (1) normalized, (2) contrast enhanced, (3) reconstructed from RBM^{50}, and (4) contrast enhancement reversed.

A principal component analysis was applied to the 1788 vectors encoded in RBM^{50}. Mapping RBM^{50} vectors onto their ordered eigenvectors is an optional final computational step of our encoder that is lossless and reversible. Although this step does not further reduce the dimensionality of data, it provides (linear) decorrelation of RBM^{50} vectors and ordering of features by variance in the data explained. The encoder including this final step is named the "decorrelated encoder". The resulting feature space is referred to as "DRBM^{50}". Dimensions of (D)RBM^{50} are referred to as "features".

4 Exploration Through Experimentation

Under the paradigm of discovery science we subject the resulting encoder to a series of smaller experiments. These collectively form the exploration phase of our study, in which we attempt to uncover information about shared visual features among the Exactitudes portraits, and individual or grouped variation thereon.

Every experiment is described following the structure: method, results and observations. In this workflow, the results of one experiment may lead to a hypothesis that is tested in a following experiment. For sake of brevity, we can only provide selected images resulting from experiments. Larger versions and collections of experimental images are available from [13].

4.1 Straight Paths Between Feature Vectors Pairs

Method. For the 50 most distant pairs of (D)RBM^{50} portrait vectors, a reconstruction of their straight connecting path in (D)RBM^{50} is made in 10 equidistant steps. It is noteworthy to mention that the most distant pairs in RBM^{50} do not necessarily have great distance in DRBM^{50}.

Results. From the resulting 50 paths, a subset is chosen based on visual disparity, and shown in Fig. 3 (top two in RBM^{50}, bottom two in DRBM^{50}).

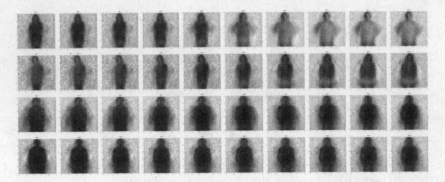

Fig. 3. Four linear connecting paths between two distant (D)RBM^{50} vectors.

Observations. The outermost reconstructions from RBM^{50} (top two) show a clear visual difference in color intensity, posture and body position. The outer reconstructions from DRBM^{50} (bottom two) have less visual differences, despite their relative large Euclidean distance. This indicates that distance in RBM^{50} is more related to visual differences, as opposed to distance in DRBM^{50}. Regardless of the position in either RBM^{50} or DRBM^{50}, each intermediate vector on a connecting line reconstructs to an image that could be assessed as human-shaped.

4.2   Random Points in Feature Space

Method. From the previous experiment one might hypothesize that every arbitrary point in (D)RBM^{50} reconstructs into an image with human shape. Here, 100 random points (c.f. a uniform probability distribution) in (D)RBM^{50} are reconstructed into images for visual review.

Results. A random selection is presented in Fig. 4.

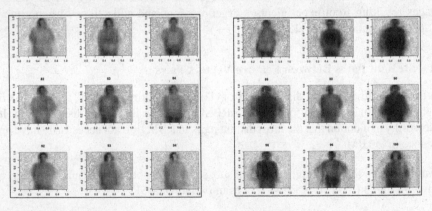

Fig. 4. Reconstruction from nine random points in RBM^{50} (left) and DRBM^{50} (right)

Observations. The hypothesis that every arbitrary point in (D)RBM^{50} reconstructs into and image with recognizable human shape is visually confirmed. It is notable that the reconstructions' color intensity differs between RBM^{50} and DRBM^{50} vectors, the latter appearing noticeably darker.

The reconstructions indicate that decorrelation changed the distribution of portrait vectors across RBM^{50}, placing darker photos throughout DRBM^{50} whilst the lighter ones are closer together. This observation might also explain the lack of visual differences in DRBM^{50} reconstructions in Fig. 3.

4.3   Activation Distributions Per Feature

Method. Activations per (D)RBM^{50} feature are summarized in box plot visualizations, over the entire dataset ($n = 1788$).

Results. Features of DRBM^{50} are ordered by the variance each one explains of the portrait collection. Thus the first explains most, subsequently decreasing until the least explanatory feature. Figure 5 (right) shows activations of the first and last three features of DRBM^{50}. In RBM^{50} there is no particular ordering of features; a random subset of 7 features and their activation distributions are shown in Fig. 5 (left).

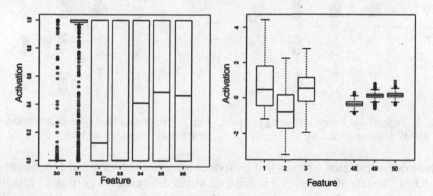

Fig. 5. Box plots of RBM^{50} (left) and DRBM^{50} (right) activations over the entire dataset.

Activations are considered an outlier if the value is outside 1.5 times the interquartile range above the upper quartile or below the lower quartile. Outlier positions are marked with a small circle in both box plots.

Observations. As expected, lower-numbered DRBM^{50} features show more variance than higher-numbered ones. Portrait activations in RBM^{50} are distributed differently because the feature order is arbitrary, and many differences exists (Fig. 5, left). For example feature 30 is fairly outspoken at zero with a few exceptions, whereas feature 34 is about similar in high and low activations.

Activations in DRBM^{50} are mostly around the center whereas activations in RBM^{50} are mostly located on the outer edges. For the latter, a potential explanation is the probabilistic nature of Restricted Boltzmann Machines [8] in combination with contrast enhancement.

4.4 Highest and Lowest Activating Portraits Per Feature

Method. As observed in experiment Sect. 4.3, feature activations in RBM50 are disordered and can be outspoken, contrasting DRBM50 in which features are ordered and activations congregate around center-values. We now order all portraits by their activation on every individual feature in (D)RBM50, revealing the highest and lowest activating portraits per feature.

Results. Figure 6 illustrates the highest and lowest activating portraits for RBM^{50} features 30 and 34. These features were selected based on results of experiment Sect. 4.3. Figure 7 illustrates the same for DRBM^{50} features 1 and 50.

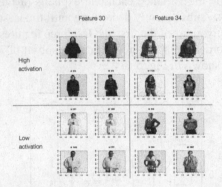 

Fig. 6. Highest and lowest activating portraits on RBM^{50} features 30 and 34.

Fig. 7. Highest and lowest activating portraits on DRBM^{50} features 1 and 50.

Observations. Figure 6 shows that outspoken activations observed in experiment 4.3 on RBM^{50} feature 30 correspond to apparent visual differences in portraits, i.c. mainly the color intensity. In contrast, less outspoken RBM^{50} feature 34 correlates with less salient visual differences. Figure 7 shows the highest and lowest activating portraits on the first DRBM^{50} feature (left), exposing a correlation with both color intensity and posture. Less outspoken DRBM^{50} feature 50 (right) correlates to less visually salient differences.

4.5 Single Feature Variations in Portrait Context

Method. To visualize effects of feature changes on portrait reconstructions, every (D)RBM^{50} feature is individually set to its lowest and highest observed value over the dataset. All other feature activations are set according to an existing portrait, providing context to the visual assessment of reconstructions.

Results. The complete results consist of the high and low feature variations, for every (D)RBM50 feature on a randomly selected subset of 10 portraits [13]. Figure 8 shows variations for features 30–35 on a single portrait.

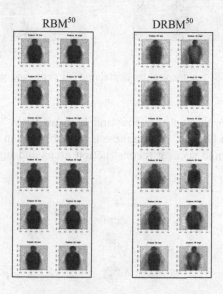

RBM^{50} DRBM^{50}

Fig. 8. Variations on one portrait, setting features 30-35 (top to bottom) to their lowest (left) and highest (right) activations, in RBM^{50} and DRBM^{50}.

Observations. RBM^{50} reconstructions in Fig. 8 lack apparent visual diversity, whereas in DRBM^{50} they show visual diversity not only across features 30–35, but also across low and high activations. E.g. DRBM^{50} feature 30 correlates to color intensity, and 34 to apparent arms position.

4.6 Single Activation Feature Vectors

Method. Encouraged by the results of experiment Sect. 4.5, here all (D)RBM^{50} features are set to their minimum possible value (zero) with the exception of a single highlighted feature being set to its maximum value (one). Reconstructions were made as such for all features once highlighted (50 in RBM^{50}, 50 in DRBM^{50}). To visually enhance differences between reconstructions, their mean reconstruction ($n = 50$) was subtracted from each individual reconstruction.

Results. A single visually salient feature was selected and presented in Fig. 9. High scoring areas in the reconstructions appear as lighter color intensities, and v.v.

Observations. Reconstructions on all highlighted features show high and low scoring areas on various body parts [13]. Visually apparent in the RBM^{50} reconstruction of Fig. 9 are the high scoring areas around the arm positions, whereas high scoring regions in the DRBM^{50} reconstruction are more "spread out". This topological salience is also

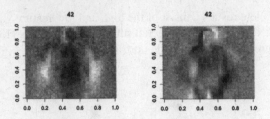

Fig. 9. Reconstructions of feature vectors containing all zeros and value 1 on feature 42, in RBM^{50} (left) and DRBM^{50} (right).

apparent from portraits that activate low and high on RBM^{50} feature 42 (Fig. 10), although it is less so in DRBM^{50} where there is arguably and in general a lesser visual relationship between high scoring reconstruction areas and high and low activating portraits.

Fig. 10. Highest and lowest activating portraits on RBM^{50} and DRBM^{50} features 42.

4.7   Portrait Distributions Over Feature Space

Method. Visualizing (D)RBM^{50} is hardly possible in a meaningful manner. However, the distribution of portrait encodings over (D)RBM^{50} can be essential information, e.g. for training a classifier. Although it captures not all information, a histogram of the Euclidian distance from the center of (D)RBM^{50} for all 1788 portrait encodings, provides a rough impression.

Results. Figure 11 shows the described histograms.

Observations. Distributions of portrait encodings across RBM^{50} and DRBM^{50} are quite different, as to be expected. Observations from experiment Sects. 4.2 and 4.3 are confirmed by these results: portrait encodings in RBM^{50} are mostly situated on the outer edges, whereas they are situated more around the center of DRBM^{50}.

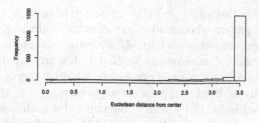

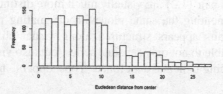

Fig. 11. Distribution of portrait encodings' Euclidian distance to the center of RBM^{50} (top) and DRBM^{50} (bottom); $n = 1788$.

4.8 Feature Pair Visualizations

Method. In an attempt to further understand the structure of (D)RBM^{50}, portrait activation values ($n = 1788$) on pairs of features are plotted, while color-coding for series ($n = 149$).

Results. Plots of portrait activations on feature pairs {1,2}, {3,4}, {5,6},..., {49,50} in RBM^{50} (Fig. 12) and DRBM^{50} (Fig. 13) feature space are made. Plots are ordered row by row, from left to right, and generated with the lowest feature number on their horizontal axis. Larger versions in [13]. Series membership is coded with random color assignment.

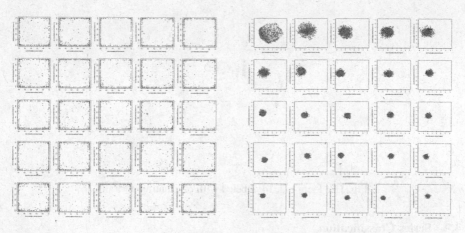

Fig. 12. Portrait activations on feature pairs in RBM^{50}.

Fig. 13. Portrait activations on feature pairs in DRBM^{50}.

Observations. Portrait encodings in RBM^{50} appear to have no particular ordering, but show a conspicuous pattern where activation values are distinctively along the space's edges (Fig. 12). Portrait encodings in DRBM^{50} form a circular clustered shape (Fig. 13). These observations are c.f. experiments Sects. 4.2, 4.3, and 4.7. Plots in Fig. 13 are clearly ordered by variance explained per feature, an effect that is also apparent from the slight horizontally oriented oval point clouds in plots.

Series are marked individually with a different color each, something that is not visually apparent among 149 series (Figs. 12 and 13). A smaller subset of series plotted onto DRBM^{50} features pair {1,2} are visually much more distinct and clearly clustered (Fig. 14). When augmenting the same plot with originating portraits (Fig. 15), the arrangement of portraits appears structured along visual characteristics. DRBM^{50} features pair {1,2} is able to not only separate series, but also visually distinct portraits within series, e.g. spacing light and dark toned 'Topshoppers' apart.

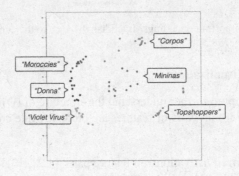

Fig. 14. Portrait activations on DRBM^{50} features pair {1,2} for six labeled series.

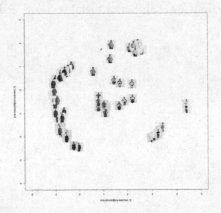

Fig. 15. Idem Fig. 14, labeled with originating portraits.

4.9 Series Classification

Method. From Figs. 14 and 15, questions arise regarding series classification from portrait encodings. A random forest classifier [7] is repeatedly trained and tested on randomly resampled training (1500) and test data (288). Results for three encodings are compared: RBM^{50}, $DRBM^{50}$, and mapping onto 50 principal components of the normalized portraits (referred to as PC^{50}).

Results. Classification correctness rates are reported in Table 1.

Table 1. Classification correctness rates after different encodings over three trials.

	RBM^{50}	$DRBM^{50}$	PC^{50}
Trial 1	0.622	0.517	0.642
Trial 2	0.615	0.531	0.677
Trial 3	0.628	0.535	0.667
Mean	0.622	0.528	0.662

Observations. All three methods are clearly successful in that they far outperform hypothetical 149-sided die classification (expected correctness 0.007). Nevertheless it appears that decorrelation of RBM^{50} features has negative effect on the ability to classify encoded portraits by series. Intuitively, this contrasts the visual distinctiveness that decorrelation appears to have added, c.f. Figures 12, 13, 14 and 15. This observation is even more striking when noting that PCA by itself (in the form of PC^{50} encoding) outperforms RBM^{50} in terms of classification, albeit only marginally.

4.10 Best and Worst Classifiable Portraits

Method. The relatively poor contribution of $DRBM^{50}$ to series classification appears to discord with prior expectations. Attempting to gain insight into the classification process, the best and worst classified series are inspected.

Results. Notably, the five best and worst classified series were the same under both RBM^{50} and $DRBM^{50}$ encoding, and illustrated in Fig. 16.

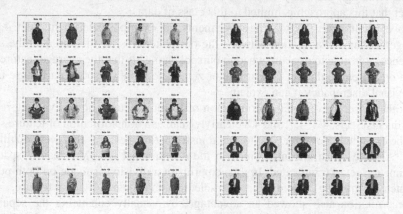

Fig. 16. Five best (left) and worst (right) classified series, each displayed horizontally.

Observations. The corresponding results for RBM^{50} and DRBM^{50} indicate that the classifier likely learns the same differences between series. This observation leaves uncertain why the classification results between both feature encodings differ. Notable also is the homogeneity within the best classified series. Although homogeneity is a characteristic of Exactitudes, it is to a greater or lesser degree present within individual series. As expected, the least classifiable series appear more heterogeneous.

5   Discussion and Future Work

In this concluding section we reflect on the experimental results, and introduce the robot curator as future potential application area for this type of research.

5.1   Analytical Reflections

Despite their popularity and interest from researchers, dimensionality reduction methods such as deep learning are commonly referred to as black boxes. This exploratory study aimed at giving an insight in the inner workings of such networks when applied to semi-structured portrait photos of the Exactitudes project. Understanding is gained using an exploratory approach, one might consider atypical: given a dimensionality reducing network, referred to as *encoder*, a series of small experiments was conducted. These experiments were of manageable size, making it possible to develop new experiments based on previously obtained results and explore encoders from different viewpoints.

This approach provided a number of noteworthy results that will be reflected upon. However, the encoder itself has a notable aspect also, namely the use of a sigmoid-shaped contrast enhancement function to force the input data into sparse binary-like vectors. Prior research did not indicate a similar approach to handle real valued data in Restricted Boltzmann Machines. Our results showed that the network could learn the binary-like data, whilst being able to reverse the transformation and reconstruct input images. Although our study only uses photos as network input, there is no reason to suspect that this approach is limited to this class of data.

Novel results emerged from the experiments. For example experiments Sect. 4.3 and 4.4 together suggest a relation between the outspokenness of a feature and the visual differences it is able to explain, wherein outspoken means that the average feature activation on the entire dataset is near one or zero. The more outspoken a feature is, the more visual discrepancy it can explain.

Furthermore, the optional decorrelation of feature vectors showed varying results. Experiments Sects. 4.3, 4.7, and 4.8 clearly visualize a transformation of feature space, as to be expected. However, due to this transformation, features that were distilled in RBM^{50} feature space are intertwined in the decorrelated DRBM^{50} feature space, as experiment Sect. 4.6 indicated. In this experiment a division is made on both posture and intensity, making it more difficult to select photos on an individual feature.

Decorrelation also showed an advantage, by visually enhancing the separation between series (Fig. 14), whilst simultaneously discriminating on visual differences (Fig. 15). These results make it worthwhile to use feature decorrelation when eyeballing visualizations for clusters or groups within the dataset. However, a distinction should be made between the human eye and classification algorithms. Experiment Sect. 4.9 shows that feature decorrelation adversely effects series classification, something that is difficult to explain from a computational viewpoint.

The same experiment also indicates that series classification on the first 50 principal components of the normalized portraits outperforms the previous methods. Although all classifications outperform random classification by a roughly 85 times higher success rate, it is noteworthy that a relatively simple technique such as principal component analysis can compete with contemporary dimension reduction algorithms. Despite the enormous popularity of dimension reduction or deep learning, it remains worthwhile to make a comparison with well-established techniques.

Furthermore, what this research has yielded in a broader sense is the approach itself. Because of its exploratory nature, the path to follow is less clear in comparison with more traditional hypothesis-driven studies. Thus, there is an increased need for structure and guidance. Performing small and manageable experiments that lead to new hypotheses and experiments in rapid succession is, to our best knowledge, novel for exploratory research into deep learning.

The method allows new insights to be fitted directly into new experiments, thus adjusting the direction of the research based on preliminary results and advancing insights. One must ensure in such exploratory research that the problem is viewed from multiple angles, something we have done to our very best in this study.

When looking back at the exploratory goal, we find this research has made a positive contribution to the understanding of deep neural networks. By means of various visualizations, a better understanding of what is learned in these networks was obtained. This understanding could not only help in the selection of relevant features but also bring light to possible shortcomings in the data set or previously unknown features that might be beneficial. Therefore a good understanding of what is learned adds considerable value in exploring and exploiting large datasets.

5.2 The Robot Curator: Artificial Curation as a Computational Creativity Task

In our introduction we identified prior research in arts and humanities as well as a range of potential future applications of our research. In this section we propose one specific area for future application: automated curation.

When creating an exhibition, choices are made about what works to include, and also in what context to present these works, for example by spatial organization. Hence curation is not just a value free reflection on works of the artists but can also been seen as a form of artistic expression in itself. The curator aims to tell a story not limited by chronological order only, may want to make the public aware of hidden relationships between works within an oeuvre or across artists, provide new ways of seeing connections and relationships or actually want to juxtapose works that each provide a different perspective to create dialogues or interventions.

Given that curation can be seen as a creative activity, it is a potential task of interest within computational creativity research. This can be approached as an artificial creativity task; both fully automated curation as well as curation augmentation, where man and machine have a more symbiotic role — comparable with artificial intelligence versus intelligence augmentation.

Our study can be seen as a simplified example of how computational learning techniques can be applied towards this aim. The various features provide new ways to organize Exactitudes series and individual portraits, or to provide a spatial ordering of series that could be leveraged for a real world exposition. For example series could be organized on a floor plan according to their topology in decorrelated feature space. Likewise, the encoder could be used to augment existing artworks with new portraits, through various generative methods given.

This is mostly barren territory for research and many directions could be explored. How can we create methods that are suitable within a semi-structured body of work by a single artist, but also generalize across artists? Moreover, there is no reason to restrict this approach to visual art only, or to artists working within a single domain. One could focus on just photography or paintings, but could also take multimodal information such as background texts, artist network structures or other background knowledge into account. Naturally, for augmented creativity approaches, the robot curator will need to make space for letting the human curator into the loop.

Acknowledgements. We acknowledge Ari Versluis and Ellie Uyttenbroek as the creators of Exactitudes and owners of all intellectual rights and privileges, and are grateful to them for sharing this wonderful collection with the world via www.exactitudes.com.

References

1. Candy, F.J.: The fabric of society: an investigation of the emotional and sensory experience of wearing denim clothing. Sociol. Res. Online 10(1) (2005)
2. Casimir, G.: Interaction of societal development and communication technology. Int. J. Home Econ. **4**(1), 3–13 (2011)
3. DeMers, D., Cottrell, G.: Non-linear dimensionality reduction. Adv. Neural Inf. Process. Syst. **5**, 580–587 (1993)
4. Doherty, P.C.: The Beginner's Guide to Winning the Nobel Prize: Advice for Young Scientists. Columbia University Press, New York (2008)
5. Gardien, P., Djajadiningrat, T., Hummels, C., Brombacher, A.: Changing your hammer: the implications of paradigmatic innovation for design practice. Int. J. Des. **8**(2), 119–139 (2014)
6. Arntzen, M.G.: Dress Code: The Naked Truth About Fashion. Reaktion Books (2015)
7. Han, J., Kamber, M., Pei, J.: Data Mining: Concepts and Techniques. Elsevier, New York (2011)
8. Hinton, G.E., Salakhutdinov, R.R.: Reducing the dimensionality of data with neural networks. Science **313**, 504–507 (2006)
9. Huffington Post. Artists Create Anthropological Photo Series (2012). www.huffingtonpost.com/2012/06/25/exactitudes-interview-art_n_1619483.html. Retrieved on: 10 Nov 2016
10. Ranzato, M.A., Poultney, C., Chopra, S., LeCun, Y.: Efficient learning of sparse representations with an energy-based model. In: Advances in Neural Information Processing Systems, vol. 19, pp. 1137–1144 (2006)

11. Smelik, A.: The performance of authenticity. ADDRESS J. Fashion Writ. Criticism **1**(1), 76–82 (2011)
12. Trocchianesi, R., Guglielmetti, I.: Design teaching and cultural companies: Languages, tools and methods toward a profitable involvement. Strateg. Des. Res. J. **5**(1), 49–57 (2012)
13. Verkoelen, S.D.: Exploring Dimensionality Reduction on Semi-structured Photos — A Closer Look at Exactitudes. Master's Thesis for the Media Technology programme, Leiden University (The Netherlands) (2015)
14. Versluis, A., Uyttenbroek, E.: Exactitudes. 010 Publishers, Rotterdam (2002)

Evolving Mondrian-Style Artworks

Miri Weiss Cohen[1]([✉]), Leticia Cherchiglia[2], and Rachel Costa[3]

[1] Department of Software Engineering, Braude College of Engineering, Karmiel, Israel
miri@braude.ac.il
[2] Department of Media and Information, College of Communication Arts and Sciences, Michigan State University, East Lansing, MI, USA
letslcherchiglia@gmail.com
[3] Department of Arts, State University of Minas Gerais, UFMG, Belo Horizonte, Brazil
rachelcocosta@gmail.com

Abstract. This paper describes a Genetic Algorithm (GA) software system for automatically generating Mondrian-style symmetries and abstract artwork. The research examines Mondrian's paintings from 1922 through 1932 and analyses the balances, color symmetries and composition in these paintings. We used a set of eleven criteria to define the automated system. We then translated and formulized these criteria into heuristics and criteria that can be measured and used in the GA algorithm. The software includes a module that provides a range of GA parameter values for interactive selection. Despite a number of limitations, the method yielded high quality results with colors close to those of Mondrian and rectangles that did not overlap and fit the canvas.

Keywords: Art- style mondrian · Genetic algorithm · Computer art

1 Introduction

Considered one of the founders of abstract painting, Piet Mondrian is known for his unique style, which he termed Neoplasticism [1, 2]. Neoplasticism reduces works of art to their essential variables, that is, to basic simple forms and colors, in the pursuit of pure abstraction. This presentation of forms serves as a substitute for a complex motif or scene that tells a story, one of the most important attributes of traditional art. The Neoplasticism conceptualism has found a way to change this by substituting scenes with built forms that serve as references to the conceptual objective of each artwork. These works are representations of a different sort in that they represent the concepts that created them. As a modernist painter, Mondrian was interested in finding clues in addition to painting for comprehending the world. He did so by configuring compositions based on geometrical figures and primary colors.

We chose to focus on Mondrian's works from the period between 1922 and 1932 due to their simplicity, which Henning [3] calls the purification of Mondrian's work. In 1926, Mondrian published an essay in the Vouloir magazine explaining his ideas about Neoplasticism [2]. In this article, he emphasized the importance of the relations and the rhythm between colors and forms. He referred to rhythm as the organization by

© Springer International Publishing AG 2017
J. Correia et al. (Eds.): EvoMUSART 2017, LNCS 10198, pp. 338–353, 2017.
DOI: 10.1007/978-3-319-55750-2_23

opposition that generates the balance, i.e., the dynamic equilibrium of each variable of the picture.

Understanding this relationship between colors and forms surpasses the capacity to explain what Mondrian did. It allows the reproduction of his paintings based on the essential idea that incited Mondrian's work. To this end, we developed a genetic algorithm based on a careful evaluation of the two variables—shapes and colors. The algorithm's results can provide clues to Mondrian's composition and offer the possibility of generating new digital art forms that could be perceived as Mondrian style. These will be assessed for their similarity and the degree to which they capture the "essence" of Mondrian's artworks.

In examining and analyzing the various works of Mondrian during the years in question (1922–1932), we referred to art researchers as well as to Mondrian's own writings in order to capture and formulate the basic essences and primary ideas of these works of art. Mondrian [2, 4, 5] interpreted the plastic medium as a flat plane or a rectangular prismatic field with lines and colors that represent the organization of the world. Filling the canvas with these two variables only is a challenge that must meet two major criteria.

The first criterion is equilibrium. Equilibrium means duality and serves as the basis of the composition. Since equilibrium is a matter of forces, the duality neutralizes opposing forces in order to build a composition based on geometry. These forces are the size and the colors of the rectangles as well as the space they occupy [6–8]. Thus, straight lines (rectangles) are somewhat delimited within the space they enclose. On the other hand, they are combined to create spaces that need to be balanced.

The second criterion is Mondrian's exclusive use of primary colors (red, blue, and yellow) and of what he refers to as non-colors (white, black, and grey). The non-colors balance the colors. The use of colors and non-colors depends on the combination of spaces, so that the colors are not symmetrical. Thus, the organization of the variables in the picture takes both positions and proportions into account. Hence, the value of the color is proportional to the area it occupies, and each color has a different value. For example, red is more powerful than yellow [8, 9].

Defining a central focal point achieves local unity by placing the various planes (rectangles) around the focal center. In some cases, the color planes have a tendency to refer to the central rectangle as the unifying element in the composition and as a dominant center with respect to placement of the other planes [5, 10].

The works have an interesting sense of composition, forming the compositional dynamics of an axis around which its various elements revolve. This dynamic placement of their spatial ambiguity conveys a tendency toward breaking down a stable centric organization. Schufreider [11] defines this deployment of elements as eccentrically related, in contrast to having a concentric relationship, which means they do not share a common center even though they are related to one another.

In another interpretation, Milner [1] defines a "dominant square without color" which the author claims the composition is "built on". This harmony of relations requires that each element maintain its own identity and its own center. On this basis, the elements may be eccentrically interrelated to one another without regard for a shared center so that their relationship is found in the interplay of their differences. We can define an "open square" which is held together by gathering its elements with respect to one another as if it were not a dominant square but rather provides some overall unity or serves as a whole that is then subsequently broken into parts.

Milner [1] defined "equivalence" as a pervasive balance through which each shape receives its due with respect to the others on the basis of a harmonic co-relation in which equivalent elements are neither subordinated to another nor merely coordinated with each other [8, 12].

In his works during this period, Mondrian completely defeats the diagonal [1]. Indeed, diagonals do not implicitly remain in the structure, leading to its structural disruption. Moreover, Mondrian excludes the diagonals as he carefully creates a formal structure in which they become irrelevant. Mondrian's strategy, of course, is to take a perfect square and fill it with imperfect rectangles (also defined by diagonals) designed to defeat the diagonals and the centricity they imply.

2   Previous Work

Many researchers worldwide have taken up the challenge of examining and analyzing Mondrian's composition. Its visual simplicity is one of the reasons it has become a target not only for art theorists [4, 8, 13–17], but also for computer scientists [14, 18–20] and psychologists [7, 21]. Because these works of art are created using two simple variables—lines (rectangles) and colors (bars)—the possibility of trying to understand the paintings beyond art theory becomes very attractive.

Investigation of Mondrian's work as a source for generating related computer images began almost fifty years ago. The initial attempt was Noll's [10] comparison of the 1917 "Composition with Lines" with a work generated by a computer. Noll's work yield computer-generated images were based on only one artwork as a source. Vaughan [12] generated quasi-Mondrian works using a BBC computer. They were called quasi-Mondrian because they were not very close to the real Mondrian paintings, despite having the same structure. Only in 1993 did McManus, Cheema and Stoker conduct the first research to use a number of Mondrian's artworks as references to generate new ones. Since then, many studies have attempted to do the same thing. Fogelman [22] achieved one of the best results. As good as these are, however, they are still very different from Mondrian's actual works. This difference results from the parameters and the evaluation of the composition, the factors that are the aim of the proposed algorithm.

The research field of evolutionary art and evolutionary aesthetic design combines evolutionary computation, computer graphics and in some cases a human advisor who participates interactively in creating the artwork. The question of whether the artwork or the design is creative remains a valid and interesting question until now. Many researchers [22–27] have studied the aesthetic criteria, the forms and the outcome images in an attempt to optimize and produce computer artwork. Many approaches have aimed at creating artificial artworks.

Lewis [6] surveys the first two decades of evolutionary artwork and some of the techniques used, beginning with one of the first attempts by Sims [28]. The art of artificial evolution by Romero and Machado [29] is a collection of studies conducted in the following decade. Some works of creating digital art are found in literature, for example, Trist et al. [30] developed an evolutionary-based software system called Shroud that tries to mimic artistic inspiration and creativity. The research by Bergen [25] goes through the stages of interactive creativity, while Liu [31] produced generative art images by means of complex mathematical functions.

Well-known artists have inspired researchers to develop evolutionary systems that attempt to produce artworks in a style similar to that of the artists' original work. These artists include evolving pop art style [32] and glitch art [26], Gauguin [33], Kandinsky [34], Esher [18] and Mondrian [18, 19, 22, 27]. de Silva Garza et al. [19, 27] proposed a GA system to produce artwork similar to that of Mondrian, but their focus was on the choice of potential artworks for the evolutionary algorithm. Their results were tested on those who are not art experts and they reported a basic correlation.

In the current study, we focus on reproducing Mondrian-style artwork without any human interaction. Our system is based on formalizing and defining the criteria and the relevant operators as defined by art researchers that have studied Mondrian's art and his writings. The goal is to create and generate a wide variety of Mondrian style digital artworks.

3 Proposed Approach

Automatically generating Mondrian evolutionary artwork with no human interaction is based on two basic principles: (a) the use of defined criteria, and (b) genetic algorithm methodology. Like many Evo Art systems [18, 23, 24, 29, 32, 35], the proposed approach does not use any human interaction except in the choice of preferred artworks for use or display.

We chose the following criteria to define our automated system. We then translated and formulized these criteria into heuristics and criteria that can be measured and used in the GA algorithm.

1. The system uses primary colors (red, blue, and yellow) and non-colors (white, black, and grey, which was considered white as well). Non-color planes are used to balance the color planes. These two planes can be defined analogically as matter and non-matter spaces.
2. The use of colors and non-colors depends on the combination of forms in each individual work and the two are not symmetrical.
3. The weight of each form is proportional to the area it occupies, but depends on weight given to each color. For example, red is more powerful than yellow Equilibrium involves this duality and forms the base of the composition.
4. The colors are frequently organized surrounding a single geometrical figure, which is primarily white [3].
5. The straight lines (slim rectangles) serve as the limits, and the use of right angles creates the forms and their combinations. The width of the lines contrasts with the empty space.
6. The composition is de-centric [11].

7. The relationship between the variables creates a rhythm [1], and placement of the variables in the picture takes positions and proportions into account. Small changes modify the entire structure and produce a different rhythm between the parts.
8. The relations between the rectangles and color spaces are based on opposing forces.
9. Mondrian's work is not about forms, but rather about relationships: "Vertical balances horizontal in both line and plane; the parts balance within the whole; color balances non-color" [1].
10. The totality of Mondrian's oeuvre is consistent in relation to each individual work [3].

4   Genetic Algorithm

Evolutionary Algorithms (EA) are stochastic search algorithms based on principles of natural selection and recombination [34]. In this sense, genetic algorithms (GA) are global, parallel, stochastic search methods. The original aim for using this type of method was to apply it to decision-making systems [36, 37].

The GA's diversity parameters undergo experimentation, which may influence the values chosen by the system and the results and detailed in [38].

A crucial issue in implementing genetic algorithms is the decision regarding how to encode the solutions as chromosomes so the algorithm will converge to good solutions [34]. Some encodings may cause the algorithm to generate infeasible and undesired solutions or may change and enlarge the search space, making it difficult to converge to reproduce Mondrian style artworks.

4.1   Input Data

Our work is based on choosing a group of 20 works of art by Mondrian produced between 1922 and 1932. These works, partially depicted in Fig. 2, will be used as the initial population for the GA. Each individual is a painting and is therefore described by one array list of strings (64-bit binary sequences) including the painting's width, length and evaluation score, as described in Sect. 4.2.

To input the paintings into the GA, we used an algorithm to manually map all shapes that composed each painting. We first opened the image file regarding the painting, then, for each shape of this painting, we used input devices to capture the color of the shape and its upper-left and bottom-right corners. Next, the algorithm would convert all information to a binary sequence and add the binary sequence to a text file. The GA algorithm used this text file as input to create the initial population.

This study could have considered a larger group of paintings. Nevertheless, the results indicate that the chosen group was sufficient to produce very good artificial artworks.

4.2   Representation and Chromosome Coding

The GA methodology requires that each solution (artwork) be represented by chromosome code. The input data (Fig. 1) was represented as a genotype of binary sequences that represents a valid phenotype. Each image can be described as a variable-sized array of 64-bit binary sequences. The following criteria were implemented:

- Each artwork is composed of a variable group of rectangles, thus a group of chromosomes.
- All shapes as rectangles, even the ones we could call lines.
- Each rectangle is represented by a color code (in RGB), the location of its upper-left corner (x and y coordinates mapped on the computer screen in pixels, relative to the limits of the canvas), its width (in pixels) and its height (in pixels).
- 24 bits are used for colors (8 red + 8 green + 8 blue), 10 bits for the x position of the upper-left corner, 10 bits for the y position of the upper-left corner, 10 bits for the width and 10 bits for the length, for a total of 64 bits.

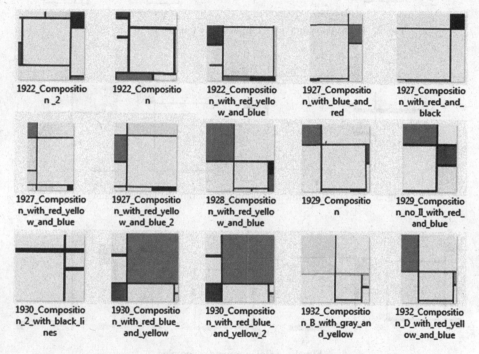

1922_Compositio n_2	1922_Compositio n	1922_Compositio n_with_red_yello w_and_blue	1927_Compositio n_with_blue_and_ red	1927_Compositio n_with_red_and_ black
1927_Compositio n_with_red_yello w_and_blue	1927_Compositio n_with_red_yello w_and_blue_2	1928_Compositio n_with_red_yello w_and_blue	1929_Compositio n	1929_Compositio n_no_II_with_red_ and_blue
1930_Compositio n_2_with_black_li nes	1930_Compositio n_with_red_blue_ and_yellow	1930_Compositio n_with_red_blue_ and_yellow_2	1932_Compositio n_B_with_gray_an d_yellow	1932_Compositio n_D_with_red_yell ow_and_blue

Fig. 1. Partial examples of Mondrian artworks used as input data (Color figure online)

4.3 Evaluation Function

The evaluation function comprises the following components: symmetry and color balance. Each component contributes to the analysis and the main aesthetic features, as outlined in Scct. 3. The aim of this stage is to translate the main observations into heuristics and measured criteria.

Symmetry: Balance/forces and Space: Diagonal vs. Axial. The algorithm searches for the biggest shape (in terms of area). If the y coordinate of its upper-left corner is zero

OR if the x coordinate of its upper-left corner is zero, the balance and force contribution are considered diagonal. Otherwise, the symmetry is considered axial. Figures 2, 3 and 4 are examples of diagonal and axial symmetries. When the algorithm is set to give higher scores to images showing diagonal symmetry, it will in fact give penalties for images that show axial symmetry (and vice versa). The score of the image decreases by 50% when penalized.

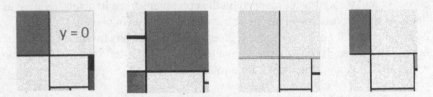

Fig. 2. Diagonal symmetry examples; y = 0 in the largest rectangle

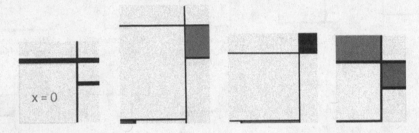

Fig. 3. Diagonal symmetry examples; x = 0 in the largest rectangle

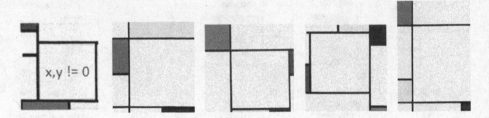

Fig. 4. Axial symmetry examples

Color balances. We set the "weight" of each color on a scale from 1–10, where the possible colors are black, blue, red, yellow and white. Then, we multiply this number by the area of the shape (calculated from the width and length of each shape). Finally, we divide this number by the total area of the image. Figure 5 depicts the deconstructed image showing the weight of each shape in terms of a chosen color weight.

Based on this approach, we divide the images into four quadrants, enabling us to determine whether the image is well balanced or not. If the sum of the weights in quadrants 1 and 4 is close to the sum of weights in quadrants 2 and 3, and the same is true for quadrants 1–2 and 3–4, the image will have a higher grade. Figure 6 illustrates this

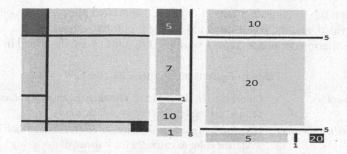

Fig. 5. Deconstructed artwork showing weight (Color figure online)

process. The position of the center of the digital artwork will determine in which quadrant the center of weight is located. Based solely in the color criteria, the image on the left will get a higher score if blue and yellow have a higher weight than red (on a 1–10 scale), because the area of the red shape is bigger than those colored in blue and yellow. The image on the right will get a higher score only if yellow has a higher weight than blue and red and if blue has a higher weight than red. The area of the yellow shape is quite small compared with the blue and red shapes. In addition, the red and blue shapes are in opposites quadrants with the yellow shape between them.

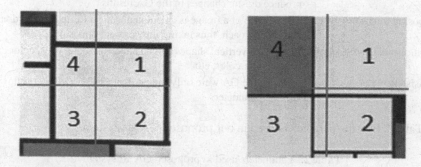

Fig. 6. Two examples of weights calculated for the color balance criteria (Color figure online)

A summary for the evaluation function can be found below:

$$F: \left(\Sigma w_1 + \Sigma w_2 = \Sigma w_3 + \Sigma w_4 \right) \&\& \left(\Sigma w_1 + \Sigma w_4 = \Sigma w_2 + \Sigma w_3 \right)$$

Where: Σw_i = sum of the weights in quadrant i = (weight of colour of shape 1 in quadrant i * area of shape 1 in quadrant i) + ... + (weight of colour of shape n in quadrant i * area of shape n in quadrant i).

A perfect image would have a score of 10. The furthest a image is from meeting the function criteria, the lower will be its score. Also, as described before, if the goal is to focus on a certain type of symmetry, all images featuring the other type will be drastically penalized. An art specialist can give scores from 0–10 to the new images produced by the system, letting the algorithm run in a step-by-step approach, or with minimal interventional (up to 60 generations at once). his approach allows more control of the images provided, and the possibility to avoid local optimal values.

The 11 criterion defined in Sect. 3 were used in our GA are described in Table 1.

Table 1. Conversion of criteria into the GA

Colors/Non-colors	Colors used in GA matched Mondrian palette for colors (red, blue, yellow) and non-colors (white, black, grey)
Use of colors/non-colors	GA takes into account all shapes in an artwork and their connection with their color to calculate the balance of the artwork
Weight	For each shape in a image, GA calculates its weight multiplying the color weight by shape area
Equilibrium	The evaluation function of the GA takes into account symmetry and color balance factors
Single geometrical figure	GA identifies the biggest shape in the image to determine the type of symmetry it has
Straight lines/right angles	GA uses only rectangles for shapes, and tries to maintain all shapes within the original canvas size
De-centric	Images are evaluated according to the balance of their shapes, not distance to a focus point
Rhythm	Small changes in color weights or shapes placement and area can produce drastic changes in the GA results
Opposing forces	The balance of a image is calculated using a Cartesian coordinate system approach, thus taking into account opposing forces
Relationships	In our GA, vertical shapes can balance horizontal ones, color can balance non-color, etc
Consistency	Input for the GA were only images from 1922-1932 that followed specific parameters

Table 2 lists the parameters used in our proposed GA art system.

Table 2. Parameters used in proposed GA art system

Representation	One chromosome for each shape within the artwork
Evaluation function	Combination of Symmetry and color balance,
Population size	20
Parent selection	Roulette or Tournament
Survival selection	Roulette or Tournament
Crossover operator	1-point crossover
Mutation operator	0.05
Elitism	Best individual

4.4 Initial Experiments

We tested the system through a series of experiment. Each experiment ran and tested each one of the criteria, as detailed in Sect. 4. Figures 7, 8 and 9 provide examples of runs and images of individuals resulting from implementation of the fitness criteria.

Fig. 7. Experimenting with images with axial symmetry

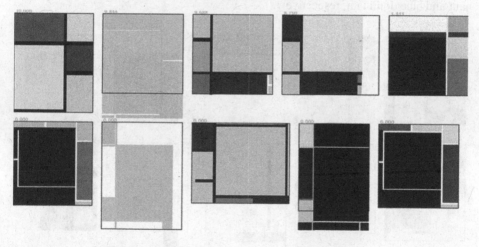

Fig. 8. Experimental images with diagonal symmetry

The images produced using the axial criterion are depicted in Fig. 7. We chose to depict only part of the whole population. These images have axial balances, as described in Sect. 4.3.

Figure 8 depicts the diagonal symmetry experiments. Note that the diagonal symmetry criterion is very dominant. These images have diagonal balances, as described in Sect. 4.3.

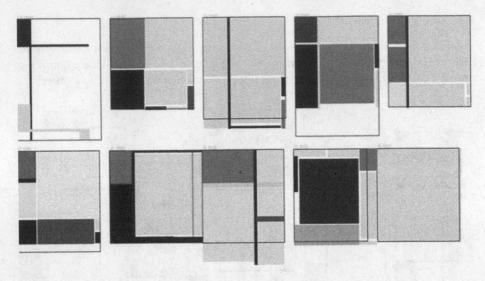

Fig. 9. Experimenting with dominant colors for the balance criterion: yellow (Color figure online)

We also experimented with the color scales by assigning a greater weight to each color separately and then studying and analyzing the effects on the total image balance, as detailed in Sect. 4. Figures 9 and 10 depicts the following color scales: yellow dominant and blue dominant, respectively.

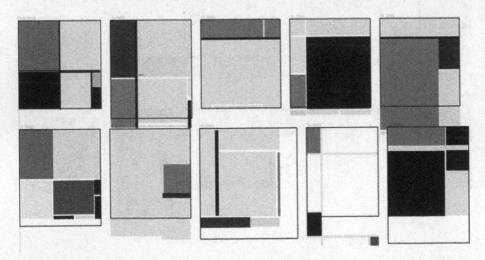

Fig. 10. Experimenting with dominant colors for the balance criterion: blue (Color figure online)

5 Results

We saw that the selection method used by the algorithm (either roulette wheel or tournament) had little impact on the quality of the pictures produced. On the other hand, the diagonal approach exhibited better results than the axial approach, both in terms of colors and in the positioning of the rectangles in the canvas. A high quality picture exhibits (1) colors closer to those Mondrian used in his paintings in terms of red, blue, yellow, white, and black, and (2) rectangles that do not overlap or have empty spaces between them, while at the same time fitting the canvas.

Fig. 11. Examples produced using the proposed system: (left) axial symmetry, tournament, 60 generations, (right) diagonal symmetry, tournament, 60 generations

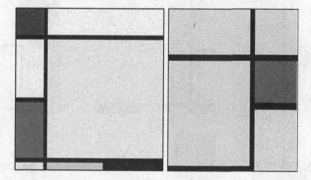

Fig. 12. Examples produced using the proposed system: (left) axial symmetry, roulette, 15 generations, (right) diagonal symmetry, roulette, 27 generations

One limitation of the proposed algorithm is that it considered all shapes in the original paintings as rectangles, instead of considering the black slim rectangles delimiting the other shapes as if they were lines. Thus, changing the position of those black rectangles in the system resulted in empty spaces between shapes or in overlapping. As a result, we can see that the high quality pictures produced do not exhibit greater change in terms of size and location of shapes when compared to the original Mondrian paintings. Considering the paintings as a set of lines that create rectangles would increase the variability of the shapes in the pictures produced. Hence, the resulting pictures would be more diverse. However, implementation of such an approach proved to be challenging at the time the research was conducted.

Another limitation deriving from the previous one is that the algorithm did not consider the fact that all slim rectangles (or lines) should be always black, because they were black in all the original paintings. Thus, the algorithm tried to change their color to red, yellow, blue, or white, resulting in pictures that did not at all resemble the original paintings (see Figs. 9 and 10). We considered addressing this issue by using the ratio of the shapes to determine if a shape should always be black, but we rejected this approach because some paintings have small black lines that would be considered regular rectangles and vice versa. This limitation could be solved by considering those lines as special shapes when coding the original paintings into the system.

Despite these limitations, we achieved some interesting results. The results are very promising. Nevertheless, some discrepancies in relation to the proposed criteria compromise the characteristic balance of Mondrian's work. These include the existence of non-primary colors in some of the samples and the extra color with weight located on one side of the image. Compared to the results of previous research, our results represent a qualitative leap with respect to visual proximity.

Figures 11 and 12 depict some examples that met our criterion produced using the proposed system.

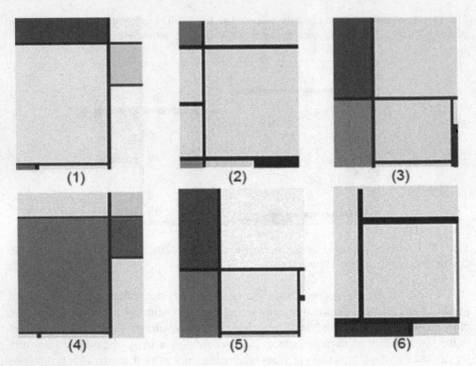

Fig. 13. Examples produced using the proposed system (images 1–6).

The examples are visually so close to real Mondrian works that only thorough analysis will lead to the above conclusions. Therefore, we conducted an experiment with students from the School of Fine Arts of the Federal University of Minas Gerais, Brazil. We showed them the images in Fig. 13 and asked them which ones they considered genuine Mondrian works. The six images are numbered from top left to top right etc. Initially, they had difficulty selecting any that was not a real Mondrian. It was only after encouraging them to make a thorough analysis that they began to respond. The majority of students considered that at least one of the images shown referred to a real Mondrian. Their basic concerns were the colors, and they tried to determine that each chosen image contained only primary colors. Almost 60% of the students identified images number 1 and 5 as real Mondrian works. Even the images with a bigger visual rejection still were identified as Mondrian's by more than 15% of the students.

These results support our conclusion regarding the visual proximity of computer-generated images to genuine works by Mondrian. This means that the genetic system criteria can be considered satisfactory and very close to those used by the artist.

6 Conclusions and Future Work

This study is valuable not only from the point of view of the developed genetic algorithm, but also because it provided insight into the logic behind Mondrian's paintings. This is a relevant finding because, as mentioned, previous works did not achieve very good results in producing Mondrian-like pictures. Future steps include addressing these limitations in a second version of our genetic algorithm and running the experiment again with the new high quality pictures produced.

References

1. Milner, J.: Mondrian. Phaidon Press, New York (2011)
2. Seuphor, M.: Piet Mondrian: Life and Work. Harry N. Abrams Inc, New York (1957)
3. Henning, E.B.: A classic painting by Piet Mondrian. Bull. Clevel. Museum Art 55(8), 243–249 (1968)
4. Hill, A.: Art and mathesis: Mondrian's structures. Leonardo 1(3), 233–242 (1968)
5. Taylor, R.: Pollock, Mondrian and nature: recent scientific investigations. Chaos Complex. Lett. 1(3), 265–277 (2003)
6. Lewis, M.: Evolutionary visual art and design. In: Romero, J., Machado, P. (eds.) The Art of Artificial Evolution: A Handbook on Evolutionary Art and Music. Natural Computing Series. Springer, Berlin Heidelberg (2008)
7. Locher, P.J.: The usefulness of eye movement recordings to subject an aesthetic episode with visual art to empirical scrutiny. Psychol. Sci. 48(2), 106–114 (2006)
8. McManus, I.C., Cheema, B., Stoker, J.: The aesthetics of composition: a study of Mondrian. Empirical Stud. Arts 11, 83–94 (1993)
9. Latto, R., Brain, D., Kelly, B.: An oblique effect in aesthetics: Homage to Mondrian (1872–1944). Perception 29(8), 981–987 (2000)
10. Noll, A.M.: Human or Machine: a subjective comparison of Piet Mondrian's "composition with Lines" (1917) and a computer-generated picture. Psychol. Record 16, 1–10 (1966)

11. Schufreider, G.: Overpowering the center: three compositions by Mondrian. J. Aesthetics Art Criticism **44**(1), 13–28 (1985)
12. Vaughan, W.: The Mondrian Maker. Supplement to Computers and Art History Group, First Issue (1985)
13. Michelson, A.: De Stijl, Its Other Face: Abstraction and Cacophony, or What Was the Matter with Hegel? (1982). Retrieved http://www.jstor.org/stable/778361
14. Blotkamp, C.: Mondrian: The Art of Destruction. Reaktion Books, London (1994)
15. Bois, Y.A., Joosten, J., Rudenstine, A.Z., Janssen, H.: Piet Mondrian 1872–1944. Leonardo Arte, Milan (1994)
16. Overy, P.: Here-I-Am-Again-Piet: a Mondrian for the Nineties. Art Hist. **18**(4), 584–595 (1995)
17. Zhang, K., Yu, J.: Generation of kandinsky art. Leonardo **49**(1), 48–54 (2016)
18. Eiben, E.: Evolutionary reproduction of Dutch masters: the Mondrian and Escher evolvers. In: Juan Romero, J., Machado, P. (eds.) The Art of Artificial Evolution: A Handbook on Evolutionary Art and Music. Natural Computing Series, pp. 211–224. Springer, Heidelberg (2008)
19. de Silva Garza, A.G., Lores, A.Z.: A cognitive evaluation of a computer system for generating Mondrian-like artwork. In: Gero, J.S. (ed.) Design Computing and Cognition '04, pp. 79–96. Kluwer Academic Publishers, Worcester, Massachusetts (2004)
20. Bentley, P., Corne, D.W. (eds.): Creative Evolutionary Systems. Morgan Kaufmann Publishers, San Francisco (2002)
21. Di Paola, S., Gabora, L.: Incorporating characteristics of human creativity into an evolutionary art algorithm. Genet. Program Evolvable Mach. **10**(2), 97–110 (2009). doi:10.1007/s10710-008-9074-x
22. Fogelman, M. (2011) http://fogleman.tumblr.com/post/11959143268/procedurally-generating-images-in-the-style-of
23. Machado, P., Correia, J.: Semantic aware methods for evolutionary art. In: Proceedings of the 2014 Annual Conference on Genetic and Evolutionary Computation (GECCO 2014), pp. 301–308. ACM, New York (2014). doi:http://dx.doi.org/10.1145/2576768.2598293
24. Heijer, E., Eiben, A.E.: Evolving pop art using scalable vector graphics. In: Machado, P., Romero, J., Carballal, A. (eds.) EvoMUSART 2012. LNCS, vol. 7247, pp. 48–59. Springer, Heidelberg (2012). doi:10.1007/978-3-642-29142-5_5
25. Bergen, S.R.: Evolving stylized images using a user-interactive genetic algorithm. In: Proceedings of the 11th Annual Conference Companion on Genetic and Evolutionary Computation Conference: Late Breaking Papers, GECCO 2009, New York, NY, USA, ACM, pp. 2745–2752 (2009)
26. Cook, T.E.: GAUGUIN: generating art using genetic algorithms and user input naturally. In: Proceedings of the 9th Annual Conference Companion on Genetic and Evolutionary Computation, London, United Kingdom (2007)
27. Silva Garza, A.G., Lores, A.Z.: Case-based art. In: Muñoz-Ávila, H., Ricci, F. (eds.) ICCBR 2005. LNCS (LNAI), vol. 3620, pp. 237–251. Springer, Heidelberg (2005). doi: 10.1007/11536406_20
28. Sims, K.: Artificial evolution for computer graphics. In: Proceedings of the 18th Annual Conference on Computer Graphics and Interactive Techniques (SIGGRAPH 1991), Vol. 25, No. 4, pp. 319–328 (1991)
29. Romero, J., Machado, P. (eds.): The Art of Artificial Evolution: A Handbook on Evolutionary Art and Music. Natural Computing Series. Springer, Heidelberg (2007)

30. Trist, K., Ciesielski, V., Barile, P.: Can't see the forest: using an evolutionary algorithm to produce an animated artwork. In: Huang, F., Wang, R.-C. (eds.) ArtsIT 2009. LNICSSITE, vol. 30, pp. 255–262. Springer, Heidelberg (2010). doi:10.1007/978-3-642-11577-6_32

31. Liu, H., Liu, X.: generative art images by complex functions based genetic algorithm. In: León-Rovira, N. (ed.) CAI 2007. ITIFIP, vol. 250, pp. 125–134. Springer, Heidelberg (2007). doi:10.1007/978-0-387-75456-7_13

32. den Heijer, E., Eiben, A.E.: Using aesthetic measures to evolve art. In: IEEE Congress on Evolutionary Computation, pp. 311–320. IEEE Press (2010)

33. Johnson, M.G., Muday, J.A., Schirillo, J.A.: When viewing variations in paintings by mondrian, aesthetic preferences correlate with pupil size. Psychol. Aesthetics Creativity Arts 4(3), 161–167 (2010). doi:10.1037/a0018155

34. Simon, D.: Evolutionary Optimization Algorithms. Wiley, Hoboken (2013)

35. Heijer, E.: Evolving glitch art. In: Machado, P., McDermott, J., Carballal, A. (eds.) EvoMUSART 2013. LNCS, vol. 7834, pp. 109–120. Springer, Heidelberg (2013). doi:10.1007/978-3-642-36955-1_10

36. Goldberg, D.E., Kalyanmoy, D.: A comparative analysis of selection schemes used in genetic algorithms. Found. Genet. Algorithms 1, 69–93 (1991)

37. Chuang, Y.-C., Chen, C.-T., Hwang, C.: A real-coded genetic algorithm with a 74-direction-based crossover operator. Inf. Sci. 305, 320–348 (2015)

38. Sivaraj, R., Ravichandran, T.: A review of selection methods in genetic algorithms. Int. J. Eng. Sci. Technol. 3(5), 3792–3797 (2011)

Predicting Expressive Bow Controls
for Violin and Viola

Lauren Jane Yu$^{(\boxtimes)}$ and Andrea Pohoreckyj Danyluk$^{(\boxtimes)}$

Williams College, Williamstown, MA, USA
ly1@williams.edu, andrea@cs.williams.edu

Abstract. Though computational systems can simulate notes on a staff of sheet music, capturing the artistic liberties professional musicians take to communicate their interpretation of those notes is a much more difficult task. In this paper, we demonstrate that machine learning methods can be used to learn models of expressivity, focusing on bow articulation for violin and viola. First we describe a new data set of annotated sheet music with information about specific aspects of bow control. We then present experiments for building and testing predictive models for these bow controls, as well as analysis that includes both general metrics and manual examination.

Keywords: Musical expression · Machine learning · Violin · Viola · Bow articulation

1   Introduction

Professional musicians seek to perform music in a way that appeals and connects to listeners [1]. In contrast, automated computer "performances" of written compositions typically lack an expressive (and thus human) quality. Computational modeling of human musical performance is difficult because of the lack of clear-cut rules for how music should be expressed. There are established guidelines and conventions for performance, but the performer ultimately decides how to incorporate these guidelines and interpret music convincingly [2].

To go from a written score to expressive music, one must perform a series of steps. At its simplest, this creates a pipeline, so to speak (see Fig. 1). From the score, a musician must decide what should be expressed. This includes aspects such as expressive volume and speed. The musician then needs to decide how to achieve those elements. For instruments such as the violin and viola, this requires deciding how the bow should be used. Similar steps can be taken by a computer to go from a given score to a plan of gestures. All of this then needs to be synthesized into a convincing sound for the instrument being modeled. In reality, this pipeline has a cycle where the musician uses the performance to then modify the plan for expression, but given the complexity of the problem at hand, we focus on one iteration of this loop.

© Springer International Publishing AG 2017
J. Correia et al. (Eds.): EvoMUSART 2017, LNCS 10198, pp. 354–370, 2017.
DOI: 10.1007/978-3-319-55750-2_24

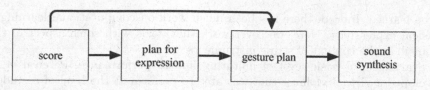

Fig. 1. The pipeline of steps that starts with a musical score and ends with an expressive performance. This pipeline is based on the one described in [3].

The specific goal of our work is to use machine learning algorithms that, when presented with a sample of annotated violin or viola music for a given era, learn models of bow articulation for similar pieces. These models can then be used to automatically annotate new pieces such that when those pieces are "performed" by a computer, they will sound as though they were produced by a human. This is represented by the top arrow from "score" to "gesture plan" in Fig. 1. Though this seemingly skips "plan for expression," we hope to implicitly capture that in the "gesture plan." The aspects of bow control we consider are bow position, bow distance from the bridge, and whether the bow should be off the string at some point during the duration of a note.

Articulation is relatively difficult to infer from sheet music alone; the notation for articulation is often ambiguous if even present [2]. Interpreting the music can include considering details about the piece's history. For example, string instruments in the Baroque Era used a different bow, so performances of Baroque music might reflect the limited capabilities of those early bows [4]. Sheet music also rarely contains instructions for how to achieve a given articulation. For violin and viola, achieving the desired sound requires varying factors such as weight from the arm holding the bow and how quickly to move the bow. One might also consider the practicality of different techniques.

The main contributions of this paper are: a data set for machine learning of musical expressivity, predictive models that seek to "perform" pieces on their own through annotating sheet music, and the design and execution of experiments to evaluate the learning of annotating models from data. Our analyses of the experimental results go beyond examining standard machine learning evaluation metrics and take into account what exactly a model is predicting.

In the next section, we discuss related work. Section 3 describes the data sets we have developed. Section 4 discusses experimental setup and results. Section 5 summarizes our conclusions and outlines directions for future work.

2 Related Work

The physics behind sound production as it relates to string instruments is well understood [5,6], but defining musical expressivity lacks the same rigor. Many researchers agree that performing a piece musically entails deviating from the score [1,7–9], but there is less agreement on what those deviations should be. While there is no repository of written clear-cut rules, there are implicit rules to be learned. Indeed, there has been much work on computational learning of musical expressivity. Here, we cover work that deals with some aspect of this problem in the context of string instruments.

Starting at a basic level of imitating human performance, Percival et al. developed a virtual violinist that was able to perform at the level of a student who had studied with the Suzuki method for a year [10]. They learned a different Support Vector Machine (SVM) model [11,12] for every string/dynamic combination, for example, *forte* on the G string. Sheet music, string being played, finger position, and distance between the bridge and bow were used as inputs for predicting a good bow force [10]. Since Percival et al.'s goal was to model a beginning-level violin student, there was no attempt at interpreting the music. A robotic performance was acceptable in this case, in contrast to our work.

Other researchers have aimed to imitate professional musicians, who are more concerned with expressive performances. Thippur et al. [3] modeled the mapping of musical terminology for a gesture plan (e.g. "up bow *piano détaché*") to the sequence of physical gestures using Gaussian processes [13]. In the pipeline represented in Fig. 1, this is within the arrow between "plan for expression" and "gesture plan." These models learned from motion-capture recordings of bow velocity, bow-bridge distance, and bow force from two professional violinists playing sequences of *détaché* notes on each of the four strings with varying dynamic levels (*f*, *mf*, and *p*). In total, six models were learned—one for each pair of bow direction (up vs. down) and dynamic. When given the pitch, duration, dynamic, bow direction, and bowing style as input, the appropriate model would be used to predict bow velocity, bow-bridge distance, and bow force.

To evaluate their models, Thippur et al. used listening tests. They synthesized two sets of violin notes: one based on the values for bow velocity, bow-bridge distance, and bow force in the original training set and one based on the bow control values predicted from the models [3]. The violin sounds were synthesized using a bowed-string model [14], but the focus of the test was the naturalness of the bow motion, not the realness of the violin sound. Eight string players participated in this test—one professional and seven advanced amateurs. They were presented with two pitches (C4 and G5 played on the G and D string, respectively) and two dynamics (*mf* and *f*) for a total of four cases. When asked "how natural is the bow motion that produced the note you heard?" most listeners rated the predicted motions (and thus the sounds produced by them) as somewhere between "artificial" and "like a human" though by varying amounts [3]. Our work expands beyond the scope of this work by starting with a score and considering additional bow strokes and dynamics.

Marchini et al. modeled expressivity on a professional level in string quartets. For training, they used both audio recordings and motion capture from a single professional string quartet performing Beethoven's String Quartet No. 4. They played this three times with different expressive intentions: mechanical, normal, exaggerated. Marchini et al. recorded the musicians both on their own and as a group. They used a Polhemus Liberty wired motion capture system to record bowing-motion data. They then aligned the audio and motion data with the score (via a dynamic programming algorithm). They also extracted features about the score. These included information about the notes in a single instrument's part as well as information about one instrument's notes relative to the other notes being played. These features were used as inputs for regression models to predict volume, note lengthening, vibrato extent, and bow velocity for each instrument. These four aspects of performance are all implicit in sheet music.

They tried three different machine learning regression algorithms: Weka's implementation [15] of model trees, linear Support Vector Machines (SVMs) [12], and k-nearest neighbor (kNN) with $k = 1$ as a baseline and distance based on Euclidean distance [16]. Trees performed the best, followed by SVM and then kNN. One limitation noted by the authors was the lack of sufficient data for training and evaluation [8]. Marchini et al.'s work is closest to ours. However, they studied string quartet performance, while we consider solo violin or viola, and they studied a variety of interpretative aspects to performance while we focus specifically on using the bow.

3   Developing Data Sets for Training and Testing Models of Expressive Bow Control

Our goal is to learn models that can label written musical compositions with appropriately expressive bow controls. In order to apply machine learning methods, we require pieces labeled with measurable aspects of bow control. At the start of our project, there was a dearth of suitable, publicly available data. Violin-specific sets, in general, contained very little data. One, for example, contained only 40 s of the bow being moved back and forth with no variation in the bow stroke. While the data set from [8] may have been suitable, it contained only one movement of a piece and focused on different bow controls from ours. Here we present a new data set we developed (available at www.cs.williams.edu/~andrea/publications.html).

3.1   Considerations

Several decisions had to be made before we could develop a data set for machine learning. Considerations included whose performances would form the base for the data set, what repertoire, how to encode information, and what aspects of bow control to examine. Here we detail those considerations and explain our decisions.

We reached out to college students studying violin or viola and performing in a semi-professional orchestra. In total, we employed four violinists and two violists. The use of advanced violin and viola students rather than professionals was primarily due to access. However, since these were students who had taken private lessons in college and performed alongside professional musicians, our conjecture was that these musicians would be skilled enough to provide expression in their performance that a predictive model could capture. For the data collection process, we relied on video recordings of musicians rather than data collected from sensors. This was to avoid the potential negative impact that sensors can have on the way a musician plays.

MusicXML was used for encoding sheet music. MusicXML uses XML syntax to encode both written sheet music and synthetic audio files (e.g., MIDIs). MusicXML had been used in previous work, and also provided the benefit of being readable through software for editing sheet music.

The choice of music was based entirely on the current repertoire of the musicians who recorded videos. This was to ensure the musicians would put substantial thought into how they were using their bows (among general considerations of expression) when performing. The pieces in our data set are the first movement of Bach's Cello Suite No. 2 played an octave higher on viola, the first movement of Bach's Cello Suite No. 6 transposed to G major for viola, the second movement of Sibelius' Violin Concerto in D Minor, and the second movement of Wieniawski's Violin Concerto No. 2.

We chose to model three aspects of bow control: (1) **Position**: where, along the length of the stick, the bow makes contact with the string. (2) **Distance from bridge**: how far away the bow is from the bridge. (3) **Off the string**: whether the bow is on the string for the duration of a note or lifted off at some point.

The defined value ranges for these are shown in Table 1. For bow position, since the bow can make contact with the string anywhere along the horsehair, we use continuous values. We chose a range of $0 - 4$ because it is large enough to capture differences without necessitating fine-grained decimals. This makes labeling easier. Similar reasoning explains our choice of values for bow-bridge distance. Since the bow is either touching the string or not, "off the string" is represented by binary values: $1 =$ off the string, $0 =$ on the string.

Table 1. Bow control values.

Property	Labels	Description	
Bow position	$0 - 4$	$0 =$ frog	$4 =$ tip
Bow-bridge distance	$0 - 2$	$0 =$ next to bridge	$2 =$ on the fingerboard
Off the string	$0, 1$	$0 =$ on the string	$1 =$ off the string

Our choice of bow controls arose from three main considerations. One consideration was consistency with the literature. The distance between the bow and the bridge, as well as bow speed and bow force, had been used in previous work. The second consideration was what could be most precisely captured in a video. Although much work in this field has included recording the force and speed of a bow, it would be impossible to watch a video and produce a precise numeric value for either. As a proxy for bow speed, we use bow position. Third, personal experience was a factor in choosing bow controls. Though not studied in previous work, whether or not the bow is lifted from the string is a consideration for effective expressive performance (A.L. Neu, personal communication, September 16, 2014).

3.2   Protocol

Generating the data set involved three distinct but related tasks: recording videos, generating MusicXML, and annotating (i.e., labeling) the sheet music with bow controls. For recordings, the violinists and violists were instructed to record videos of themselves playing, with a clear view of the bow in terms of its position along the stick and its distance from the bridge.

For most of the pieces, there was no publicly available MusicXML. As a result, we employed seven students to generate MusicXML files using software for editing sheet music. The source material was either the sheet music used by the performer or sheet music found online. Almost all printed markings were kept: pitch, rhythm, articulation markings, dynamics, indications of bow direction and any expressive text. Rehearsal letters and suggested fingerings were not kept. For chords, the non-melody note durations were changed to match that of the melody note. For passages of music with more than one voice, the notes were modified to look like chords with the rhythm of the main melody. These were both done to simplify parsing. The changes largely affected the training data for Sibelius, and thus may have negatively impacted the results, but we believe the generally poor results for Sibelius can be primarily attributed to the size and diversity of instances in the training set.

After the sheet music and video were ready, the bow controls (labels) observed in the video were encoded in the MusicXML as lyrics, with each verse representing one property (see Fig. 2). There was a label for every sixteenth note or shortest note, whichever was larger for the particular piece. Multiple labels for a note were separated by a slash ("/").

Fig. 2. Labeled sheet music.

We relied on only one person to interpret each video. Ideally we would have had three and tested their agreement, but because of our timeline and the difficulty of finding people with enough musical knowledge we used a single encoder. After the labels had been recorded in the MusicXML, the files were then parsed to create the appropriate file formats to be used as input for the various machine learning models.

3.3   Featurization

Building a predictive model requires giving the machine learning algorithm different features of the data to consider. For example, to determine whether or

Table 2. Formula from [8] for calculating the strength of a beat b for common time.

Strength	Case
4	b is the first beat of the measure
3	b is the up beat (third beat) of the measure
2	b is at a remaining quarter division of the measure (second or fourth beat)
1	b is an eighth division of the measure
0	all other cases

4 0 1 0 2 0 1 0 3 0 1 0 2 0 1 0

Fig. 3. Example of the formula described in Table 2 applied to a measure.

not a note should be played off the string, the learning algorithm might want to know the duration of the note and if there are any articulation markings. Features we extracted from the MusicXML files include:

Pitch the pitch of the note in hertz. In the case of a note with multiple pitches, the highest pitch is used, since that is usually the melody note.

Charge how many steps away in the circle of fifths the note is from the tonic of the piece [8].

Note place since every sixteenth note was labeled, some notes were split into multiple instances. This feature indicates where in the note a label is. The possible values for this feature are `begin`, `middle`, `end`, and `entire`.

Duration the duration of the note. This is expressed relative to the smallest note value in the piece. If a piece's shortest note is an eighth note, an eighth note's duration is expressed as 1, a quarter note's duration as 2, etc.

Tenuto whether or not the note has a tenuto marking.

Accent whether or not the note has an accent marking.

Staccato whether or not the note has a staccato marking.

Ornament whether or not the note has an ornament. This includes all ornaments except grace notes, due to the MusicXML encoding.

Chord whether or not more than one note is to be played at a time. This includes two melody lines being played at once.

Bow direction this value is `up` or `down` when there is a $\vee$ or $\sqcap$ explicitly on the note. Otherwise, the value of this feature is `none`.

Slur where in the slur the note is: `start`, `middle`, or `stop`. If there is no slur associated with the note, the value of this feature is `none`.

Beat which beat of the measure the note starts on.

Beat strength the relative strength of the beat. Table 2 shows the calculation for $\frac{4}{4}$ time and Fig. 3 shows the result, but this calculation is done for all meters. The formula is based on the one described in [8].

Dynamic the last explicitly notated dynamic. This is a numeric value provided by the MusicXML: 124.44 for *ff*, 106.46 for *f*, 88.89 for *mf*, 71.11 for *mp*, 54.44 for *p*, and 36.67 for *pp*. The number is *not* adjusted when there is any kind of volume-modifying marking, such as a *crescendo* or *diminuendo*.

Wedge whether or not the note contains the beginning of a *crescendo* or *diminuendo*. The values for this feature are `cresc`, `dim`, and `none`, and only the first note of a *crescendo* or *diminuendo* receives a value other than `none`. This is because a text marking of these does not indicate an end point.

Previous dynamic the dynamic of the previous note.

Previous intervals these are the number of half steps between either the current note and the previous note or two preceding notes.

Previous labels labels from previous instances are included to capture the sequencing nature of the task. When learning a bow control, e.g., bow position, previous values for only that bow control are used as features.

Features of the next note since interpreting music often involves looking ahead to see where the music is going, information about the next note is included: pitch, charge, note place, wedge, duration, tenuto, accent, staccato, ornament, chord, bow direction, beat, beat strength, dynamic, and slur.

4 Experiments and Evaluation

Our goal is to use machine learning techniques to predict expressive bow articulation. Because of the variation between specific musical genres and composers, this requires learning multiple models. In this section we discuss machine learning experiments for building and evaluating predictive models. We begin by discussing the formulation of model-building as a supervised machine learning problem as well as evaluation methodology. We then present experimental setups and results pertaining to the three aspects of bow control—bow position, bow-bridge distance, and "off the string".

4.1 Problem Formulation

We frame the task of learning a bow-articulation model as a supervised classifier learning or regression model. The goal of classifier learning is to build a classifier. For example, the categorization of emails as legitimate or spam can be framed as a classification task. Given a set of previous instances (training data)—i.e., emails and whether or not they are spam—a learning algorithm aims to create a model that both fits the training data and can classify new instances.

In the case of predicting, say, off the string, given a training set of labeled instances—where an instance is a vector of features describing a note in a score as well as whether or not it should be played off the string—can we build a model that will be able to label previously unseen notes as on or off the string? The training instances here might be instances from the beginning of the piece or another piece of a similar composer or style. Regression is similar; the main difference is that the value to be predicted is real-valued, rather than discrete.

This formulation allows for utilizing powerful learning algorithms. There are, however, two noteworthy characteristics of this domain that distinguish it from traditional classifier-learning problems. First, the instances used for prediction (the notes in a piece of music) are not independent of each other. Second, in order for predicted labels, i.e., bow controls, to make sense, they must be physically realizable in the context of playing violin or viola. As a result, our experiments differ from typical classifier-learning experiments in two important ways: (1) the formulation of training and test sets and (2) the methods of evaluation.

A standard method of evaluation for supervised classifier learning is performing a ten-fold cross-validation. This consists first of randomly partitioning the data set into ten equal-sized parts (hence, *ten*-fold). Then one set is used for testing and the remaining sets are used for training. This is repeated for each of the ten equal-sized parts. Cross-validation is used to compensate for data sets that are not large enough to simply split in half; it keeps sets as large as possible to ensure error estimates are robust, minimizes overlap between sets, and maintains proportional representations of classes in subsets of data [17].

Unfortunately, cross-validation is not appropriate to our experiments. The sequential nature causes predictions of instances in isolation to be meaningless. A set of predicted labels should make sense *together*, from the beginning of a piece to the end. So instead of cross-validation, we train on one piece and test on another when appropriate. Otherwise, we use the beginning of a given piece as the training set and the rest of the piece as the testing set to serve as a proxy for learning from one piece and applying the model to another similar piece.

4.2   Evaluation Metrics

Many classification and regression experiments rely on metrics such as accuracy and correlation coefficient for evaluating a learned model. Looking at these metrics alone, however, does not give a full enough picture of the predictions made. For example, when looking at whether or not the bow is off the string, if most of the true labels are "on the string" and the classifier defaults to always predicting this, the accuracy will appear high, even though the model is weak.

Confusion matrices give a fuller picture by showing whether or not a classifier is actually predicting all classes. Table 3 shows a confusion matrix for a binary classification task with the classes 0 and 1. The row corresponds to the true class and the column indicates the predicted class. The goal, in terms of maximizing accuracy, is to have all the nonzero numbers be along the major diagonal.

Table 3. Example of a confusion matrix.

0	1	← classified as
557	2	0
18	40	1

Unfortunately, confusion matrices cannot evaluate sequences of instances. Therefore, in addition to observing confusion matrices and looking at metrics such as percent correct and correlation coefficient, we also evaluate our predictive models by carefully examining their outputs. We put the predicted labels back into MusicXML so we can examine the labels with the sheet music. Manual evaluation of the output enables us to pinpoint areas where the learned models fall short. Additionally, relying on statistics alone assumes that the only correct result is one that mimics the original data. Because interpreting music is not exact, however, there can be predictions that contradict the original labels but still produce a plausible performance. Manually examining the output also allows us to evaluate validity of the musical interpretation on its own.

4.3   General Experimental Setup

This section introduces the general setup for our experiments. Unless otherwise noted, all results presented were gathered using the setup described here.

Training and Test Data. The data sets for these experiments were constructed from our four labeled pieces: the first movement of Bach's Cello Suite No. 2 played an octave higher on viola ("Bach No. 2"), the first movement of Bach's Cello Suite No. 6 transposed to G major for viola ("Bach No. 6"), the second movement of Sibelius' Violin Concerto in D Minor, and the second movement of Wieniawski's Violin Concerto No. 2. We parsed the MusicXML for each of these four pieces into instances for the data sets, representing each sixteenth note (or eighth note for Wieniawski) as a vector of the features described in Sect. 3.3.

Ideally, we would train on one piece of a given composer or era and test on another such piece. Below we present such experiments for our two Bach pieces. In all other cases, we used the first 25% of a piece for training and the rest of it for testing. The 25–75 split was determined empirically.

Algorithms. All of the experiments presented here used Weka's REPTree implementation with default parameter settings. REPTree is a decision tree learner that can handle both classification and regression tasks. It uses information gain and variance as the splitting criteria for building the tree. For pruning, it uses reduced-error pruning with backfitting [15]. Empirically, it performed better than other (non-ensemble) algorithms tested. The learned trees are also easy to interpret, and examining the models built was helpful in interpreting the results.

Some of our experiments employed bagging with REPTree as a base algorithm, again using Weka's implementation with default parameter settings. Bagging is an ensemble-learning method that samples instances with replacement and learns models for different sets of instances. These models are then combined to create the final model [15]. For classification, the new instances are classified by having the underlying classifiers vote. For regression, the results of the underlying models are averaged.

364   L.J. Yu and A.P. Danyluk

Table 4. Correlation coefficients for bow position, with and without bagging, for eight and sixteen previous labels.

Piece	w/o bagging		w/bagging	
	8	16	8	16
Bach No. 2	0.43915	0.3213	0.51385	0.54428
Bach No. 6	0.8027	0.72086	0.7957	0.84185
Sibelius	0.08808	0.06649	0.0692	0.05229
Wieniawksi	0.89156	0.83105	0.90839	0.90764

As mentioned above, we use the labels from previous instances as features for the current test instance. To fully simulate a model labeling a piece from beginning to end, after each test instance is labeled with a bow control, we use that label as a feature for the next note instance.

4.4   Results

Here we present a subset of our experimental results and analyses, demonstrating the viability of learning models of bow control. Due to space limitations, we present a few in detail and summarize others. We organize the analyses by the aspect of bow control to be modeled: bow position, distance from the bridge, and bow on/off the string. Lastly, we present experiments where we trained on one piece and tested on another.

Bow Position. These experiments were run on all four pieces. For each, we trained on the first 25% of the piece, and tested on the remaining 75%. We considered four, eight, and sixteen previous labels as part of the featurization. We also experimented with using both eight previous labels and the next note's features. We found that using only four previous notes was not sufficient, but sixteen produced little to no improvement over eight. Incorporating lookahead to the next note's features also failed to improve results.

Bagging over trees provided an improvement over REPTrees alone (see Table 4). An instance of a common improvement is shown in Fig. 4. The top row has the original labels, the middle row contains predictions without bagging, and the last row shows predictions with bagging. The last two predicted labels of the first note in the middle measure are both 2.18, even though previous labels all show an increasing bow position, correctly indicating a down bow. In the bottom measure, the last two labels are 1.98 and 2.5. The improvement with bagging is even more prominent for the next note. The middle measure has 3.46 for all three labels on that note while the bottom measure shows a decreasing bow position that accurately indicates an up bow. Note that the Wieniawski correlation coefficient alone would not be sufficient to tease out such differences.

For the featurization with eight previous labels and no lookahead features, the primary features used by the decision trees for Wieniawski and Bach No. 6

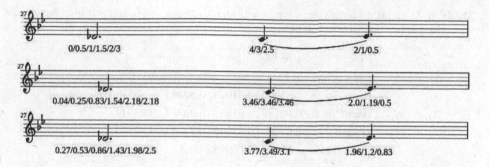

Fig. 4. A comparison of predictions for Wieniawski. The top measure is the original, the middle measure has predictions from using eight previous labels without bagging, and the bottom measure has predictions from using eight previous labels with bagging.

were the previous labels, note-place, and slur. That is what we would expect, as those features are most responsible for determining bow direction. Other features in the tree included pitch, beat, dynamics, and previous intervals.

It is unsurprising that the models performed best with Bach No. 6 and Wieniawski. Wieniawski has many long notes slurred together over the piece, and much of Bach No. 6 consists of three eighth notes with either all three notes slurred together or the first two slurred. The Sibelius, on other hand, contains a wide variety of patterns in terms of notes and slurs; training on the first 25% does not give enough information about the rest of the piece. Bach No. 2 also contains more such variety than Bach No. 6 and Wieniawski. To overcome such a problem, we might instead create training sets by sampling a series of measures from each segment of a piece with a distinct pattern.

Distance from the Bridge. In learning models for distance between the bow and bridge, we ran experiments similar to those for bow position: we used all four pieces with the same 25–75 split for training and testing. To facilitate manual examination, we discretized the labels for bow-bridge distance in Bach No. 6 (which started with more granular labels than the other pieces) so that the only possible labels were $\{0, 0.5, 1, 1.5, 2\}$. Even in this case, the learned model was still performing regression, rather than classification. As above, we varied the input features, trying four, eight, and sixteen previous labels as part of the featurization. Again, we found that using only four was not sufficient. Interestingly, bagging over trees did not provide the same improvement that we saw for bow position (see Table 5).

For some pieces, the learned models were generally successful at capturing expressive approaches. In analyzing the actual output of the models, one trend we observed was that the predicted labels were farther from the bridge during quieter sections and closer to the bridge during louder sections. In some cases, the predicted increase in difference was actually greater than that in the original labels. Though the predictions did not exactly match the original labels,

Table 5. Correlation coefficients for bow-bridge distance, with and without bagging, for eight and sixteen previous labels.

Piece	w/o bagging		w/bagging	
	8	16	8	16
Bach No. 2	0.83967	0.84023	0.82344	0.83458
Bach No. 6	0.92786	0.9135	0.93334	0.93644
Sibelius	0.6148	0.53052	0.74443	0.52495
Wieniawksi	—	0.76916	—	0.74446

one might argue that the predictions were more expressive than the original; increasing the distance from the bridge by a greater amount would create a bigger contrast in dynamics.

For all four pieces, the learned REPTrees first considered the previous label before looking at other features. This is both unsurprising and concerning. Since the distance from the bridge will often stay constant for a section of music, it makes sense that the learning algorithm would want to use the previous label to predict the current one. In the extreme case, however, this strategy can default to the same value for the entire piece. This is, in fact, what we observe in the model learned for Wieniawski using eight previous labels.

Off the String. Next we consider models that predict whether or not the bow should be lifted from the string at some point during a note's duration. Of the four pieces, only the Bach No. 6 was suitable for these experiments.[1] As before, the piece was split 25–75 for training and testing, and the same combinations of previous labels and lookahead features were considered. In this case, four previous labels proved to be sufficient. All experiments were run with the REP-Tree algorithm, both with and without bagging. Bagging resulted in roughly the same accuracy percentages but predicted "on the string" more frequently, so we summarize only results without bagging in this section.

Guessing a default of "on the string" would yield an accuracy of 94%, which is precisely what the model trained on eight previous labels did. The model trained with only four previous labels, however, was equally accurate and discriminated between "on" and "off" the string, as shown in the confusion matrix in Table 6.

Looking at the output against the sheet music, it is clear the model learned one pattern that could account for lifting the bow. If there were two slurred eighth notes followed by another eighth note, the last label of that group might be a 1, or "off the string" (see Fig. 5). For a section with this repeated pattern, the lifting of the bow every third eighth note could be considered reasonable. However, applying this pattern inconsistently does not make sense. Unfortunately, the model's output is, in fact, inconsistent.

---

[1] The percentages of "off the string" notes in Wieniawski, Sibelius, and Bach No. 2 are 0, 2, and 3, respectively. As a result, learned models only predicted "on the string."

Table 6. Confusion matrix for "off the string" for Bach No. 6 with four previous labels.

on	off	← classified as
1689	46	on
59	57	off

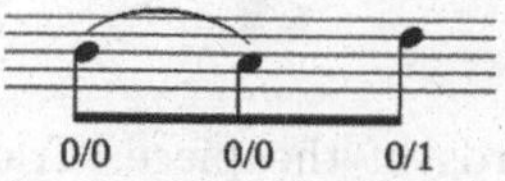

0/0   0/0   0/1

Fig. 5. A common pattern in which the model would predict "off the string." Even if this deviated from the original labels, it would be a plausible interpretation, if applied consistently throughout the piece.

Training on One Piece, Testing on Another. Ideally, we would like to train on several similar pieces and then apply the learned model to others of the same nature. Here we present experiments in which we trained on one Bach piece and tested on another. In the first set of experiments, we trained on Bach No. 6 and tested on Bach No. 2. In the second set we did the opposite. Featurization remained the same as for previous experiments.

Results for training on Bach No. 6 and testing on Bach No. 2 are shown in Table 7. Because bagging outperformed REPTree on its own, we present only results from bagging. The confusion matrix for the better "off the string" model is shown in Table 8.

The results for testing on Bach No. 2 are generally as good as or better than the results for Bach No. 2 in previous experiments. The predicted bow positions in these experiments did not perform much better or worse than in previous experiments, but we were able to learn a model for "off the string" (recall that this was previously impossible for Bach No. 2 due to the class imbalance).

Table 7. Correlation and accuracy: training on Bach No. 6; testing on Bach No. 2.

Bow control	Previous Labels	
	8	16
bow position	0.54699	0.56967
bow-bridge distance	0.83022	0.83097
off the string	97.70%	98.77%

The results for training on Bach No. 2 and testing on Bach No. 6 are shown in Table 9. Due to the class imbalance, we did not use Bach No. 2 to learn a model for predicting "off the string". Training on Bach No. 2 produced worse results for Bach No. 6 than splitting Bach No. 6 for training and testing. For bow position, this may be due to Bach No. 6 consisting primarily of two bowing

Table 8. Confusion matrix for "off the string" for training on Bach No. 6 and testing on Bach No. 2 using sixteen previous labels in the featurization.

on	off	← classified as
710	2	on
7	12	off

patterns in the first two-thirds of the piece. Training on Bach No. 2 means training on patterns that are far more different from the test instances. On the other hand, the learned model for bow-bridge distance successfully captured the movement away from the bridge for softer moments and closer to the bridge for louder moments.

Table 9. Correlation and accuracy: training on Bach No. 2; testing on No. 6.

Bow control	Previous Labels	
	8	16
bow position	0.53506	0.4307
bow-bridge distance	0.83096	0.80442

Though Bach composed both pieces, they are rather different. Bach No. 2 is slower and more melancholy; Bach No. 6 is more upbeat and energetic. That being said, the results are quite good and indicate potential for using pieces of one composer or style to learn a model for other similar pieces.

5 Summary

In this paper we have described a data set consisting of human-provided bow control annotations for violin and viola repertoire, as well as experiments in learning models that can annotate sheet music with appropriate bow controls for expressive performance. We argue that results of such experiments must be evaluated not simply based on standard machine learning metrics, but with careful manual inspection of results.

Though our learned models leave room for improvement, our results provide a clear proof of concept that models of expressivity and bow control can be learned. For example, in music with long notes or slurred notes, a learning algorithm can capture the changing direction of the bow at appropriate times, which we saw with bow position predictions for Bach No. 6 and Wieniawski. With predicting bow-bridge distance, there were instances in which the models predicted moving the bow either closer to the bridge for a louder sound or farther from the bridge for a quieter sound at appropriate moments. Lastly, with predicting "off the string" instances, the learned model was able to capture some common patterns for which lifting the bow from the string makes sense.

In addition, the results from training on Bach No. 6 and testing on Bach No. 2 suggest that there is potential for learning bow articulation models on sample pieces for a composer and then applying those models to other pieces from the same composer. Ideally, these learned models should capture a wide variety of expressive bow controls, and so having training pieces that are written by the same composer but different in style is best. At the same time, there needs to be a large number of examples of the patterns in the training set to reinforce the learning process.

There are many avenues to explore in future work. These include: producing more data sets; improving the accuracy of the training labels by utilizing multiple annotators or more precise methods of data capture; experimenting with the addition of new features, such as melodic contour; exploring different parameterizations of the learning algorithms; experimenting with new algorithms that more explicitly take into account time dependency (such as reinforcement learning algorithms); and producing a more readable output to produce an educational tool for aspiring musicians.

References

1. Woody, R.H.: Learning expressivity in music performance: An exploratory study. Res. Stud. Music Educ. **14**(1), 14–23 (2000)
2. Randel, D.M. (ed.): The Harvard Dictionary of Music, 4th edn. The Belknap Press of Harvard University Press, Cambridge, London (2003)
3. Thippur, A., Askenfelt, A., Kjellström, H.: Probabilistic modeling of bowing gestures for gesture-based violin sound synthesis. In: Proceedings of Stockholm Music Acoustics Conference 2013, Stockholm, Sweden (2013)
4. Donington, R.: String Playing in Baroque Music. Faber Music Ltd., London (1977)
5. Cremer, L.: The Physics of the Violin. MIT Press, Cambridge (1983). translated by Allen, J.S
6. Schelleng, J.C.: The bowed string and the player. J. Acoust. Soc. Am. **53**(1), 26–41 (1973)
7. Juslin, P.N.: Five facets of musical expression: A psychologist's perspective on music performance. Psychol. Music **31**(3), 273–302 (2003)
8. Marchini, M., Ramírez, R., Papiotis, P., Maestre, E.: The sense of ensemble: a machine learning approach to expressive performance modeling in string quartets. J. New Music Res. **43**(3), 303–317 (2014)
9. Neocleous, A., Ramírez, R., Pérez, A., Maestre, E.: Modeling emotions in violin audio recordings. In: ACM Workshop on Music and Machine Learning (ACM-MML), Fironzo, Italy, pp. 17–20 (2010)
10. Percival, G., Bailey, N., Tzanetakis, G.: Physical modeling meets machine learning: Teaching bow control to a virtual violinist. In: Sound and Music Conference, Padova, Italy, July 2011
11. Cortes, C., Vapnik, V.: Support-vector networks. Mach. Learn. **20**(3), 273–297 (1995)

12. Cristianini, N., Shawe-Taylor, J.: An Introduction to Support Vector Machines and Other Kernel-Based Learning Methods. Cambridge University Press, Cambridge (2000)
13. Rasmussen, C.E.: Gaussian processes in machine learning. In: Bousquet, O., Luxburg, U., Rätsch, G. (eds.) ML -2003. LNCS (LNAI), vol. 3176, pp. 63–71. Springer, Heidelberg (2004). doi:10.1007/978-3-540-28650-9_4
14. Demoucron, M.: On the Control of Virtual Violins: Physical Modelling and Control of Bowed String Instruments. Ph.D. thesis, KTH, Sweden (2009)
15. Hall, M., Frank, E., Holmes, G., Pfahringer, B., Reutemann, P., Witten, I.H.: The WEKA data mining software: An update. SIGKDD Explor. Newsl. **11**(1), 10–18 (2009)
16. Cover, T.M., Hart, P.E.: Nearest neighbor pattern classification. IEEE Trans. Inf. Theory **13**(1), 21–27 (1967)
17. Alpaydin, E.: Introduction to Machine Learning, 3rd edn. MIT Press, Cambridge (2014)

Author Index

Printed in the United States
By Bookmasters